Animal Biomarkers
as Pollution Indicators

Chapman & Hall Ecotoxicology Series

Series Editors

Michael H. Depledge
Professor of Ecotoxicology, Institute of Biology, Odense University, Denmark

Brenda Sanders
Associate Professor of Physiology, Molecular Ecology Institute, California State University, USA

In the last few years emphasis in the environmental sciences has shifted from direct toxic threats to man, towards more general concerns regarding pollutant impacts on animals and plants, ecosystems and indeed on the whole biosphere. Such studies have led to the development of the scientific discipline of Ecotoxicology. Throughout the world socio-political changes have resulted in increased expenditure on environmental matters. Consequently, ecotoxicological science has developed extremely rapidly, yielding new concepts and innovative techniques that have resulted in the identification of an enormous spectrum of potentially toxic agents. No single book or scientific journal has been able to keep pace with these developments.

This series of books provides detailed reviews of selected topics in Ecotoxicology. Each book includes both factual information and discussions of the relevance and significance of the topic in the broader context of ecotoxicological science.

Already published

Animal Biomarkers as Pollution Indicators
David B. Peakall
Hardback (0 412 40200 9), about 290 pages

Forthcoming title

Toxicity Testing in Environmental Biology
V. Forbes and T. Forbes
Hardback (0 412 43530 6), about 208 pages

Animal Biomarkers as Pollution Indicators

David Peakall

National Wildlife Research Center
Canadian Wildlife Service
Ottawa
Ontario
Canada

with a contribution on immunotoxicology
by Michel Fournier and co-workers
Université du Québec
Montréal
Québec
Canada

Half of the royalties of this book are being
donated by Environment Canada to
the World Wildlife Fund (Canada)

CHAPMAN & HALL

London · New York · Tokyo · Melbourne · Madras

Published by Chapman & Hall, 2–6 Boundary Row, London SE1 8HN

Chapman & Hall, 2–6 Boundary Row, London SE1 8HN, UK

Chapman & Hall, 29 West 35th Street, New York NY 10001, USA

Chapman & Hall Japan, Thomson Publishing Japan, Hirakawacho
Nemoto Building, 7F, 1-7-11 Hirakawa-cho, Chiyoda-ku, Tokyo 102,
Japan

Chapman & Hall Australia, Thomas Nelson Australia, 102 Dodds Street,
South Melbourne, Victoria 3205, Australia

Chapman & Hall India, R. Seshadri, 32 Second Main Road, CIT East,
Madras 600 035, India

First edition 1992

© 1992 Chapman & Hall

Typeset in 10/12 Times by Graphicraft Typesetters Limited, Hong Kong
Printed in Great Britain by T.J. Press (Padstow) Ltd, Padstow, Cornwall

ISBN 0 412 40200 9

A catalogue record for this book is available from the British Library

Library of Congress Cataloging-in-Publication data available

To
J.A. Keith
For his support, advice
and friendship

Contents

Series foreword

Ecotoxicology is a relatively new scientific discipline. Indeed, it might be argued that it is only during the last 5–10 years that it has come to merit being regarded as a true science, rather than a collection of procedures for protecting the environment through management and monitoring of pollutant discharges into the environment. The term 'ecotoxicology' was first coined in the late sixties by Prof. Truhaut, a toxicologist who had the vision to recognize the importance of investigating the fate and effects of chemicals in ecosystems. At that time, ecotoxicology was considered a sub-discipline of medical toxicology. Subsequently, several attempts have been made to portray ecotoxicology in a more realistic light. Notably, both F. Moriarty (1988) and F. Ramade (1987) emphasized in their books the broad basis of ecotoxicology, encompassing chemical and radiation effects on all components of ecosystems. In doing so, they and others have shifted concern from direct chemical toxicity to man, to the far more subtle effects that pollutant chemicals exert on natural biota. Such effects potentially threaten the existence of all life on Earth.

Although I have identified the sixties as the era when ecotoxicology was first conceived as a coherent subject area, it is important to acknowledge that studies that would now be regarded as ecotoxicological are much older. Wherever people's ingenuity has led them to change the face of nature significantly, it has not escaped them that a number of biological consequences, often unfavourable, ensue. Early waste disposal and mining practices must have alerted their practitioners to effects that accumulated wastes have on local natural communities; for example, by rendering water supplies undrinkable or contaminating agricultural land with toxic mine tailings. As activities intensified with the progressive development of human civilizations, effects became even more marked, leading one early environmentalist, G.P. Marsh, to write in 1864: 'The ravages committed by Man subvert the relations and destroy the balance that nature had established'.
destroy the balance that nature had established . . .'.

But what are the influences that have shaped the ecotoxicological studies of today? Stimulated by the explosion in popular environmentalism in the sixties, there followed in the seventies and eighties a tremendous increase in the creation of legislation directed at protecting the environment. Furthermore, political restructuring, especially in Europe, has led to the widespread implementation of this legislation. This currently involves enormous numbers of environmental managers, protection officers, technical staff and consultants. The ever-increasing use of new chemicals places further demands on government agencies and industries required by law to evaluate potential toxicity and likely environmental impacts. The environmental manager's problem is that he needs rapid answers to current questions concerning a very broad range of chemical effects and how to control discharges, so that legislative targets for *in situ* chemical levels can be met. It is not surprising, therefore, that he may well feel frustrated by more research-based ecotoxicological scientists who constantly question the relevance and validity of current test procedures and the data they yield. On the other hand, research-based ecotoxicologists are often at a loss to understand why huge amounts of money and time are expended on conventional toxicity testing and monitoring programmes, which may satisfy legislative requirements, but apparently do little to protect ecosystems from long-term, insidious decline.

It is probably true to say that until recently ecotoxicology has been driven by the managerial and legislative requirements mentioned above. However, growing dissatisfaction with laboratory-based tests for the prediction of ecosystem effects has enlisted support for studying more fundamental aspects of ecotoxicology and the development of conceptual and theoretical frameworks.

Clearly, the best way ahead for ecotoxicological scientists is to make use of the strengths of our field. Few sciences have at their disposal such a well-integrated input of effort from people trained in ecology, biology, toxicology, chemistry, engineering, statistics, etc. Nor have many subjects such overwhelming support from the general public regarding our major goal: environmental protection. Equally important, the practical requirements of ecotoxicological managers are not inconsistent with the aims of more academically-orientated ecotoxicologists. For example, how better to validate and improve current test procedures than by conducting parallel basic research programmes *in situ* to see if controls on chemical discharges really do protect biotic communities?

More broadly, where are the major ecotoxicological challenges likely to occur in the future? The World Commission on Environment and Development estimates that the world population will increase from *c.* 5 billion at present to 8.2 billion by 2025. 90% of this growth will occur in developing countries in subtropical and tropical Africa, Latin America and Asia. The introduction of chemical wastes into the environment in these regions is likely to escalate dramatically, if not due to increased industrial output, then due to the use of pesticides and fertilisers in agriculture and the disposal of damaged, unwanted or obsolete consumer goods supplied from industrialized countries. It may be many years before resources become available to implement effective waste-recycling programmes in countries with poorly developed infrastructures, constantly threatened by natural disasters and poverty. Furthermore, the fate, pathways and effects of chemicals in subtropical and tropical environments have barely begun to be addressed. Whether knowledge gained in temperate ecotoxicological studies is directly applicable in such regions remains to be seen.

The Chapman & Hall Ecotoxicology Series brings together expert opinion on the widest possible range of subjects within the field of ecotoxicology. The authors of the books have not only presented clear, authoritative accounts of their subject areas, but have also provided the reader with some insight into the relevance of their work in a broader perspective. The books are not intended to be comprehensive reviews, but rather accounts which contain the essential aspects of each topic for readers wanting a reliable introduction to a subject or an update in a specific field. Both conceptual and practical aspects are considered. The Series will be constantly added to and books revised to provide a truly contemporary view of ecotoxicology. I hope that the Series will prove valuable to students, academics, environmental managers, consultants, technicians, and others involved in ecotoxicological science throughout the world.

Michael Depledge
Odense, Denmark

Foreword

In recent years, public interest in environmental pollution has grown very rapidly, an interest that has been promoted by extensive coverage in the media. Such a high level of publicity has been something of a mixed blessing. On the one hand, it has heightened public awareness of important issues, and so has helped to generate political pressure to sponsor research. On the other hand, these issues have sometimes been presented in a sensational, simplistic and even misleading way, with unfortunate consequences. Issues of relative unimportance have sometimes been given considerable attention, while more important ones have been largely ignored. The dangers of such a tendency when it comes to the allocation of the very limited resources available for research are all too obvious. There is, inevitably, a pressure to do short term, superficial studies to meet immediate political needs, rather than take on longer-term and more expensive research that would deal with problems at a more fundamental level. This is unfortunate, because many pollution problems are only soluble in the longer term by a more fundamental approach. The definition of environmental problems, the study of effects of pollutants in the environment and the discovery of ways of solving them require scientific work of high calibre involving many disciplines.

In the context of increasing research effort in the field of pollution, a new discipline, 'Ecotoxicology', has emerged which is concerned with the study of harmful effects of chemicals upon ecosystems. The emphasis upon effects of chemicals on *ecosystems* is a feature which distinguishes it from classical toxicology. Changes in both population numbers and in genetic composition are relevant here. A particularly difficult problem is establishing the effects of chemicals in the environment, either upon individual organisms, or at the level of populations or communities. Although some progress has been made in relating levels of pollutants to population numbers, a complex situation exists in the field. Here organisms are exposed to many different pollutants and other stress factors, and it can be very difficult to establish whether correlations represent causal relationships. A different approach to the problem has been the development of models for risk assessment which include data from the laboratory (e.g. toxicity data and physical properties of chemicals). Until now this approach has been of very little value in predicting effects of chemicals at the level of individuals and populations. The complexity and variability of the environment place a serious limitation on what can be achieved by such means. The use of biomarkers is an exciting development in ecotoxicology which can resolve some of the problems outlined above. Biomarkers can provide sensitive and specific measures of exposure – and sometimes of toxic effects – using samples obtained from the field by non-destructive sampling procedures.

The ability to measure characteristic biochemical responses to a chemical in the environment can provide the evidence necessary to establish causality and so provide a link between exposure and effect. Furthermore, it is possible to relate biochemical response to consequent effects at the population level. The technology for doing this should not be difficult to acquire. Already great progress has been made in exploiting some of the new techniques of molecular biology and biochemistry to measure the effects of drugs and environmental chemicals on human beings. The production of simple diagnostic kits which can be used to measure the environmental impact of chemicals has obvious attractions. Unfortunately, despite its high promise, this area of research (like other fundamental aspects of ecotoxicology) has until now received little support or much attention in the media. There is a need for authoritative texts to provide a secure foundation upon which decisions can be based.

Dr David Peakall's book is, thus, extremely welcome and timely. He has had a long and distinguished research career in the field of wildlife toxicology, and a great deal of international experience in wildlife conservation and related pollution problems. Having been trained as a chemist, and being also a very knowledgeable ornithologist, he has been able to develop fundamental research in an ecological context. He is well known for his work on the effects of pollutants on birds, originally at Cornell University, and later at the Canadian Wildlife Research Center where he is Chief of Wildlife Toxicology. The present work gives an up-to-date account, in depth and breadth, of the use of biomarkers in ecotoxicology. It is logically organized and lucidly presented. It also has the advantage, all too rare these days, of being a well-integrated single-author work by a leading expert in the field. It should be of great value to people working in ecotoxicology. It should also provide valuable guidance in the development of ecotoxicological research in the years ahead.

Colin Walker,
School of Animal and Microbial Sciences
University of Reading, UK

Preface

This monograph is an attempt to look, in a practical way, at the 'biomarkers' available to assess the health of organisms in the environment. The definition of biomarker used here is that put forward by the National Academy of Sciences: 'A biomarker is a xenobiotically-induced variation in cellular or biochemical components or processes, structures, or functions that is measurable in a biological system or sample' (NRC, 1987). Typically these biomarkers are changes in the activity of enzymes or in the level of a specific biogenic compound. Events such as the formation of adducts by covalent linkage to DNA and rearrangements of DNA in sister chromatid exchange are also considered as biomarkers. Changes to biomarkers are caused by a wide variety of stresses, natural and man-made. The changes caused by natural stresses are considered briefly in order that changes caused by toxic chemicals can be accurately identified.

The study of biological effects caused by toxic chemicals is complementary to studies on the occurrence of these chemicals in the environment. The classical approach to establish the hazard of toxic chemicals to wildlife is to determine the amount of a chemical present and then compare that value with those found to do harm in experimental animals. The levels of each of the widely used organochlorines associated with avian mortality have been determined by workers at the US Fish and Wildlife Service. Thus we know that brain levels of DDT in the range 60–100 mg/kg (Stickel and Stickel, 1969) are associated with mortality and that this can be used to diagnose that death was caused by this particular pesticide. Furthermore, the ratio of lethal levels to those found in wild populations gives an indication of the safety margin that this population has with regard to this chemical. There are two major difficulties with this approach: the first is the large number of chemicals that can be involved, and the second and even more critical problem is that of mixtures of chemicals. In the case of industrial chemicals it is rare for one compound to be predominant. The persistent organochlorines, for example, have a strong tendency to cross-correlate. Pesticide spray programmes are one of the few cases where a specific chemical is predominant. The second major difficulty is that experimental studies are carried out only on a very limited range of species and the inter-species extrapolation involves a good deal of uncertainty. This classical approach to wildlife toxicology is considered further in Chapter 2. Lest I sound too negative, let me say that this approach has furnished the basic knowledge base for wildlife toxicology.

Inter-species variation is a major theme throughout this book. The number of species in the world has been variously estimated at from 2 to 5 million. Many of these are rare and poorly described, and, regrettably, appreciable numbers of them are passing into extinction. Nevertheless, the number of species that can be studied for toxicant effects is more than enough to cause great difficulty in making comparison between species. In this book I have largely confined myself to the vertebrates, feeling that I lack the expertise in both the ecology and the physiology outside this subphylum. My own research has been almost exclusively on the effects of toxic chemicals on birds, but I have attempted to include work on other classes of vertebrates, particularly mammals and fish. Work on birds and fish constitutes the largest database for wildlife toxicology, whereas in laboratory studies on the effects of toxicants, mammals predominate. One of the fundamental difficulties with mammalian toxicological studies is that the effects are usually considered solely in terms of the dose given to the animal. Since accurate dietary levels are rarely known under field conditions, effects seen have to be related to residues found in the animal.

The core of this book is a consideration of the major biomarkers that are available. At the beginning of each section there is a brief overview of the biochemistry and physiology of the system under consideration. Natural factors that influence the measurements such as age, sex, seasonal variations, etc. are then discussed briefly. The summary of experimental studies that follows is necessarily highly selective. The literature on this subject is so vast that it is impossible to collect it all, let alone read and digest it. Even a comprehensive review of reviews on mixed function oxidases would be a formidable undertaking. I have tried to select those papers which have the greatest interest from a wildlife viewpoint. Nevertheless in selecting from so large a field there is a considerable degree of arbitrariness. I am reminded of the words of the great naturalist Gilbert White written 200 years ago in his country parish of Selborne on his own observations, 'which are, I trust, true in the whole, though I do not pretend to say that they are perfectly void of mistakes, or that a more nice observer might not make many additions'. To those who feel that their work should have been included I tender my apologies.

Chapter 3 looks at biomarkers of the nervous system, the inhibition of acetylcholinesterase and other esterases, the catecholamines and finally the neurotransferases and other indicators of neurotoxicity. Reproduction is considered in Chapter 4, which looks first at field investigations, secondly at using embryos, thirdly on effects on the hormone systems, and finally at receptors. The techniques available for examining adduct formation of pollutants with DNA and the effects on DNA structure are considered in Chapter 5. The mixed function oxidases system, one of the key defences against xenobiotics, is discussed in Chapter 6. A series of other well-established biomarkers is examined in Chapter 7: the thyroid hormones and function, retinol, porphyrins and other effects on haem and finally serum enzymes. Effects of toxic chemicals on behaviour are looked at in Chapter 8 from the viewpoint of their relationship to parallel biochemical changes. The final chapter on specific biomarkers discusses the immune system.

One of the difficulties and conversely one of the great interests of this topic is the interrelationship between the different biomarkers. To the writer it causes real problems. Should the relationship between induction of cytochrome P450 system and retinol levels be considered under mixed function oxidases or under retinol? From the scientific viewpoint, these inter-relationships are both important and fascinating. Our knowledge of how closely interrelated many physiological processes are is expanding rapidly. In a commentary of the new biology of receptors, Lefkowitz *et al.* (1989) refer to the 'magnificent seven' superfamily of receptors. This highly conserved structure includes not only the receptors for the steroids, but also those for the structurally and functionally dissimilar thyroid hormones. In the field of toxicology, by far the most detailed studies have been carried out on the Ah receptor. One requirement for the study of a receptor is the identification of a specific inhibitor or antagonist. In the case of the Ah receptor this antagonist is 2,3,7,8-tetrachlorodibenzo-p-dioxin. The discovery by Poland and co-workers (Poland *et al.*, 1976) of highly specific binding of dioxin, a compound known to be one of the most toxic of environmental contaminants, to the Ah receptor and the evidence that this was the receptor for the induction of aryl hydrocarbon hydroxylase opened the door to the combination of toxicology and receptor biochemistry.

The thesis that is put forward here is that the systematic use of appropriate biomarkers would serve to indicate the state of health of wildlife living under any particular set of environmental conditions. A clinical approach, comparable to that used in human medicine, is proposed. The availability of a battery of suitable physiological and biochemical tests covering the major physiological systems of the organism would make it possible to judge which, if any, of these functions were abnormal. Establishing the relationship between an alteration in a biomarker and a specific health effect in the organism, or population of organisms, is the next step. In the case of some biomarkers, such relationships are already reasonably well established – for example, the inhibition of acetylcholinesterase (AChE) or aminolevulinic acid dehydratase (ALAD) – whereas for others the link to definable health effects is less well established. Nevertheless, in an approach analogous to that used in human clinical medicine, an observation that, in a given wildlife population, one or more biochemical/physiological indices were outside the normal range would be cause for concern. Indicators of change at the biochemical level would trigger investigations of possible consequences at the level of the whole organism or population, such as changes in behaviour, reproductive success, population size, or community structure. Thus, while accurate predictions of population change based solely on biochemical measurements are not possible at this time, such measurements will at the very least indicate where more in-depth studies should be undertaken. At best, they can provide an early warning of effects that, although not yet manifest, will likely occur unless mitigative steps are taken. Conversely, routine population surveys and ecological field studies can reveal changes that should subsequently be investigated at the biochemical or physiological level.

The concept of dose-response is a basic one in toxicology, but it is formulated for controlled laboratory studies in which an organism is exposed to a range of doses of a single chemical. In nature, organisms are rarely exposed to only a single chemical and are subjected to many other stresses that are normally absent in laboratory studies. Under these conditions it is unlikely that a single unique dose-response relationship exists that can relate a biomarker to the dose of a chemical. It is more likely that there is a family of dose-response relationships, each corresponding to a different combination of interacting chemical and environmental stresses.

One of the most difficult decisions facing environmental lawmakers today is 'how much clean-up is necessary'. It is a highly emotional issue, with a wide range of opinions from the most radical environmentalists to the most conservative industrialists. The pressures to do something are immense, as are potential costs. A decision to do too little may cause irreversible damage to the environment, while a decision to do more than is necessary may waste large sums of money which could have been better used elsewhere.

The first basic question that society has to come to grips with is, 'How much damage are we prepared to tolerate?' Are we going to take the view that not a sparrow shall fall, or are we content if conditions are such that, despite losses of individuals, the population survives? At one extreme there will not be an outcry if a few aquatic invertebrates die within a few metres of the end of the pipe. At the other end of the scale, an event such as the destruction of most of the biota of the Rhine and the recent environmental terrorism in the Persian Gulf resulted in a worldwide reaction. The problem is both geographical and species related. It is not only the scale, but also what species are involved, that concerns us. Our concerns increase from algae to mammals. In human medicine the section of the Hippocratic oath 'for the good of the sick to the utmost in your power' relieves the physician of the most difficult of decisions, when to give up. In protecting the health of the environment there are no such absolutes. 'How much damage are we prepared to tolerate?' is a question for society to answer. Admittedly the answer cannot be given in absolute terms, nor will there ever be unanimity on this issue. But in order for society to make reasonable environmental decisions there needs to be a broad consensus on what is an acceptable answer to the question. Once it is answered it will be possible to design protocols to meet the standards required. Unanswered, it leaves the regulators in an impossible position.

The second basic question is 'How much proof is enough?' There is considerable inertia in decision-making. Industry is understandably reluctant to alter its production methods or to install expensive clean-up equipment; researchers want to continue their research, and assessors are only too well aware of the gaps and inconsistencies of the data that could be exposed in a legal challenge. Nevertheless decisions must be made. No decision is itself a decision.

Pragmatically, the approach of using an appropriate suite of biomarkers in biomonitoring programmes would help to determine whether, for a particular species and location, the environment is healthy, and would enable us better to defend the advice we provide concerning environmental issues. Defendible positions cannot always be arrived at solely on the basis of contaminant residue data. The problems inherent in predicting or assessing the biological impacts of complex mixtures of chemicals, and of unknown or transient chemicals in the environment, may be avoided by incorporating biomarkers in the assessment process. I hope that this monograph will make a contribution to this process, but, in the words of Pooh (A.A. Milne, *The House at Pooh Corner*, Chapter 6), 'a Thing which seems very Thingish inside you is quite different when it gets into the open and has other people looking at it.'

Acknowledgements

The process of writing even a small book is a salutary one. Certainly it requires a lot of assistance and it is difficult to be sure that everyone has been included. To Professor Peter Peterson, who hosted me at the Monitoring Research and Assessment Centre at King's College, London, and to his helpful and patient staff I tender my grateful thanks. The reference list gives an idea of the amount of library work needed. I would like to thank Bev Kouri (National Wildlife Research Centre of Canada) and Joan Lovenach (King's College). To Colin Walker, University of Reading, who offered valuable advice (both on wildlife toxicology and wine) and friendship during my stay in England. To Peter Stanley, Ministry of Agriculture, Fish and Food, for making the facilities of his organization freely available to me. To Sean Kennedy, Pierre Mineau, Lee Shugart and Colin Walker for reviewing specific sections of the book. To Susan Peakall for her patient editing. To Wendy Scott for her advice and encouragement. To the members of the Toxicology Research Division of the Canadian Wildlife Service, Sean Kennedy and Tony Scheuhammer, who took over my duties in my absence. To Jim Learning for his cheerful assistance on all computer-related problems. Finally to Anthony Keith, Director of the Wildlife Toxicology and Surveys Division of CWS, for much more than making professional leave available to me.

1   *Scope and limitations of classical hazard assessment*

There is general agreement that hazard is a function of exposure and toxicity. In the simplest terms if there is no exposure there is no hazard and if there is no toxicity there is no hazard. To be accurate, there is no such thing as zero toxicity; as the 16th-century Swiss physician Bombastus von Hohenheim, better known by his boastful nickname Paracelsus, expressed it: 'sola dosis facet veninum'. Furthermore the advances of analytical chemistry have made zero exposure rare. Nevertheless, despite these limitations and despite the disagreement of what the exact function of the relationship is, the concept that hazard = f[exposure.toxicity] is a useful one. Here I look at the two parts of the equation, first singly and then at the combination.

I would like to suggest that the time has come to take a hard look at the proportion of resources used for measurement of residue levels in the environment compared to those used to study the effect of these, and other chemicals, on the environment.

1.1 The exposure side of the equation

Analytical chemistry has made and continues to make an immense contribution to environmental science. The demonstration of the biomagnification of DDE in ecosystems and the finding of PCBs in the environment, which was rapidly followed with evidence of the global nature of their distribution, are two early triumphs. More recently, one could mention the congener-specific analysis of PCBs, polychlorinated dibenzofurans (PCDFs) and polychlorinated dibenzodioxins (PCCDs), which have been essential to the development of wildlife toxicology.

Nevertheless, there are limitations. Although a wide variety of chemicals have been detected in environmental samples (nearly 500 compounds of 19 chemical classes in Great Lakes fish (Passino and Smith, 1987)), environmental chemistry is still virtually limited to two major classes of chemicals, the polyhalogenated aromatic hydrocarbons (PHAHs) and the heavy metals. Even within the PHAHs the identification of those compounds present may be far from complete. Wesén *et al.* (1990) in a note in *Ambio* state that of the solvent-extractable, organically bound chlorine compounds in cod, only 10–15% have been identified.

With heavy metals this problem does not arise. The analytical difficulties here are associated with problems of contamination of samples and the speciation of organically bonded metals. Background contamination is important in measurements of levels of metals in air and water that can be critical to our knowledge of the transport of these elements. It is less critical to those measurements associated with toxicological investigations since analysis at these extremely low levels is not required.

1.1.1 Surveys and monitoring programmes

It is important to be clear about the distinction between surveys and monitoring. One of the definitions of *survey* in the *Oxford Dictionary* is 'The act of viewing, examining or inspecting in detail for some specific purpose'. In the present context surveys can be defined as measurements over a geographic range and/or range of species within a limited period of time.

The same authority defines monitoring as 'To observe, supervise or keep under review, especially for the purpose of regulation or control'. For environmental work, monitoring can be defined as the repeated measurement in the same place and on the same substrate at fixed intervals of time.

Initially the need for surveys was clear. When the global nature of the pollution by PHAHs was realized in the mid- to late 1960s with the discovery of DDT and its metabolites in the fauna of Antarctica (Sladen *et al.*, 1966) and shortly afterwards of PCBs in sea birds from several remote areas (Risebrough *et al.*, 1968), there was an obvious need to map the degree of contamination so that key areas could be investigated for effects. The bioaccumulation of PHAHs in ecosystems was soon established (Woodwell *et al.*, 1967) as leading to harm to top predators, in some cases even those living in remote regions (Cade *et al.*, 1971). Surveys are a necessary precursor to monitoring, giving basic information on where and what to sample. Regrettably some major programmes – essentially glorified surveys – have been continued without the fixed-time and temporal conditions. The type of programme that I have in mind is the analysis by an institute of road-kills, beached sea birds or other readily available material. The output is usually in the form of tables of data, often in government reports of limited availability, from which little information on temporal or geographical trends can be obtained.

The basic concept of monitoring, the need to know, on an ongoing basis, the levels of xenobiotics in the environment is attractive and easy to grasp.

The next level of question, too rarely asked, is, 'Why do we need to know?' or, to put it in a more practical form, 'What are we going to do with the information when obtained?'

The rationale for wildlife toxicology can be divided into three broad categories. These are: situations where there is reason to believe that the health of wildlife is or may be endangered; where there are concerns for the health of humans consuming wildlife; and when such studies are of use in determining environmental quality. While these categories are not mutually exclusive, and ideally a single programme will serve more than one purpose, it is convenient to discuss these broad categories in turn.

1.1.2 Wildlife health considerations

An example of the first category is the measurement of PHAHs in the eggs of peregrine falcons. The residue levels in eggs were soon related to reproductive success. This was important evidence of a widespread effect of the persistent organochlorine pesticides which was occurring far from the initial point of use. The banning of these materials in much of the Northern Hemisphere has resulted in improvement in the populations in many areas (for a comprehensive review see Cade *et al.*, 1988). Currently concerns are limited to a few specific populations. A viable population has not yet been established in eastern Canada and it is necessary to check on the residue levels of potential prey in the selection of release sites and to follow this up with measurements on any infertile eggs laid. In California measurements of eggshell thickness and residue levels in membranes of eggs has been used to decide which eggs are brought into the laboratory to be incubated artificially; after hatching, the young are returned to the wild. At present the coastal population in California is maintained only by such manipulations. The work on the peregrine falcon, which can be considered as one of the classic studies in wildlife environmental studies, was a reactive response to a population decline that occurred over almost the whole Holarctic (Hickey, 1969).

There is a need to be more proactive to the possible effect of toxic chemicals on endangered species. One approach is to map pesticide use and major industrial activity of concern and overlay this information with the ranges of species of concern. An example of this is the overlap of the range of burrowing owl with agricultural land in the Canadian prairies. The densities of burrowing owls on farmland, where pesticide use is heavy, is much greater than on range land. Investigations showed that the use of one particular pesticide, carbofuran, was causing considerable mortality of this rare species (Fox *et al.*, 1989). In other cases, where the likelihood of pollutants being important is low, there is still a case for a small, one-time survey to establish that this indeed is the case. As a specific example, addled eggs of the piping plover from the upper Great Lakes have been examined and found to have low levels of PHAHs. At this point it was possible to close the books. In most cases a holistic approach is needed for endangered species, with one facet, albeit often a small facet, being the effects of toxic chemicals.

1.1.3 Human health considerations

In general, human health concerns are outside the scope of this book. One point of interaction occurs when wildlife is used as a significant source of food for humans. By far the largest monitoring programmes are those on fish. Safety levels for the major PHAHs and mercury have been set by both national and international agencies and fish is routinely analysed. Failure to meet these set levels can lead to banning of commercial sale and to the issuing of advisories on sport fishing. That these programmes are still needed is demonstrated in the Great Lakes by the act that fish advisories are still in effect in some areas in all of the lakes. Consumption advisories affect 22 species of fish from Lake Ontario, and even in Lake Superior some species tested above limits at all sites (Rathke and McRae, 1989). Since considerable resources are required to run this type of programme, it is important that the data generated be used as widely as possible.

Data on fish and other lower trophic level organisms with that in gulls for PCBs and mercury are compared in Figure 1.1. The levels of PCBs increase from 0.01 ppm in plankton to 50 ppm in herring gulls, giving a concentration factor of 5000. Since the concentrations in water are in the low parts per trillion range the total biomagnification factor is about 25 million. The equivalent figures for mercury for plankton and herring gulls are 0.004 and 0.540 ppm, and the total biomagnification is about 1000. Another example of cooperation is the data generated on the levels of mercury in fish by the Ontario Ministry of the Environment. These data, which are generated for human health concerns, have been used in investigations of the release of heavy metals occurring when lakes are acidified by acid rain (Scheider *et al.*, 1979 and McMurtry *et al.*, 1989).

Other major programmes are focused on the levels of organic compounds in water. These are defended on the grounds of the need to know that water supplies are safe. The scientific basis for relying on analysis of water for this purpose is weak. The concentration factors of PHAHs from water to aquatic invertebrates are so large that the safety levels of PHAHs in water are usually below detection levels. The statistical handling of data containing a large number of observations where the chemical is below the detection level makes meaningful decisions on water quality virtually impossible. A more reliable method is to use surrogates, for example, caged invertebrates such as mussels, which should be capable of producing better data at much lower cost. A filter feeder has already done the concentrating work for us. Techniques for the use of caged systems have been worked out. Young *et al.* (1976) demonstrated the bioaccumulation of PHAHs in caged mussels

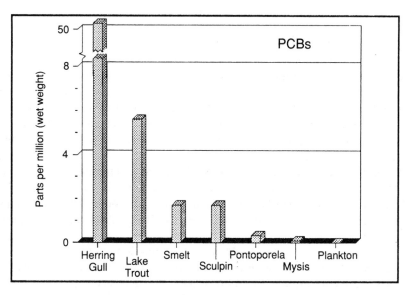

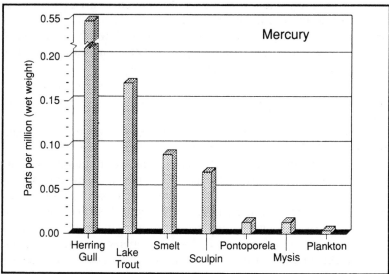

Figure 1.1 Biomagnification of PCBs and mercury in Lake Ontario food chains. After Environment Canada (1991).

in contaminated areas of the coast of California. The levels in the mussels followed the general pattern of contamination, although exact water concentrations were not determined. Davies and Pirie (1978) demonstrated that measurements on caged mussels accurately reflect the mercury concentration in the surrounding water in the Firth of Forth. The use of such bioindicators has the additional advantage of providing information on the degree of bioavailability rather than the total amount of the pollutant. The capacity of these bioindicators to average out pollution levels is an important advantage, but the degree to which they do so varies greatly with the pollutant being studied. Half-lives of metals in molluscs are generally long (cf. 96–190 days for cadmium, Borchardt (1983)) whereas those of PAHs are much shorter (cf. 25–50 hours for naphthalene, Widdows *et al.*, 1983).

There are other areas of concern. The levels of contaminants in animals killed by hunters – cadmium in deer, and PHAHs and mercury in ducks, for example – are often higher than those normally found in commercial foods. These concerns are of particular importance for those sub-populations, often native peoples for whom these animals form a major proportion of their diet.

In addition there are more specialized subjects. A good example is the contamination of the food-chain in the Arctic. In recent years it has been shown that the long-range transport of pollutants into the Arctic is a serious problem (Norstrom *et al.*, 1988). The bioaccumulation through the food-chain has led to Inuit in the high Arctic having levels of PCBs four to five times those of the general population of Canada (Dewailly *et al.*, 1989). In this case the evidence from surveys shows that a monitoring programme should be put into place.

1.1.4 Environmental quality

Monitoring of pollutant levels in wildlife, even when no adverse effect on the wildlife itself is known or suspected, can be used as an index of environmental quality. One monitoring programme, initially set up at the time when marked adverse effects were seen and continued subsequently as a monitor of environmental quality, is the Great Lakes Herring Gull Egg Program. The systematic use of the herring gull goes back to 1974 and currently consists of annual sampling at 13 colonies throughout the Great Lakes (see review by Mineau *et al.*, 1984). The temporal trend data for the levels of DDE for eight of the monitoring sites throughout the Great Lakes are shown in Figure 1.2. At most sites there has been a sharp decline from the levels found in 1978–1980 to those found in the rest of the 1980s. At some sites, notably those on Lakes Erie and Ontario, the levels throughout the 1980s have been remarkably constant. At others, those in Lakes Michigan and Superior, levels have decreased over this period. With other organochlorines, PCBs, mirex, HCB and TCDD, similar trends have been seen.

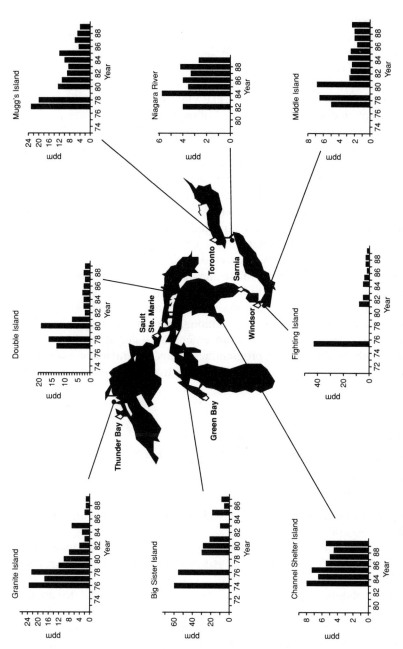

Source: Canadian Wildlife Service, Environment Canada. The scale for each bar chart is different.

Figure 1.2 Temporal trends of DDE in Great Lakes herring gull eggs.

The pattern of the changes, with the exception of dieldrin, is similar, although there are considerable differences in the absolute values. Rapid decreases occurred at most sites, following the bans and restrictions put into place on these compounds in the mid-1970s. These decreases terminated in the early 1980s and since then levels have remained essentially constant (Environment Canada, 1991).

The temporal trends for four organochlorines, DDE, PCBs, dieldrin and TCDD in lake trout collected from Lake Ontario are given in Figure 1.3. The pattern of changes is less marked. Although the highest values for DDE are in the late 1970s, there has been little reduction. With PCBs, there was a sharp and significant decrease between 1984 and 1985, followed by essentially constant values since that time. The highest values for dieldrin were also in 1978 and 1979, but there has been little change since. No distinct trend can be seen in the data for TCDD.

These monitoring programmes enable us to quantify the degree of success of legislative action. It resulted, initially, in a substantial decrease, but it is also now becoming quite clear that a new equilibrium is being reached. From the viewpoint of wildlife toxicology this means that we are now looking at a chronic situation, where the levels of contaminants are still substantial even though they are below the levels that caused widespread reproductive failure of fish-eating birds in the early to mid-1970s. From a regulatory point of view it means that although we have obtained substantial benefit from earlier actions, further improvements are going to require different actions. Another value of these programmes is that they maintain a watching brief on the possibility of increased leaching from toxic chemical dumps around the Great Lakes.

1.1.5 Design of monitoring programmes

Selection of indicator species The criteria for selection of indicator species were first discussed by Moore (1966) and his criteria, given below, have stood the test of time.

1. They should be easily available, i.e. widely distributed, relatively abundant and easy to collect.
2. Analysis can be made of single animals or of bulked samples. If monitoring is to be based on the chemical analysis of organs of single animals, each animal should be sufficiently large.
3. Ideally, it should be possible to ascertain the age of the indicator animal by inspection. To indicate contamination levels occurring just before collection, young or short-lived animals are necessary; where one needs to know about accumulative build-up of residues, long-lived animals are required.
4. Species which, in a preliminary survey, show relatively high contamination

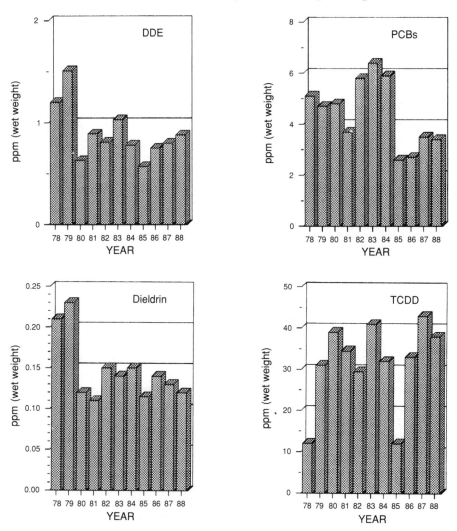

Figure 1.3 Temporal trends of PCBs, TCDD, DDE and dieldrin in lake trout.
After Environment Canada (1991).

are most suitable. Species which contain very high residue levels are
unsuitable indicator organisms since an increase in environmental contamination is likely to have toxicological effects which will affect the
significance of the samples. On the other hand, species with low levels
are also unsuitable since the identification of residues is less reliable and
because a decline in environmental contamination may result in levels
being too low to record at all.
5. If measurements of local change are required, indicator species must be

sedentary in their habits. On the other hand, where changes in the contamination of a large area are to be measured with limited resources, species with extensive ranges may give a more reliable guide to the general situation. For either purpose the range of the indicator species must be known.

One point that needs to be made clear is that the selection of an indicator species should be made in the context of the problem to be examined. The idea of a single indicator species for an entire country or other geographical unit is appealing, but unrealistic. One reason for the necessity of different indicator species is clearly stated in Moore's last criterion. Another is the type of chemical problem being investigated. For stable, bioaccumulating chemicals, studies of species at the top of food chains have advantages. For the impact of acid rain, studying invertebrates low in the food web may be more appropriate.

Coefficients of variance, sample size   To design a monitoring programme effectively, one should be able to answer the question, 'What does one wish to detect by the programme?' in mathematical terms. If the question can be answered in such terms as 'to measure a 25% change of a specific chemical with 95% confidence', then it is possible, if the distribution of values is known from a preliminary experiment, to calculate the necessary sample size.

There are different sources of variation: those caused by variation of the analytical method itself (analytical variance), and those caused by the variation of the levels in the samples (sample variance). The first source of variation can be calculated from the results of repetitive analysis of the same sample. Gilbertson *et al.* (1987) give figures for this variance based on analysis of a pool of Lake Huron herring gull eggs which was used for quality assurance in the Great Lakes Program. The values obtained for DDE, dieldrin, HCB and PCBs were, respectively, 10.4%, 11.9%, 22.6% and 15.3%. This analytical variance gives, for this method in this laboratory, the minimum degree of variation that can be obtained.

Using this analytical variance Gilbertson and co-workers calculated the sample size estimated to detect a 20% change in residue levels at an 80% confidence level for the data on the Canadian Wildlife Service East Coast Sea Bird Monitoring Program. The sample size needed in the case of puffin eggs was calculated to range from 13 for dieldrin to 34 for HCB.

One strategy for reducing the number of analyses necessary to achieve the requirements of the programme is to use pooled samples. This approach may be used to study either samples from the same or closely related sites, or samples collected over a period of time from the same site. The underlying assumptions are that there are no restrictions on the number of samples that can be collected, that the collecting of additional samples does not appreciably increase collection costs and that analytical costs are high in comparison to

Table 1.1 Comparison of relative efficiencies of pooling strategies for residue analysis

No. of mixtures = No. of analysis	No. of eggs per mixture	Total no. of eggs	Relative efficiency %
10	1	10	100
30	1	30	309
10	3	30	283
8	3	24	213
7	4	28	214
6	5	30	242
5	5	25	198
5	1	5	40

After Gilbertson *et al.* (1987).

collecting costs. Under these conditions, which are usually met, the use of pooled samples is an effective strategy. By using a number of pools, one is able to obtain an index of variation at the same time as one increases the degree to which the environment is sampled. Calculations of the increase in statistical power obtained, based on data of DDE levels in storm petrel eggs collected in New Brunswick, are given in Table 1.1. There are, of course, conservation considerations. In the case of the sea birds cited above, the colonies are large, from thousands to, in some cases, hundreds of thousands of pairs. The fraction of egg-producing young that survive to adulthood is small, and eggs collected early in the breeding season are usually replaced. However, the greatest waste of all is to have a monitoring programme that does not provide the answers required.

The coefficient of variation for DDE and PCBs in fish is much higher than those recorded in birds. Based on the work of Zitko *et al.* (1974) for herring and yellow perch and Skåre *et al.* (1985) on flounder and cod, Gilbertson *et al.* (1987) calculate coefficients of variation from 57 to 108% for DDE and from 78 to 500% for PCBs. These figures show that larger sample sizes (analysis of a series of pools is a possible method without increasing analytical costs) and tightening of the design of the monitoring programme (closer control over the size and lipid content of fish used, for example) would be necessary to permit detection of anything but the grossest changes in levels. Collection of large sample sizes is unlikely to be a problem since commercially caught specimens are generally used. Analytical costs are another matter.

1.2 The toxicity side of the equation

1.2.1 Broad studies of lethal levels

The broadest survey published is surely that of Kenaga (1981). Using the screening data compiled by the Dow Chemical Company, he compared the

Table 1.2 Comparison of the range of lethal concentrations of chemicals to plants and animals

Organism	Sample size	% of chemicals causing 100% mortality	
		0.01–0.09 ppm	0.1–0.99 ppm
Daphnia	33 909	0.6	2.4
Composite of fish species	35 305	0.14	1.3
Algae	49 082	0.006	0.02
Composition of aquatic plants	27 781	0	0.1
Composition of terrestrial plants	131 596	0.006	0.17

comparative toxicity of 131 596 chemicals to five different species of terrestrial plant seeds and then extended that comparison to daphnia, fish and aquatic plants. Within the terrestrial plants soybean was generally the most sensitive, although no one or two species were a good indicator of toxicity for the other species. The comparison of the range of lethal concentrations of chemicals to various plants and animals from Kenaga's paper is given in Table 1.2. Detailed comparisons cannot be made from such a broad survey, but it does point out the relative sensitivity of daphnia and the vast amount of data collected by industry which is not readily available.

One of the standard toxicity tests, now falling somewhat into disrepute, is the LD_{50} test (section 9.2). A comparison of the LD_{50} values for the rat and starling for a wide range of chemicals based on the work of Schafer (1972) is shown diagrammatically in Figure 1.4. It will be seen that in a great majority of cases the starling is more sensitive than the rat; in the case of 5 chemicals out of 71 this sensitivity is more than 100-fold, with a maximum value of 500. The values for two closely related species of birds, the starling and the red-winged blackbird, are shown in Figure 1.5. Again there is clear skewing of the data, with the red-winged blackbird being appreciably more sensitive than the starling. Here the spread of the data is considerably less, with most of the ratios between within one and all falling between two orders of magnitude.

In contrast, Tucker and Haegele (1971), measuring the LD_{50} of six avian species to 16 pesticides, found that although the species varied widely in their sensitivity to specific compounds, the average sensitivity of any one species to the 16 pesticides was not statistically different from any other. Three of the species used were closely related, and the other three much more distantly, but these studies did not include either the starling or the red-winged blackbird. Examination of the toxicological data by pairs of species failed to show any phylogenetic relationships. Schafer and Brunton (1979) expanded their work to cover six avian species and 36 pesticides. The list of species used had three in common with those used by Tucker

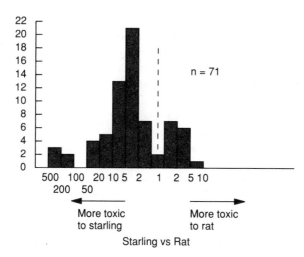

Figure 1.4 Comparison of LD$_{50}$s; rat to starling. After Schafer (1972).

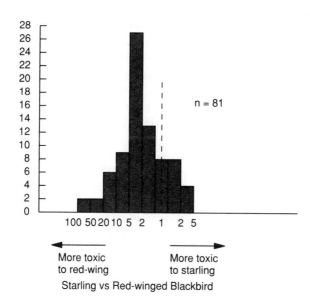

Figure 1.5 Comparison of LD$_{50}$s; starling to red-winged blackbird. After Schafer (1972).

and Haegele. Schafer and Brunton found statistical differences, but these are dependent on the marked insensitivity of the starling.

A comparison of the data for the mallard and bullfrog is shown in Figure 1.6. This inter-order comparison showed more differences than the intra-order comparisons already discussed. In most cases the bullfrog was much less sensitive than the mallard. Particularly striking is the insensitivity of the bullfrog to organophosphate and carbamate pesticides. Indeed this is more striking than an examination of Figure 1.6 would indicate since several of these pesticides with greater than 2000 or 4000 mg/kg values have been used as actual LC_{50} values are not available.

The wide extrapolation from *Daphnia magnia* to the rat was made by Crosby *et al.* (1966). They concluded that while anatomical, physiological and biochemical differences preclude any simple, direct interspecies relation of responses, 'among the many compounds for which daphnia toxicity data now exists, no instance has been observed in which a substance showing significant acute toxicity to mammals has failed to be highly toxic to the aquatic animal also'. The basis for this statement is unclear. In their paper data are given for 14 pesticides, but their statement implies a wider database. The converse, however, was not true. There are several substances highly toxic to daphnia that have low toxicity to mammals.

The toxicity of 19 widely different chemicals to 13 algal species was examined by Blanck *et al.* (1984). The inter-species variation in sensitivity varied over three orders of magnitude, but no particularly sensitive or in-sensitive species could be identified – exactly the same finding as with avian species if one excludes the Rasputin-like starling. Examining algal test batteries, Blanck *et al.* (1984) found that a predictive value within two orders of magnitude with 95% confidence required a three-membered test battery and that if a 99% confidence level was required the number had to be increased to five.

All the studies examined so far are related solely to dose, but the time factor of exposure is also important. The simplest extrapolation of the acute LD_{50} test is to increase the time. A plot of survival time against concentration of DDT for rainbow trout is shown in Figure 1.7. This type of plot can be extrapolated to infinite time to give a minimum value for chronic lethal exposure. Naturally this value assumes that the concentration of the pesticide remains constant over this period, a condition that will never, well, hardly ever, occur in the real world.

1.2.2 Tissue levels associated with mortality or indices of harm

One of the major difficulties in relating laboratory experiments to environmental conditions is that laboratory experiments usually give only the dosage employed, whereas in environmental studies only the final tissue residue level is available. While, of course, tissue levels can be measured in laboratory

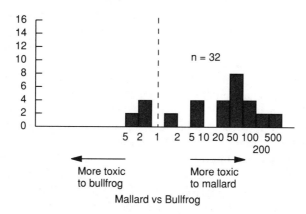

Figure 1.6 Comparison of LD_{50}s; bullfrog to mallard. After Tucher and Haegele (1971).

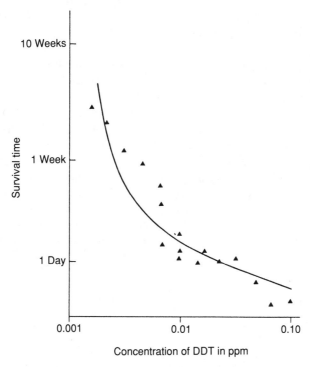

Figure 1.7 Survival time against concentration of DDT for rainbow trout.

experiments, this is done only in a small minority of cases. For hazard assessment to humans the approach of using dosage only is entirely reasonable. Market basket surveys of contaminants in food and measurements of concentrations in air and water are the main basis for calculating human exposure to chemicals. Only in some isolated cases, lead levels in blood for example, are residue levels used. In contrast, only in special cases, such as observing the rate of feeding on granular pesticides, is it possible to calculate for wildlife the dietary intake of a contaminant.

One limitation of the residue level approach is that one does not know under what conditions the residue was acquired. A tissue level of x mg/kg may result from a long-term exposure to low concentrations, caused by a single large exposure some time ago followed by a period of elimination, or any combination of fluctuating exposure regimes. The same type of difficulty exists for any physiological measurement: an abnormal enzyme level may represent a steady-state situation, but it could equally represent either the initial or recovery phase of exposure. Usually, diagnostic work in the field is based on residue levels. A series of papers by the Stickels and their co-workers (Stickel and Stickel, 1969; Heinz and Johnson, 1981; Stickel *et al.*, 1984) at the Patuxent Wildlife Research Center established that the levels associated with mortality have been determined for the common persistent PHAHs.

Criteria documents, such as those developed by the World Health Organization, have given safety levels for such pollutants as mercury, cadmium and lead. These are based largely on dietary intakes. In fact, in human medicine the only critical organ values that have been proposed seem to be those for the cadmium concentration in the renal cortex (Foulkes, 1990). There are a number of problems with identifying the critical concentration of a chemical in the target organ. For example, if the target is the nervous system the effects may occur at a much later stage than the initial exposure that causes the damage. In the case of a toxicant that affects the immune system there is the problem of the diffuse nature of the target. In a recent review of the concept of critical levels of heavy metals in target tissues, Foulkes (1990) looks at the relationship of dose to tissue levels of rats dosed with lead. It is clearly shown that the level in organs does not increase in a linear fashion with the dose administered and that the relative concentrations in different organs varies with dose.

1.3 Hazard assessment

Initially hazard assessment was relatively easy to make. Following the spraying of elm trees with DDT in New England, American robins were found dead (Wurster *et al.*, 1965). The residues in their brains were compared with those in other passerine species which had died after experimental exposure and found to be similar (Stickel and Stickel, 1969). Once concern was extended to pollution over wider areas and to industrial chemicals in addition to pesticides, the problems rapidly increased. Briefly stated, the problems are:

1. The increase in the number of species that are involved.
2. The increase in the number of chemicals involved.
3. The fact that mixtures of chemicals, of varying composition and concentration, are involved.

In most real-world studies we are dealing with mixtures. Most studies of the effects of mixtures of chemicals have been carried out by treating the mixture as a single entity. The mixture, for instance, an industrial effluent, river water, or crude oil, is put into the test system and the result observed. The mixtures can be extremely complex. Chemical problems are nothing new on the Rhine. A hundred and sixty years ago Coleridge wrote,

> I counted two and seventy stenches,
> All well defined, and several stinks!
> Ye Nymphs that reign o'er sewers and sinks,
> The river Rhine, it is well known,
> Doth wash your city of Cologne;
> But tell me, Nymphs, what power divine
> Shall henceforth wash the river Rhine?

In recent years, water from the Rhine has caused elevated occurrence of chromosomal aberrations (Prein *et al.*, 1978), chromosome breakage (Koss *et al.*, 1986), tumours and porphyrin levels (Slooff and De Zwart, 1983) in fish compared to those from cleaner waters. These studies are valuable since they give information on the effects caused by the complex mixture of chemicals in the river. However, nothing is learned about the effects of the various components of the mixtures.

Two major approaches for elucidating the effects of individual chemicals in a mixture are by fractionation or combination. The mixture can be fractionated, the fractions tested, and the results compared with the effects of the original mixture. Combinations can be made of specific chemicals to mimic the mixture. Both approaches can be expensive, and in both the mixture would need to be analysed by such techniques as gas chromatography/mass spectrometry. Even with this approach there is no proof that all compounds would be identified. Using the first approach, the mixture would need to be fractionated, re-analysed, tested and so on. In the second, while the analytical back-up is not so extensive, the number of combinations of chemicals can rapidly become impractical. Nevertheless, there are situations which call for detailed investigations. These include the following:

1. Areas in which significant adverse effects are seen that are contaminated by multiple sources, and it is necessary to pinpoint which specific sources are the major contributors;

2. When the knowledge of which components of an effluent mixture are causing the problem may enable a clean-up procedure or reformulation to solve the problem rather than invoking a complete ban;
3. When the knowledge of interactions may enable synergistic reactions to be predicted. In some cases this has been done deliberately to enhance the toxicity of pesticides. In these cases the underlying mechanism is based on a knowledge of the mixed function oxidase system (Wilkinson, 1968). Detoxification occurs in two steps; first modification of the molecule to render it more polar, followed by conjugation so that the material can be excreted (section 5.3). Slowing either process can enhance toxicity. A recent example is the enhancement of malathion toxicity to the red-legged partridge by fungicide prochlorax (Johnston *et al.*, 1989). This amplification of malathion toxicity is attributed to an increased activation of malathion to its active metabolite, malaoxon.

Several years ago a panel concluded, 'Simply stated, our current understanding of the fundamental scientific principles of chemical interactions and methodology is insufficiently developed to permit the required assessments' (Stich *et al.*, 1982). Major advances have not been made since that time. A less basic approach is to use tests that can be carried out rapidly enough to make it feasible to test a large number of environmentally relevant chemicals. This technique has been used for a wide range of chemicals found in the Great Lakes (Passino and Smith, 1987). These workers provided a hazard ranking for 19 classes of chemicals representing many of the 500 compounds identified in trout from the Great Lakes, using a combination of acute toxicity testing with daphnia combined with the levels found in the fish.

In writing a book, it is hard to resist the temptation of quoting oneself. Discussing the problems of extrapolations of single-species studies to population, communities and ecosystems, my long-time friend and colleague of the US Environmental Protection Agency, Dick Tucker, and I concluded: 'Hazard assessment has a degree of imprecision that scientists dislike and that calls for tough political decision which politicians dislike. Nevertheless, the subject has to be tackled since the refusal of scientists and politicians to handle the problem can only lead to major environmental problems' (Peakall and Tucker, 1985). 'Si non fa, non falla' does not apply.

2   *Biomarkers of the nervous system*

Two major classes of pesticides, the organophosphates (OPs) and carbamates, act directly on the nervous system via the inhibition of esterases. The most important of these is acetylcholinesterase (AChE) on which many studies, from detailed molecular biology to large-scale field studies, have been carried out. While the effects of OPs via cholinesterase inhibition are transient, there are longer-term neurological effects mediated through the neurotoxic esterases (NTE). The effects of pollutants on biogenic amines, which act as neurotransmitters to the autonomic nervous system, have also been investigated, but here the focus has been on PHAHs and heavy metals.

2.1 Esterase inhibition

2.1.1 Classification of esterases

The initial attempt at classification was made by Aldridge (1953), who divided them into 'A'- and 'B'-esterases. A-type esterases are not inhibited by diethyl p-nitrophenyl phosphate and hydrolyse p-nitrophenyl acetate at a higher rate than p-nitrophenyl butyrate. The B-type are readily inhibited by diethyl p-nitrophenyl phosphate and hydrolyse p-nitrophenyl butyrate at the same or higher rate than p-nitrophenyl acetate. Thus, for the 'B'-esterases, which include the cholinesterases (acetyl and butyryl), neurotoxic esterases (NTE), and carboxylesterases, dephosphorylation is slow, and they are readily inhibited by OPs. In contrast, with A-esterases the regeneration to the free enzyme is rapid. The basic classification is diagrammed in Figure 2.1.

The position is made more complex by the fact that a number of enzymes that hydrolyse esters can also hydrolyse non-ester bonds as well. A good deal of the difficulty in the classification of esterases lies in their purification, or rather the lack thereof. Attempts in a number of laboratories to solubilize and purify NTE have been discouraging (Johnson, 1990). The carboxyesterases also have not yet been completely characterized, (Walker and Mackness, 1983) but have the character of a 'B'-esterase.

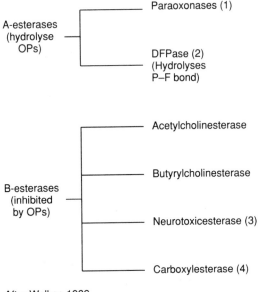

After Walker, 1989.
(1) Mackness and Walker, 1988
(2) Cohen and Warringa, 1957
(3) Johnson, 1975
(4) Mentlein *et al.*, 1987.

Figure 2.1 Classification of esterases.

From the viewpoint of environmental toxicology the two most important esterases are AChE and NTE. Butyrylcholinesterase, the precise physiological role of which is unknown (although it is regarded as a marker enzyme for glial or supportive cells and other non-neuronal elements) has sometimes been studied in parallel with AChE.

2.1.2 Mode of action

Acetylcholinesterase The basic mechanism of the toxic action of the organophosphate and carbamate insecticides is inhibition of the AChE activity of nervous tissue. This causes an accumulation of acetylcholine at the nerve synapses and disruption of nerve function. The outline of this process is shown in Figure 2.2. Recovery of AChE activity in organisms that survive the acute effects is dependent on the spontaneous, but slow, dephosphorylation of the inhibited site and on the synthesis of new AChE. Mechanistic differences, which explain the marked differences in species sensitivity to OPs, have been found by Kemp and Wallace (1990). They compared the kinetic constants of inhibition of rat, chicken and trout brain acetylcholinesterase and found that trout brain AChE possessed both less steric tolerance and weaker nucleophile binding than the other two species.

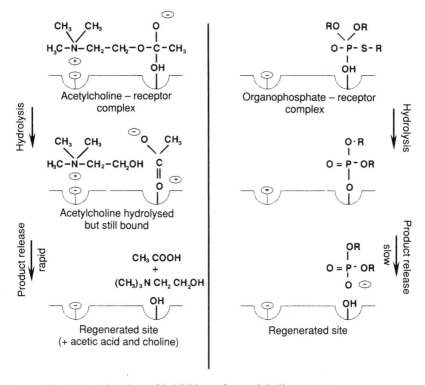

Figure 2.2 Mode of action of inhibition of acetylcholinesterase.

Neurotoxic esterase Neurotoxic esterase, or neuropathy target esterase, referred to in either case as NTE, has been extensively studied by Johnson and co-workers (major reviews are Johnson, 1975, 1982 and 1990).

The basic outline of the process of interaction of OPs with NTE which can lead to organophosphorous-compound-induced delayed neurotoxicity (OPIDN) is outlined in Figure 2.3. Although it is assumed that the NTE protein, which is tightly bound to neuronal membranes, does serve a physiological function, the normal role of NTE is unknown. The relationship of binding of OPs to NTE leading to OPIDN is discussed in section 2.1.3.

2.1.3 Factors affecting activity

Species variation The activity of brain AChE and A-esterases (those hydrolysing nitrophenyl acetate) was determined in a considerable number of mammalian and avian species by Westlake *et al.* (1983). While a wide range of wildlife species was examined – 17 mammalian and 47 avian – the sample sizes for each species were small, usually only one to five individuals. The mammalian species means ranged from 4.75 to 27.5 mm/min/g for AChE, with most species lying within the range 10–12. Species showing

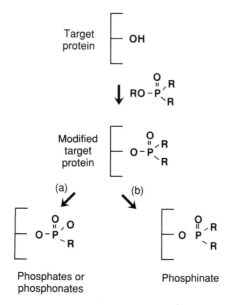

Figure 2.3 Schematic representation of the interaction of OPs with NTE.

low levels were cat, fox and polecat ferret, and those showing high levels were mole and grey squirrel. The levels of A-esterase were similar to those of AChE for most species; the two outliers, the fox and roe deer, are only represented by a single individual. Among avian species the activities varied from 15 to 57 units. The enzyme activity in all species of ducks and geese studied was at the low end; that of the chicken was high in contrast to other members of the Phasianidae. In all three species of Alcidae the activity was high and in all nine members of the Passeriformes studied the activity was towards the high end of the scale. Hill (1988) has given values for 48 species of wild birds from North America. In this study the sample size for each individual species was larger, mainly in the range of 8 to 20. Overall, Hill found a threefold variation. The distribution was similar to that reported by Westlake *et al.* (1983) with low values for duck and geese and high values among the Passeriformes.

Both the A-esterase activity and the aryl-esterase activity were measured in the blood of 11 species of mammals representing 5 different orders, and 14 species of birds representing 7 different orders by Mackness *et al.* (1987). In the mammals examined the highest values of aryl-esterase were reported for the badger and pig and the lowest for the rabbit and feral cat. Among the avian species the highest values of aryl-esterase were found in swan and goose; the three Passeriformes were also high, while the activity in quail and chicken was at the low end. The three species of Alcidae represented were all towards the lower end. The overall range of activity was similar for the two classes of organisms. However, as with the studies of Westlake *et al.* (1983), the sample size for each species was small (one to seven). The most significant finding was that, in contrast to the mammals, the avian species examined showed an almost complete lack of any A-esterase activity.

Mineau (1991) analysed the inter-species data sets available for the toxicity of organophosphates and carbamates to wild bird species. Considerable data sets exist for the following ten species of five families: e.g. Anatidae (mallard), Phasianidae (quail, pheasant and chukar), Columbidae (rock dove), Icteridae (red-winged blackbird, starling, and grackle), and Ploceidae (house sparrow and red-billed quelea). To attempt to reduce the variability, definable groups of chemicals – phosphates, phosphorothioates and methyl carbamates – were used and the data sets analysed, using the observed LD_{50}s or their ranking. The among-species covariance matrix was calculated and the first two principal components were derived. In almost all cases there was no similarity of response between closely related species. The only case of similarity was between the three Phasianidae for the methyl carbamates, but no similarities were seen with this group of chemicals for either the Icteridae or the Ploceidae.

Kinetic studies by Wang and Murphy (1982), covering four classes of animals (mammals, birds, fish and amphibians), suggest that sensitivity to OPs is a combination of binding affinity and rate of phosphorylation of the enzyme. These workers found a good correlation between the observed *in vitro* concentration required for 50% inhibition and that calculated from the kinetic data.

The species variation of NTE has been summarized by Johnson (1975). Phylogenetic comparisons are made difficult by the fact that some measurements of activity were made on specific parts of the brain (frontal cortex, spinal cord, cerebellum) and others on the whole brain. Using the limited data set available on the whole brain, the ranking is mammals $\geq$ fish > amphibians. No activity was found in the two crustaceans studied (lobster and crayfish). Another important criterion is whether or not a single dose causes neuropathy. This variation in sensitivity has been summarized by Johnson (1982). Man and some other primates are sensitive to a single dose, which is also true for the chicken, pheasant and duck. In contrast the partridge and quail have not been found sensitive to a single dose in some cases. Among the other mammals rodents seem particularly resistant. Francis *et al.* (1983) found that a range of OPs and carbamates failed to cause OPIDN in several strains of mice. In this respect the effect on OPs on NTE is quite different from that on AChE. The test animal of choice for NTE is the mature chicken.

Age and Sex   Grue *et al.* (1981) found a strong linear relationship between brain ChE activity and age of nestling starlings. The activity increased from 5 to 15 µmol/min/g from 4 to 18 days of age, by which time they were 70%

of adult values. In a subsequent study the activity was studied in fledglings up to 60 days of age (Grue and Hunter, 1984). The value was close to adult levels by 30 days. A linear increase of brain AChE activity has also been demonstrated in three species of the Ardeidae – great and snowy egrets and black-crowned night herons (Custer and Ohlendorf, 1989). The most comprehensive data set, that for the snowy egret, showed a doubling of AChE activity from the time of hatching to day 16. In contrast to the altricial species, those precocial birds studied, mallard (Hoffman and Eastin, 1981) and laughing gull (White *et al.*, 1979), show that the increase of AChE activity occurs largely during the embryonic stage of development. Ludke *et al.* (1975) found plasma ChE values were higher in two-week-old than in four-week-old quail. In adult house and harvest mice, Shellhammer (1961) found a negative correlation of brain AChE with body weight, inferring that activity decreased with age. Similar decreases have been reported in the laboratory rat (Bennett *et al.*, 1958). For NTE both young cats and hens have been found to be insensitive to a single dose of several OPs whereas the adults of both species are sensitive (Johnson, 1975).

No difference in the AChE levels between sexes was found for the starling (Grue and Hunter, 1984). No differences was seen in brain AChE levels between sexes in young quail, but plasma levels of adult females were significantly lower than in males (82–84%). Hill (1989) found no difference between the sexes in brain levels of AChE either in controls or when treated with dicrotophos or carbofuran. In agreement with the findings of Grue and Hunter for starlings, the plasma levels of females were lower than the males, and marked differences in the response of both sexes to the two pesticides were noted. Male quail exposed to 3 mg of dicrotophos had plasma AChE activity levels of 73–76 units compared to 480–579 units in females. The effect of carbofuran was also marked, with the activity of AChE in males being only a quarter of that in females. Additionally, storage of plasma at −25 °C caused a decrease of AChE activity in plasma from females, but not in that from males. In contrast, Dieter *et al.* (1976) found no significant differences between sex and age class (immature/adult) in plasma activity of ChE in canvasback ducks.

No statistically significant differences were found between the sexes in harvest and house mice (Shellhammer, 1961). A rapid increase of AChE activity was found in the embryos of the Hensel toad after day 5 of development, reaching a peak by day 15, after which some decrease in activity occurred (de Llamas *et al.*, 1985).

Diurnal and Seasonal Thompson *et al.* (1988) showed considerable diurnal variation of the activity of carboxylesterases in starlings whereas butyrylcholinesterase did not show a clear pattern of diurnal variation. No seasonal change was found in three species of shore bird examined on their wintering ground (Mitchell and White, 1982). Seasonal variation of the plasma levels of butyrylcholinesterase in four avian species was studied by

Hill and Murray (1987). In bobwhite, the activity was highest in the winter and lowest in the autumn. In the starling, values in the spring were much lower than during the other seasons. No significant differences were found in either red-winged blackbirds or common grackles. Shellhammer (1961) reported seasonal fluctuations in harvest and house mice, with the highest levels in the winter.

Variation within the brain Hart and Westlake (1986) found marked differences in AChE activity in different parts of the brain of the starling. Levels in the midbrain were over five times higher than those in the cerebellum. This finding shows the importance of using whole brain homogenates – or at least carefully prepared half-brains – in work on AChE activity levels. NTE is widespread in human and chicken nervous tissue, with no special high-activity regions (Johnson, 1975).

The problems of storage were studied by Ludke *et al.* (1975). They found that birds stored at 2 °C showed no loss of brain ChE activity. At ambient temperature, material was suitable for analysis for up to 48 hours. Freezing causes a decrease of AChE activity and is unsuitable unless control material is treated in the same manner.

Diet/Cold In the quail, Rattner (1982) found only modest changes in AChE activity were caused by decreased food intake or by exposure to elevated temperatures.

Other contaminants Dieter (1974) found an increase of plasma ChE in quail exposed to DDE or PCBs; the relationship between enzyme activity and the log of toxicant concentration was essentially linear. For both compounds an increase of 80% above control was found at a dietary level of 100 ppm.

Chambers *et al.* (1979) found an erratic pattern of activity change in striped mullet exposed chronically (10 months) to crude oil, although the only statistically significant results were a threefold increase after two weeks and a 20% decrease after two months.

Chronic exposure of rats to cadmium, mercury, and lead caused no effect on the activity of AChE, although some decreases of the concentration of acetylcholine in the brain stem were found (Hrdina *et al.*, 1976). No changes of acetylcholine levels were found by Silbergeld and Goldberg (1975) in mice exposed to lead. Modak *et al.* (1975) found a significant decrease of AChE in some regions of the brain of rats exposed to lead.

2.1.4 *Direct mortality caused by cholinesterase inhibitors*

Direct mortality of wildlife, especially birds, has been frequently reported following the use of OP pesticides. A listing of published incidents in North America and Europe has been made by Grue *et al.* (1983). Very few incidents of poisoning of mammals have been reported, probably due to lower exposure and to their being, in general, less conspicuous than birds. The scale of the incidents has varied greatly, from a few individuals to an estimated 2.9 million songbirds during the forest spray operations in eastern Canada in 1976 (Pearce *et al.*, 1976). In North America, four pesticides – diazinon, fenthion, parathion and phosphamidon – have been responsible for three-quarters of the incidents, whereas in Europe the chief culprits, usually as seed dressings, are carbophenothion and chlorfenvinphos.

While published records are the most reliable, they represent only a small portion of the available information. Peakall and Bart (1983), reviewing the impact of forest spraying on birds, estimated that 90% of the data resided in the 'grey' literature. Grue *et al.* (1983) cited four incidents involving diazinon, whereas over 60 reports have been made to the USEPA of bird kills associated with this pesticide. These involve over 20 species, and numbers of individuals in specific cases range up to over 1000. Hardy *et al.* (1987) stated that 2500 suspected wildlife incidents have been investigated by the Ministry of Agriculture, Food and Fisheries in England and Wales since 1964. Of these cases, 45% were identified as involving agricultural chemicals, again with birds predominant.

The possibility has been put forward that appreciable mortality of birds occurs regularly as part of normal agricultural use of pesticides. The following calculations are based on field studies of carbofuran effects on birds in corn, sorghum and soybean fields (USEPA, unpublished data). The mortality rate is calculated by dividing the number of carcasses found by the acreage searched in the study. Based on avian mortality rate and the acreage of the crop, it is possible to estimate the potential avian mortality. The estimates involve several assumptions: first, that avian density and distribution in the study areas are typical; second, that observed mortality is typical; and finally, that the number of carcasses found is the actual number killed. Hopefully, the errors in the first two assumptions even out as studies are carried out in several different areas. The last assumption clearly leads to under-estimation of total avian mortality.

A geometric mean value of 0.3 avian mortalities per acre was calculated from the cornfield data study data submitted to the USEPA. Using this value, it is possible to estimate avian mortality in the United States:

Estimated Annual Avian Mortality results
from Carbofuran in the United States

Crop	Acres treated	Total mortalities
Corn	4 500 000–5 500 000	1 350 000–1 650 000
Sorghum	640 000–2 040 000	192 000–600 000
Soybeans	210 000–280 000	63 000–84 000

While the figures should not be considered exact since even small adjustments to the figure of 0.3 mortalities/acre would make sizable differences to the total, the figures do suggest that pesticide usage on these crops could be a significant mortality factor.

Secondary poisoning of raptors has been reported. Six bald eagles in the Chesapeake Bay area have been killed by feeding on poisoned birds and mammals. This figure includes poisoning of immatures by adult birds bringing contaminated prey back to the nest. It has been estimated that the eagle kills due to carbofuran constitute 6% of the breeding population in this area (US Fish and Wildlife Service, unpublished data). The effect of another secondary poisoning threat to this species, that of lead, is considered elsewhere (section 6.3.8).

Since most of the OP and carbamate pesticides are rapidly metabolized and eliminated, chemical analysis of tissues has not been widely used in the diagnosis of poisoning by these compounds. Measurements of AChE in brain and plasma have been the main means of monitoring exposure and are diagnostic of death. Reduction of AChE activity by 20% – which for a reasonable sample is roughly two standard deviations – has been put forward as the criterion for exposure to these classes of chemicals (Ludke *et al.*, 1975). Acute inhibition of 80% and chronic inhibition of 50% of brain AChE has been associated with direct mortality (Ludke *et al.*, 1975; Hill and Fleming, 1982).

2.1.5 *Summary of experimental studies*

There has been a vast number of studies on the effect of cholinesterase inhibitors – OPs and carbamates – and those discussed below are only a few which seem to the author to be of the greatest interest from an ecotoxicological viewpoint. One of the most detailed has been the series of studies carried out by the US Fish and Wildlife Service on the starling, involving both captive and free-living birds.

The effects of OP famphur on nestling free-living starlings were studied by Powell and Gray (1980). Daily doses of 0.3, 1 and 3 mg/kg were given. The highest dose caused almost complete mortality within eight hours; only one nestling died in each of the other two groups. No difference in the increase of body weight compared to the controls was seen in the two lower dosage groups. The brain ChE was reduced to 44% of control in the 0.3 mg/kg group and to 28% in the 1.0 mg/kg group.

The effect on the care of nestlings by female starlings treated with an OP (dicrotophos) at a dosage found to reduce the AChE to 50% of control values was studied by Grue *et al.* (1982b). The females were dosed when the nestlings were 12 days old and parental care was examined for the following day. In order to standardize conditions, the males were killed and the brood size adjusted to four at the beginning of the experiment. It was found that the OP-dosed females made significantly fewer sorties to feed their young and remained away from their boxes for longer periods. Nestlings of treated parents lost significantly more weight than the controls. The LD_{50} of dicrotophos for free-living five-day-old nestling starlings was about half that of 15-day-old and adult birds (Grue and Shipley, 1984). This finding is in agreement with the lower activity of AChE in younger nestlings. Growth of nestlings was severely depressed following OP exposure.

The post-fledgling survival of starlings exposed to dicrotophos was studied by Stromborg *et al.* (1988). The young were dosed at 16 days of age and a mortality of 18.5% occurred among the experimental birds. One young per brood was collected at 18 days of age for AChE activity measurements; the reduction was to 46% of control values. Age at fledging, post-fledgling survival, flocking behaviour and habitat use were not affected by OP treatment.

The activity budget of captive male starlings, exposed to a dose of dicrotophos that caused 50% inhibition of AChE, was studied by Grue and Shipley (1981). Within two to four hours of dosing, the behaviour of the treated males was markedly altered. The treated birds spent significantly more time perched and significantly less time flying, singing and displaying (Figure 2.5). Effects on foraging were less marked, but consumption of food from hanging feeders was significantly decreased. No significant effects were seen at either 26–28 or 50–52 hours after exposure. The behavioural changes shown in these studies have implications for monitoring studies carried out shortly after exposure to OPs, especially those based on singing males.

A detailed study of the reproduction of the red-winged blackbird, a close relative of the starling, was carried out by Powell (1984) in two meadows sprayed with fenthion compared to two control meadows. No significant effects were seen on frequency of nest abandonment, clutch size, hatching success or fledgling success. The only significant effect found was lower growth rate of young nestlings in one area, but overall growth rate to fledging was not affected, despite the demonstration that one principal food item, noctuid larvae, was significantly reduced by the insecticide. The AChE levels of a sample of eight- to ten-day-old nestlings showed no difference from controls.

Studies on the effects of methyl parathion on the incubation behaviour and nesting success of red-winged blackbirds were carried out by Meyers *et al.* (1990). Females were dosed with methyl parathion at a dosage which was demonstrated to cause 35% inhibition of brain AChE. Although some clinical effects – ataxia, lacrimation and lethargy – were seen initially, no adverse effects were seen on reproduction. The weights of nestlings of females dosed with pesticide were the same as those of controls, indicating that there was no effect on the ability of the female to provide food. Fledging

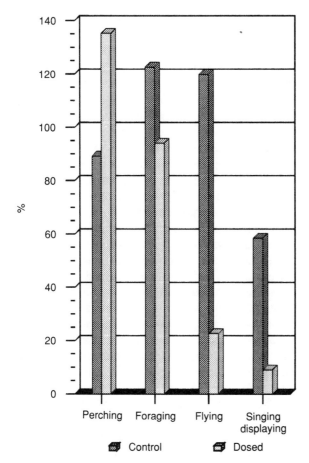

Figure 2.4 Effect of OP on behaviour of starlings. After Grue and Shipley (1981).

success of young from treated nests was normal and band returns did not indicate any difference in over-winter survival.

Hoffman and Eastin (1981) examined the effects of external exposure of mallard eggs to three OPs, using formulations and concentrations similar to those used operationally and at five times these levels. Mortality was increased by the high dose of malathion and the number of abnormal embryos by the high dose of parathion. Brain ChE was inhibited, but there was no relationship between the effects – even mortality – and ChE activity.

Rattner *et al.* (1982a) examined the effect of parathion on the reproduction of bobwhite quail. They found a decrease in body weight, egg production, luteinizing hormone (LH) and progesterone in the group exposed to 100 ppm. No significant effects were seen at 25 ppm. The inhibition of AChE was 27% and 56% respectively. Interaction of parathion and cold stress showed a marked rise (fivefold) in plasma corticosterone in the group receiving the highest dosage. Using a wider range of dosages, Rattner *et al.* (1982b) found that food intake, loss of body weight, egg production and ovary weight decreased in a dose-dependent manner over the range 50–400 ppm. The inhibition of AChE was also dose dependent, reaching nearly 80% at the highest dose. In a subsequent experiment, Rattner *et al.* (1986) linked the effects more closely with LH.

An evaluation of the embryotoxicity and teratogenicity of a wide range of chemicals of environmental concern was made by Hoffman and Albers (1984). Abnormalities among survivors were noted with the following AChE inhibitors: acephate; dimethoate; parathion and sulprofos and reduced growth with acephate, diazinon, dimethoate, parathion and sulprofos.

The changes in terrestrial animal activity following spraying with aminocarb were examined by Bracher and Bider (1982) in the Laurentian forests of Quebec. Studies were carried out over two years and in the second year one site was sprayed at the maximum operational dosages (175 g/ha). After spraying, large numbers of midges (Chironomidae), spiders (Araneae), harvestmen (Opilones), beetles (Coleoptera) and ants (Formicidae) were found dead. No vertebrates were found dead. The decrease of activity of invertebrates of the Arachnida, Lepidoptera and Mollusca was attributed to the direct toxic action of the aminocarb. There was an increase of millipedes (Diplopoda) and centipedes (Chilopoda), which may have been due to decreased competition and/or increased food availability. The authors refer to a dramatic increase of avian activity following spraying, but as the natural rate of increase is so high at that time, it is hard to be certain that this change should be attributed to the spray. The pattern of activities of short-tailed shrews, ermines, and snakes was quite different on the two sites and again it was difficult to pinpoint any changes due to the spray. Patterns for the amphibians showed no significant changes.

Murphy and Cheever (1968) examined the AChE and carboxyesterase activity in rats exposed to OPs. They found that for some OPs carboxyesterase activity was more sensitive to inhibition than AChE. Pretreatment with these compounds rendered the animals more susceptible to inhibition of brain AChE by malathion.

A strong correlation between the reduction of nicotinamide adenine dinucleotide, an important co-factor in the electron transport system, and the severity of teratogenic effects was found by Proctor and Casida (1975) for 36 OPs and 12 carbamates. The relationship of this finding to inhibition of esterases is unclear.

The concentration required to inhibit by 50% AChE and NTE in an *in vitro* preparation of human and hen brain was measured by Lotti and Johnson

(1978). The ratio of the inhibitory power (I_{50}s) for AChE/NTE varied greatly, from less than 0.01 to over 100.00. The ratios were similar for the two species and the ratio of the ratios varied only from 0.3 to 8.0. These workers compared these results with the available data on LD_{50}s and the dose producing severe ataxia in hens. They found a good correlation between the ratio of these two concentrations and the I_{50}s of AChE and NTE. These workers divided OPs into two classes, group A and group B. Group A compounds (phosphates, phosphonates, and phosphoramidates) inhibit NTE and the inhibited complex can undergo ageing and may cause neuropathy. Group B compounds (phosphinates, sulphonates, and carbamates) produce an inhibited enzyme complex that does not age and these compounds cause neuropathy.

The neurotoxic effects of the organophosphate insecticide phenyl phosphonothioic acid-0-ethyl-0-[4-nitrophenyl] ester (EPN) on the embryonic development of the mallard and on the adult has been studied by Hoffman and Sileo (1984) and Hoffman *et al.* (1984). In the studies on embryonic development EPN was applied topically at doses equivalent to one, three and nine times normal application levels. The activity of AChE in 18-day-old embryos (exposed on day three) was 60% of control for all dosage levels; the activity of NTE ranged from 50% in the low dose to only 19% of control in the high dose. From the lower two dosage groups some were allowed to hatch; in these hatchlings the difference between the two esterases was even more marked, with AChE at 80% of controls and NTE at 21–27%. Embryonic mortality increased and brain weight decreased in a dose-dependent manner. Examination of the embryos surviving to 18 days revealed that over a third were abnormal, but the percentage of abnormalities was not dose-dependent. The studies on the adult mallard (Hoffman *et al.*, 1984) showed that the onset of ataxia was dose-dependent. The inhibition of NTE was also dose-dependent with 10, 30, 90 and 270 ppm causing 16%, 69%, 73% and 74% inhibition respectively. The lowest dose group did not display ataxia. Inhibition of AChE followed a similar pattern, but was less marked.

In a recent review of the mechanism underlying OPIDN, Carrington (1989) concluded that 'hundreds of compounds have been tested for their ability to inhibit NTE activity both *in vivo* and *in vitro* and a very strong correlation between inhibition and the development of OPIDN has been established (Johnson, 1975; 1982). However, the fact that it has not been possible to purify the protein with which the esterase activity is associated in active form has limited studies.' Kimbrough and Gaines (1968) screened a number for the teratogenic potential of a number of OPs on rats. Slight teratogenic effects were seen with several compounds at dosages close to those causing toxic effects. Similar findings were reported by Budreau and Singh (1973), who found teratogenic effects of demeton and fenthion only at dosages close to the lethal level. This finding on the lack of teratogenic effects in rats is in agreement with this species' lack of sensitivity to inhibition of NTE.

Meneely and Wyttenbach (1989) found that although chicken and bobwhite quail embryos showed similar axial skeletal malformations and AChE inhibition when eggs were injected with diazinon or parathion, there were some striking differences in embryotoxicity and effects on the cardiovascular system and the nonaxial skeleton.

Interesting differences, illustrated in Figure 2.5, between different OPs and inter-species variations were found by Fulton and Chambers (1985) in studies on three species of frog. An advantage of this test system is that the sample size of embryos examined can be large (40–210 in this study). Nevertheless, marked teratogenic effects are not seen below concentrations that cause appreciable mortality. A similar finding was reported by Elliott-Feeley and Armstrong (1982) for the effects of fenitrothion and carbaryl on *Xenopus*.

2.1.6 Relationship to other biomarkers

A number of studies have examined the relationship between the degree of AChE inhibition and behavioural changes. In general, behavioural changes are less sensitive than AChE inhibition, with few significant changes in behaviour being identified until AChE inhibition reaches 50%. These studies are considered in more detail in Chapter 7.

The increased rate of metabolism of OPs in animals with induced MFO systems is considered in section 5.3. Hinderer and Menzer (1976) assayed for A- and B-esterases, cytochrome P450 and glutathione transferase in liver, lungs, kidneys and testes in quail. Liver microsomes contained the highest A-esterase activity and cytochrome levels. B-esterases were more generally distributed and there was little difference in glutathione transferase among tissues.

2.1.7 Use of esterases as biomarkers

Inhibition of brain AChE was put forward by Bunyan *et al.* (1968) as a diagnostic method of assessing OP and carbamate toxicity. The problems involved in the use of this technique have been considered in detail by Ludke *et al.* (1975) and Zinkl *et al.* (1979). The relationship between inhibition of brain and plasma AChE has been investigated by a number of workers (Fleming and Grue, 1981; Holmes and Boag, 1990). Unfortunately inhibition of plasma AChE can only be used for a few hours after exposure. The problems involved in the use of plasma are considered more fully in Chapter 9. While the recovery of activity of brain AChE is slower, it starts a few hours after exposure ceases. The plots presented by Holmes and Boag

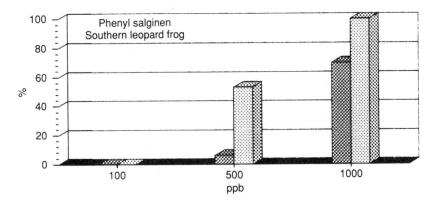

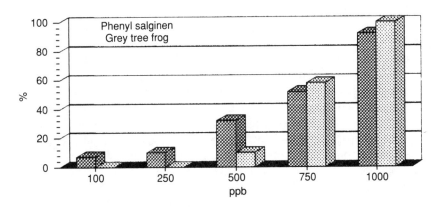

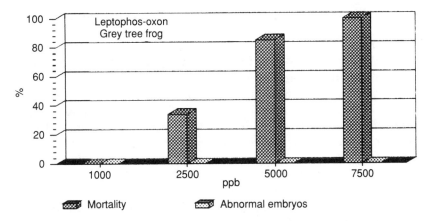

Figure 2.5 Inter-species variation of effects on frogs. After Fulton and Chambers (1985).

(1990) for zebra finches exposed to fenitrothion show a linear increase over the period of 3 to 30 hours after exposure to fenitrothion. In Atlantic salmon, Morgan *et al.* (1990) found that levels were normal within a week at a low dosage (0.004 µl/l), but that activity was still significantly depressed after this time in fish exposed to 0.16 µl/l.

Basically, two questions need to be answered. The first is, 'Was the pesticide spray the cause of death?' and the second is, 'Did the spray cause an impact on the population?'

The first question can be readily answered under ideal conditions. If the animal is fresh and can be matched exactly by controls, then inhibition in the range of 50–80% can be taken as proof of mortality from the pesticide. In practice, these conditions are rarely met; the degree of denaturation is often unknown and adequate controls are difficult to obtain (Hill and Fleming, 1982). Currently work is underway (Kennedy, in press) to develop a method based on the extent of phosphorylation of the active site which would not be affected by enzyme denaturation and would not require the use of controls since it would determine the degree to which the active sites were inhibited. Another approach has been put forward by Prijono and Leighton (1991). These workers have carried out parallel measurement of brain AChE and muscarinic cholinergic receptor (mCBR) activities. The mCBR is a component of the post-synaptic membrane to which the neurotransmitter acetylcholine binds to induce depolarization and hence transmission of nerve impulses across the synapse. Since both mCBR and AChE are involved in cholinergic neurotransmission it was considered likely that the ratio would be similar even if the absolute values varied. Two experiments were carried out. In the first the activity of AChE and mCBR was measured in brains of quail maintained at 25 °C for various periods of time up to 8 days. Some, rather modest, decrease of activity of both AChE and mCBR occurred, but the ratio remained constant at 1.2–1.3: 1. In the second experiment, quail were exposed to diazinon. The activity of AChE was initially reduced by 63%, but fell to 84% of initial activity by eight days, whereas the activity of mCBR, which does not bind to OPs, followed the same course as it had in the control birds. It is proposed that the normal activity of AChE can be calculated from the activity of mCBR.

The use of AChE inhibition as a diagnostic tool to determine mortality caused by OPs and carbamates is well established. It has been used widely both to indicate and predict environmental impact. As an example of work on agricultural lands, one can cite the extensive research of Bunyan, Stanley, Westlake and co-workers (Bunyan *et al.*, 1981; Jennings *et al.*, 1975; Stanley and Bunyan, 1979; Westlake *et al.*, 1978; 1981a,b). Their studies, recently reviewed by Hardy *et al.* (1987), involved laboratory studies, field trials and monitoring bird populations. Inter-species differences, even between different species of geese (Westlake *et al.*, 1978), show the importance of using target species. Much of the work of these workers was carried out on relatively open land and was triggered by a number of incidences of mortality of large numbers of birds.

There is considerably more difficulty in assessing the impact of forest spraying where the proportion of dead animals found is likely to be too low for this to be an effective means of assessing the actual mortality. Most studies have been carried out on birds. Small forest mammals are secretive and difficult to study and forest spray is less likely to penetrate their habitat than that of canopy-dwelling songbirds. This is confirmed by the studies of Buckner and McLeod (1975). Amphibians have been less frequently studied, but those studies that have been carried out indicate an effect caused by DDT, but not with OPs and carbamates (Pearce and Price, 1975).

Census techniques have been extensively used to assess the impact of forest spray on songbirds (Peakall and Bart, 1983). This approach is hardly a biomarker and is thus outside the scope of this book. For a critical and comprehensive review of avian census techniques, the reader is referred to Ralph and Scott (1981).

The difficulties of answering the second question, 'Did the spray cause an impact on the population?', have been considered by Mineau and Peakall (1987). In addition to the problems already considered above, the following need to be considered:

1. The degree of AChE depression varies with time after exposure to the pesticide. It is usually difficult to collect enough samples at any one time, but serious error can be involved if part of the collection is made after substantial enzymatic recovery has occurred.
2. Behavioural changes in exposed birds may cause collection biases. Observations suggest that heavily exposed individuals seek shelter and are less likely to be collected.
3. If the spraying is carried out on comparatively small blocks there is the problem of post-spray immigration of birds from outside the spray area. Rapid replacement of songbirds holding territory following collection has been demonstrated (Stewart and Aldrich, 1951).

These considerations suggest that there will be a strong collection bias towards underestimating the degree of inhibition of AChE. The data on degree of inhibition at various dosages of fenitrothion (Figure 2.6) support this contention. A better response is found when the percentage of individuals that have inhibition of 20% or more is plotted against the dosage. The data for fenitrothion are shown in Figure 2.7. The point at which the curve starts to rise steeply corresponds to the dosage found from census and survey work to be acutely hazardous to birds (Peakall and Bart, 1983), even though the degree of inhibition, about 30%, is well below that known to cause mortality.

Inhibition of AChE will remain the reference line for measuring the impact of the two widely used classes of pesticides, OPs and carbamates.

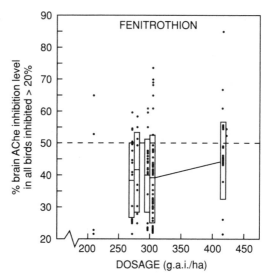

Figure 2.6 Plot of % AChE inhibition against dosage of fenitrothion. After Mineau and Penkall (1987).

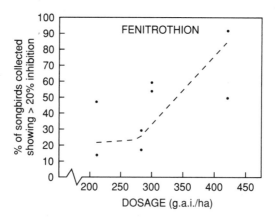

Figure 2.7 Plot of % birds with more than 20% inhibition against dosage for fenitrothion. After Mineau and Penkall (1987).

Improved methodology is likely to make these diagnoses more accurate, although the difficulties of assessing the impact on populations will remain. There is a case for using a wider variety of esterases. Murphy and Cheever (1968) found that for some OPs carboxyesterase activity was more sensitive to inhibition than AChE. Carrington (1989) concluded that NTE has already proven to be a useful marker for assessing the potential neurotoxicity of organophosphorous compounds. It also represents a good working hypothesis for the study of the aetiology of OPIDN. It would be valuable to examine

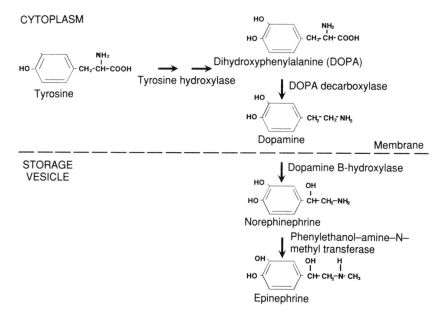

CYTOPLASM

Tyrosine

Tyrosine hydroxylase

Dihydroxyphenylalanine (DOPA)

DOPA decarboxylase

Dopamine

Membrane

STORAGE VESICLE

Dopamine B-hydroxylase

Norephinephrine

Phenylethanol–amine–N– methyl transferase

Epinephrine

Figure 2.8 Biosynthesis of catecholamines.

in more detail the interrelationship with the automonic nervous system (section 2.2). In at least one case, the effects of acephate on rats (Singh and Drewes, 1987), the changes of catecholamines are more marked than those of AChE.

2.2 Biogenic amines

2.2.1 Introduction

The major catecholamines – epinephrine (adrenaline), norepinephrine (noradrenaline), and dopamine – mediate a variety of responses through their interaction with specific receptors in responsive cells. The biosynthesis of these compounds is outlined in Figure 2.8. Norepinephrine (NE) is the primary transmitter in the autonomic nervous system, while epinephrine acts both as a neurotransmitter and as a hormone reaching the target cells via the blood stream. A metabolite of NE that has been extensively studied is 3-methoxyl-4-hydroxyphenylglycol (MHPG) (Figure 2.9). Dopamine (DA) and dihydroxyphenylalanine (DOPA) are intermediates in the synthetic pathway (Figure 2.10) and activate receptors – both stimulatory and inhibitory – in the brain.

The biogenic amine serotonin (5-HT) is a powerful vasoconstrictor and a stimulator of smooth muscle contraction, and may act as a neurohormone. Serotonin is formed by the decarboxylation of 5-hydroxytryptophan and is

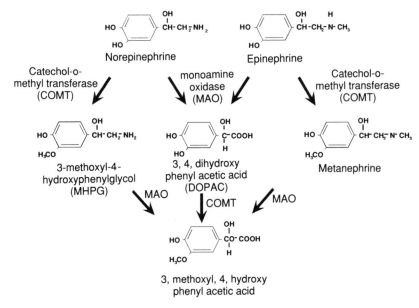

Figure 2.9 Metabolism of epinephrine and norepinephrine.

converted by monoamine oxidase to 5-hydroxyindoleacetic acid (5-HIAA), which is excreted in the urine (Figure 2.11). Serotonin is also involved in the cyclic AMP cascade in the sensory neurons associated with short-term memory. Dudai (1989) in his 'Neurobiology of Memory' refers to the involvement of this compound in the evocative phrase 'a puff of serotonin'.

Brain tissue is rich in gamma-aminobutyric acid (GABA), which is implicated in the transmission of nerve impulses and is considered to be involved in the increase of membrane permeability to the chloride ion.

2.2.2 Summary of experimental studies

Numerous studies involving a number of PHAHs, OPs and heavy metals have been carried out. Many of these have been done to examine the mechanism of action of the pesticides and metals when seizures and other neurological damage occur. Attempts to use these data from a comparative toxicological viewpoint are bedevilled by the wide variety of different brain separations and by the problem of which biogenic amines and their metabolites were measured. In an attempt to tabulate the mammalian data on the effects of PHAHs, I counted effects on eight biogenic amines and their metabolites, to say nothing about free amino acids, in 16 different areas of the brain. Only a brief summary of the effects of the major PHAHs is given. A much more limited number of studies have been carried out from the viewpoint

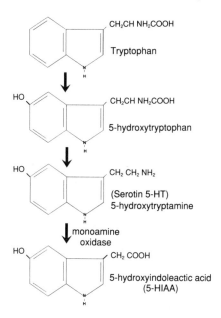

Figure 2.10 Synthesis and metabolism of serotonin.

of using biogenic amines as biomarkers of environmental agents. These studies are considered in more detail below.

Effects of PHAHs The effect of DDT on the rat over an oral dose range of 25–100 mg/kg and over several time intervals (2–24 hours) on the levels of biogenic amines and amino acids was studied by Hudson *et al.* (1985). They found marked dose-related increases of 5-HIAA and MHPG, and some increase of DOPA, whereas the parent compounds 5-HT, NE, and DA were not affected. Dose-related changes in free amino acids were found with aspartate and glutamate, but GABA was not affected. In the time sequence the most marked effects were seen on 5-HIAA, MHPG, DOPA, aspartate, glutamate and GABA, with, in all cases, a maximum elevation at 12 hours. The much more marked effect of DDT and also of chlordecone on the metabolites 5-HIAA and MHPG compared to those on 5-HT and NE was found by the studies of Tilson *et al.* (1986).

A comparison of whole brain and regional brain changes of NE, 5-HT and DA levels in rats fed a diet containing 50 ppm dieldrin for one, two, four and eight weeks was carried out by Wagner and Greene (1978). Only small inconsistent changes were seen in the whole brain experiments. Much more marked changes were found when specific regions of the brain were investigated. These workers found partial depletion of NE in all regions initially, followed by a recovery to control values. In the case of 5-HT, marked increases were noted in the medulla and pons, whereas there were decreases in the striatum and hippocampus.

The effect of dieldrin on the levels of NE, 5-HT and DA in the brains of mallards exposed to dieldrin was studied by Sharma (1973). Three levels of dietary dieldrin were used (4, 10 and 30 ppm). This investigator divided the brain into two equal halves by cutting along the medial line. Half of the brain was used for determination of the biogenic amines and half for the residue levels of dieldrin. The decreases were dose dependent, but only that of 5-HT was significant at the 10 ppm level, whereas decreases to 20–30% of control values were found for NE, 5-HT and DA at the highest dosage. The level of GABA was not affected at any of the levels used.

The effects of dieldrin on the whole brain levels of 5-HT and 5-HIAA levels in trout, chickens and hamsters were examined by Willhite and Sharma (1978). Single, injected doses of 10 mg/kg caused a modest, but significant, increase of 5-HT in all three species, whereas 5-HIAA was decreased in the trout and unaffected in the other two species. At 25 mg/kg the response was mixed – a decrease of 5-HT in the chicken, a slight increase in the trout, and no effect in the hamster. The only effect on 5-HIAA was a decrease in the hamster. The activity of the enzyme monoamine oxidase was significantly decreased in the chicken and the hamster, but was not affected in the trout.

The dose response of NE and DA to DDE, dieldrin, and PCBs has been studied in the ring dove (Heinz *et al.*, 1980). These workers used the same approach as Sharma (1973), splitting the brain and using half for biogenic amines and half for PHAH determinations. Their major findings are illustrated in Figure 2.11. The pattern observed is very similar for all three PHAHs. The brain residue levels at the end of eight weeks on diet suggest that the middle dosages – which caused significant decreases of both amines – correspond to those that might reasonably be expected in a contaminated environment. Only in the highest dieldrin dosage did levels approach those that have been associated with mortality.

The effects of methoxychlor on the whole brain levels of 5-HT, 5-HIAA and tryptophan in the flagfish over a range of concentrations and times were examined by Holdway *et al.* (1986). The only significant changes were decreases of 5-HT levels during the first 24 hours after exposure at the high exposure levels (1.9, 3.1 and 5.1 mg/l); at longer time periods (up to 14 days) no effects were seen. No changes in the concentration of 5-HIAA were seen through the experiment and the only significant change in tryptophan was at the highest dose for the longest period.

Evidence for the stimulatory action of GABA being inhibited by the cyclodiene insecticides has been put forward by Ghiasuddin and Matsumura (1982) and Lawrence and Casida (1984). Using a receptor assay, the latter workers examined the competitive binding of the various isomers of hexachlorocyclohexane, toxaphene and other cyclodienes and found within each series that mammalian toxicity was closely related to strength of binding.

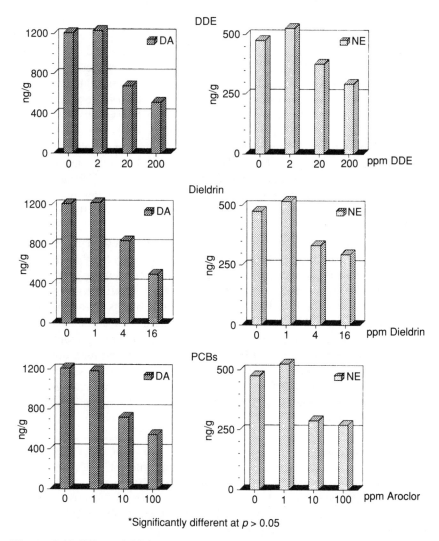

Figure 2.11 Effect of DDE, dieldrin and PCBs on the levels of DA and NE in the brain of the ring-dove. After Heinz *et al.* (1980).

In contrast, DDT, mirex, and kepone were not inhibitors of binding. Bloomquist *et al.* (1986) also found correlation of binding of cyclodienes to mortality, but found lindane was a less potent inhibitor of GABA-dependent chloride uptake than was found by previous workers. Bloomquist and co-workers also examined the effects of the pyrethroids and concluded that the mechanism of action of these insecticides was not mediated by binding to the GABA receptor.

Effects of organophosphates The effects of dosing rats with 3 mg/kg of dichlorvos for ten days on the levels in the cerebral hemisphere, cerebellum and brain stem were examined by Ali *et al.* (1980). At the end of the period dopamine was reduced by 32–37% in all areas; NE was reduced by 50% in the cerebral hemisphere and to a lesser extent in other areas, while maximum reduction of 5-HT occurred in the brain stem. No data are presented on the effect on AChE, but the dosage is high enough to expect considerable inhibition. The same evaluation can be made of the studies of Aldous *et al.* (1982) using leptophos. These workers describe the compound as a potent AChE inhibitor, and again the dosages used would be expected to cause severe inhibition. Effects were seen on the turnover rates of NE and DA at cumulative dosages of 75 mg/kg, but the authors consider that this may be artefactual, as a cumulative dose of 225 mg/kg showed no difference from controls.

The sensitivity of the response of the biogenic amines compared to AChE to treatment with an organophosphate was demonstrated by Singh and Drewes (1987). These workers exposed rats to levels of 1 or 10 mg/kg acephate for a period of 15 weeks. The lower dose did not affect AChE activity in either blood or brain, but the blood levels of epinephrine and NE were markedly increased and the brain level of dopamine decreased. Free amino acid levels were not affected, although the activity of the enzyme glutamic acid decarboxylase was decreased at both dosages. Even at the high dose level brain AChE levels were not significantly reduced, although serum AChE activity was reduced to 60% of control values. Epinephrine levels were increased further by the higher dose, but the other catecholamines did not show further changes.

Effects of heavy metals The best studied of the heavy metals is lead, which is hardly surprising in view of the long-standing interest in lead encephalopathy. Lead, at levels that affected motor activity in rats (5 mg/ml in drinking water), was found to increase levels of NE in the forebrain, but those of dopamine were not altered (Silbergeld and Goldberg, 1975). Although the same finding is claimed by Golter and Michaelson (1975), using a daily oral dose of 0.1 ml of 2% solution of lead acetate, a significant increase of NE was found at only one of the six time periods examined. Direct comparison of the dose received in the two experiments cannot be made without knowing the amount of water drunk by the individual rats in the first experiment. The lack of any data on residue levels in these, and many other experiments, makes comparison with environmental studies difficult.

Operant behaviour in rats was affected by exposure to a daily lead intake of 27 or 81 mg/kg body weight over a three-week period (Sobotka *et al.*, 1975), but no changes were observed at 9 mg/kg. Inhibition of both AChE and butyrylcholinesterase was found, and there was a marked dose-response inhibition of ALAD. No consistent changes were found in NE, DA, 5-HT and 5-HIAA. The effect of higher dosages of lead (100 and 400 mg/kg body weight) on the regional distribution of biogenic amines in the brain was studied by Kumar and Desiraju (1990). The rats on both these dosages showed a decrease in body weight and in the weight of the brain. Significant elevations of NE were found in the hippocampus, cerebellum, hypothalamus and brain stem. The levels of dopamine altered significantly in different directions, with elevation in the hypothalamus and decrease in the brain stem. Serotonin was elevated except at the lower dose in the motor cortex and brain stem. Changes in the levels of GABA were small.

The effects of 45-day oral exposure of rats to cadmium (0.25 and 1 mg/kg/day), methylmercury (0.4 and 4 mg/kg/day), and lead (0.2 and 1 mg/kg/day) were examined by Hrdina *et al.* (1976). Exposure to both cadmium and methylmercury (higher dose only) caused a decrease in brain-stem 5-HT, whereas lead had no effect. NE levels were increased by the lower dose of methylmercury and decreased by the higher dose of lead. All these effects were reversed by 28 days on a clean diet, except the reduction of 5-HT by cadmium. No effects were seen on AChE activity, although some alterations in the concentration of acetylcholine were observed.

The effects of manganese, in combination with iron and copper, on the biogenic amines in the brains of mice were examined by Chandra *et al.* (1980). The dosages of metals used were low (4 mg/kg Mn; 1 mg/kg Fe and Cu) and resulted in brain levels approximately three times those of controls, except that the combination of copper and manganese resulted in a 10-fold increase in the level of copper. The levels of dopamine were increased by all metals and metal combinations; NE by all except manganese alone. 5-HT was reduced by most combinations.

Effects of other compounds The effects of the synthetic pyrethroid, permethrin, on the levels of biogenic amines in the various regions of the brain of rats, were studied by Hudson *et al.* (1986). The levels used, an oral dose of 90 or 180 mg/kg, were sufficient to cause tremors in the rats. Increases, particularly marked in the brain stem and striatum, were found for the metabolites 5-HIAA and MHPG in the hypothalamus and brain stem. DA was not affected and NE levels were altered only at the higher dosage of permethrin. Bloomquist *et al.* (1986) found that deltamethrin was not a strong inhibitor of the GABA receptor.

2.2.3 Factors influencing activity

Regions of the brain There are marked differences in the levels of the various biogenic amines in various parts of the brain. In many experiments the levels in the different parts of the brain are measured separately. Comparative studies are made more difficult by the differences in the separations used by different investigators. These two points can be illustrated by examining the 5-HT levels given for the rat by Tilson *et al.* (1986) and for the ferret by Bleavins *et al.* (1984). The values for the former are given as hippocampus 2.82, brain stem 3.88, hypothalamus 4.09, and caudate nucleus 6.62 pmol/mg; and the latter as cerebellum 123, cerebral hemispheres 175, hypothalamus 687, medulla 755 and midbrain 901 ng/g. Only one measurement was common to both and even then the difference in units has to be calculated before a comparison can be made.

Sex and Age Wagner and Greene (1978) found no differences in the levels of NE, DA and 5-HT between the two sexes for rats on a whole brain basis, although there were some sex differences in the response to DDT when examinations were made on a regional basis. Control levels of the same compounds in the various regions of the brain of ferrets also showed little difference between sexes, although the higher values of NE and 5-HT in the cerebral hemispheres of the females appear to be significant. Again there were some statistically significant differences in the responses of the levels of three biogenic amines to HCB (Bleavins *et al.*, 1984). The level of serotonin in female flagfish was significantly higher than in males (Holdway *et al.*, 1986). There are marked differences between the levels of NE, DA and 5-HT in young and adult female mink (Bleavins *et al.*, 1984). Particularly striking are the much higher levels of all three compounds in the hypothalamus in the young animals.

Inter-species variation Only a few differences in the levels of NE, DA and 5-HT were found in different parts of the brain of ferrets and mink, the only marked differences being the higher levels of NE and DA in the hypothalamus of ferrets (Bleavins *et al.*, 1984). The whole brain levels of 5-HT were measured for golden hamster (4.65 μmol/g), chicken (6.35) and rainbow trout (1.30) by Willhite and Sharma (1978). No detailed studies of inter-species variation, such as have been carried out for cholinesterases (section 2.1.3) are available for the biogenic amines.

2.2.4 Relationship to other biomarkers

The recent discovery that serotonin and the structurally unrelated muramyl peptides, which are well-known immunomodulators, both bind specifically and competitively to certain cells of both the immune and nervous systems has been reviewed by Silverman and Karnovsky (1989). Several interactions between 5-HT and the immune system have been found. Serotonergic pathways in the central nervous system act on the hypothalamic-pituitary-adrenal axis to regulate suppressor activity of B- and T-lymphocytes. 5-HT released from mast cells mediates the delayed-type hypersensitivity response, and 5-HT also alters the effect of other regulators on macrophages.

While it is becoming increasingly evident that the immune and nervous systems interact extensively, the biochemical events following the binding of 5-HT or muramyl peptides are, as yet, poorly understood. Silverman and Karnovsky consider that 'common chemical signals and receptors provide a basis for a biochemical language which the two systems can use to communicate within and between themselves.' Despite the considerable differences in chemical structure, there is increasing evidence that 5-HT and muramyl peptides can share the same biological binding sites. It seems likely that there are several types of serotonin receptors. The sequence of events that follows the binding of 5-HT and muramyl peptides will be of considerable interest. The possibility raised by these workers that it would not be surprising if the brain came to be regarded as the most important organ of the immune system is a challenging concept.

2.2.5 Use of biogenic amines as a biomarker

The data on PHAHs show that changes can be expected from this class of compounds, and the work on the GABA receptor indicates the possibility of differentiating between the various compounds. The comparison of these results with those found for the Ah receptor (section 5.6) would be interesting.

The inhibition of esterases (section 2.1) is clearly the reference line for the effects of OPs. Nevertheless, the work of Singh and Drewes (1987), which indicates that, under some circumstances, alterations of biogenic amines can be more sensitive than AChE, is worthy of further investigation.

Experiments on heavy metals are often difficult to interpret because of the difficulty of translating the dosing regime into something that is environmentally meaningful. Nevertheless, the indication is that changes in biogenic amines are not a sensitive indicator for this group of pollutants.

There are a number of factors that make studies of biogenic amines difficult. Studies directed towards environmental monitoring have largely been based on whole brain determinations and frequently do not include determinations of important metabolites such as 5-HIAA and MHPG. The problem of cellular heterogeneity of nervous tissue needs to be carefully considered. Several different types of neurons and gila are usually found together, and when brain tissue is homogenized, cells of different physiological and biochemical make-up are intermixed. Another difficulty is that some important neurochemicals, such as the monoamine neurotransmitters, are present at low concentrations. Some chemicals, especially those associated with the energy sources, change rapidly after death. Interdisciplinary approaches – biochemical, histological and behavioural – can help to overcome some of the specific problems by giving a wider picture of events occurring in the brain.

3

Biomarkers of the reproductive system

Much of the work of the effects of pollutants on the reproductive system has focused on the most sensitive life stage, for example, the OECD (Organization for Economic Cooperation and Development) early life cycle test in fish. Clearly this approach has merit in the testing of chemicals, as reproductive tests can be much more sensitive than those relying on acute mortality. A comparison of such data is given for heavy metals on fish in Table 3.1.

Another approach is to study the effect of chemicals on the total reproductive capacity over the lifetime of an individual. The total reproductive capacity is a function of four factors: (1) rate of fertilization from individual matings; (2) frequency and timing of matings; (3) reproductive lifespan; and (4) ability of carrying the conceptus to term. All but the last factor apply to both sexes.

Formulae can be developed to sum these effects, but their ecological significance is open to doubt. In a wide range of avian species most individuals die without ever passing their genetic material on to a successfully breeding next generation. Recruitment into the breeding population relies to a remarkable extent on a small number of successful individuals (Newton, 1989). The accumulative production, plotted from the data given in Newton's own chapter on the sparrowhawk, is given in Figure 3.1. From these data it can be calculated that 50% of the young are produced by 20% of the adult females and the 50% less productive females are responsible for only 18% of the young. Similar figures were found for a number of other species, including a small data set for the red-billed gull. For a wide variety of species it is concluded that the breeding lifespan is a more important determinant than variation in egg production and young survival. These considerations, coupled with other density-dependent mortality, make estimation of the effect of reproductive impairment on natural populations difficult.

Nevertheless, some of the most significant effects of environmental pollutants have been on reproduction (Table 3.2.). The vulnerability of the reproductive process was clearly demonstrated by the widespread mortality of the fry of lake trout exposed to DDT (Burdick *et al.*, 1964) and the

Table 3.1 Comparison of LC_{50} (96 hr) and LOEC* for heavy metals in fish

Metal	LC_{50} (96 hr)/LOEC	Species
Cadmium	130	Fathead minnow
	255	Bluegill
	312	Flagfish
Chromium	9	Fathead minnow
	168	Brook char
Copper	4–13	Fathead minnow
	6	Brook char
	27	Bluegill
Lead	34	Brook char
	41 and 146	Rainbow trout
Mercury (inorganic)	560	Fathead minnow
(methyl)	83	Brook char
Nickel	37	Fathead minnow
	60	Rainbow trout
Zinc	1.5	Brook char
	4 and 51	Fathead minnow
	8	Guppy
	29	Flagfish

* LOEC: lowest-observed-effect concentration based on full life cycle, partial life cycle, or early life stage tests
After Atchison *et al.* (1987).

reproductive failure of ranch mink fed fish from the Great Lakes (Hartsough, 1965). One of the best-documented cases is DDE-induced eggshell thinning where reproductive failure caused by egg breakage resulted in population declines of several species of predatory birds over wide geographical areas (Risebrough, 1986). The time sequence of events surrounding the discovery and elucidation of DDE-induced eggshell thinning in the peregrine falcon is given in Table 10.1. A number of other species were also affected. Complete reproductive failure of the double-crested cormorant in Lake Huron occurred in 1972 due to egg breakage, although no mortality of adults was observed (Weseloh *et al.*, 1983).

In this chapter, although the consideration is mainly at the biochemical level – hormones and genetic material – studies carried out on the effects of environmental toxicants on the breeding cycle and on embryos are briefly discussed in order to indicate where biomarkers can be used. Emphasis is placed on studies which have both a laboratory and field component.

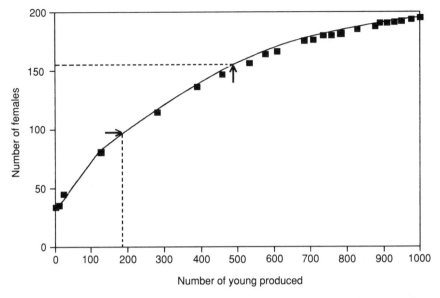

Figure 3.1 Plot of number of young produced against number of females. After Newton (1989).

3.1 Studies of the breeding cycle

3.1.1 Introduction

Studies which observe the effect of chemicals on an organism throughout its reproductive cycle can clearly give a great deal of information. There are, however, severe limitations, especially those involving vertebrate species. Inter-species variation is a serious problem as the number of potential target species is large and the number of species available for testing is small. These tests require that the animals be exposed to toxic chemicals for prolonged periods, especially if multi-generation studies are undertaken. These studies are expensive, in terms of both time and money.

Studies on invertebrates are largely outside the scope of this book. Studies on the reproductive cycle of daphnia were included in 'Minimum Premarket Data' for new chemicals put forward by the OECD, and are part of the chemical registration scheme of the European Economic Community. The cladoceran *Daphnia magna* has a short life cycle, reaching reproductive maturity within eight days, and a single mature individual can produce as many as 20–30 young per day. Thus a relatively inexpensive test completed in 10 days is a chronic test from the viewpoint of the species. A vast number of data is generated by the requirements of these schemes, but, regrettably, most of it is unavailable because of confidentiality. The number of data in existence is demonstrated by the Kenaga (1981) review, which

Table 3.2 Major pollutant-related effects on reproductive processes

Species	Major findings	Biomarker studies
Mink	Complete reproductive failure of ranch mink fed Great Lakes fish (1). Population declines around the Great Lakes (2). Laboratory studies show sensitivity to PCBs and dioxins (3–5).	None carried out. Examination of reproductive impairment by dioxin equivalents should be made.
Seals	Decreased reproductive success and population declines in the Baltic and Wadden seas (6, 7) Experimental studies on Wadden Sea population suggest correlation with PCBs (8).	Marked changes in retinol and some changes in thyroid (9). Pathological changes in the Baltic population (10).
Raptoral birds	Decreased eggshell thickness leading to reproductive failure and widespread population declines (11, 12) Marked inter-species variation (13).	Caused by DDE (and closely related pesticide dicofol). Inhibition of Ca-ATPase (14) and effects on calmodulin (15).
Fish-eating birds	Decreased reproductive success, embryotoxic, behavioural effects and congenital abnormalities (2, 16, 17).	Good inverse relationship between reproductive success and dioxin equivalents based on AHH induction (18, 19).
Fish	Mortality of fry in hatcheries (20) and in field situations (21). Population effects seen in trout in New Brunswick and salmon in the Great Lakes (21, 22).	None carried out.

References: (1) Hartsough (1965); (2) Environment Canada (1991); (3) Aulerich and Ringer (1977); (4) Aulerich *et al.* (1985); (5) Aulerich *et al.* (1988); (6) Helle *et al.* (1976); (7) Reijnders (1980); (8) Reijnders (1986); (9) Brouwer *et al.* (1989a); (10) Jensen *et al.* (1979); (11) Ratcliffe (1967); (12) Hickey (1969); (13) Peakall (1975); (14) Miller *et al.* (1976); (15) Lundholm (1987); (16) Gilman *et al.* (1977); (17) Fox *et al.* (1978); (18) Kubiak *et al.* (1989); (19) Hoffman *et al.* (1987); (20) Burdick *et al.* (1964); (21) Elson (1967); (22) Mac (1988).

refers to daphnia tests on nearly 34 000 chemicals carried out over the years by a single chemical company. The possibility of examination of the reproductive data contained in the daphnia test, which takes this little creature through several reproductive cycles, and relating it to the structure-activity of a wide range of chemicals is an exciting one. Studies on the breeding biology of vertebrates may be carried out under laboratory con-ditions, either on the basic experimental species or on captive wildlife species, by means of experimental studies in the field or by observation of untreated and unmanipulated individuals in the wild. Obviously both the lack of control over the variables and the realism increase as we move from experi-ments on a well-defined genetic stock in the laboratory to observations in the environment.

3.1.2 Laboratory experiments

Availability of test species The number of readily available test species is quite small, being limited to about half-a-dozen each for mammals, birds and fish and to even smaller numbers of other classes of organisms. Additional species have been bred in captivity for specific purposes. For a SCOPE (Scientific Committee on Problems in the Environment) meeting in 1981, I compiled a list of ten additional avian species which had been bred in captivity for toxicological experiments, although not in all cases could the young be reared successfully (Peakall, 1983). Today, ten years later, the list is shorter, consisting of only six species.

The fundamental reason for the limited number of species available is the cost factor. These specialized species are particularly costly to maintain, since all of them are labour intensive to rear successfully. Obviously, there is no consensus on this point, but there is a strong body of opinion that these studies are no longer cost-effective. Certainly, at few of the institutes where such colonies were established have the proponents been able to hold the budget cutters at bay.

Multi-generation studies These are important when considering the effects of highly stable contaminants such as PHAHs and heavy metals. For both classes of compound, multi-generational effects have been observed. Peakall *et al.* (1972) found that, while the first generation of ring doves was unaffected by a diet of 10 ppm Aroclor 1254, reproduction in the second generation was severely reduced, with only a tenth of the fledging success of the control population. Residue levels in eggs were not appreci-ably different in the first and second generation. Carnio and McQueen (1973) found that 15 ppm of DDT caused marked reduction of fertility of eggs in the second and third generation of quail, with the number of fertile eggs per female decreasing from just under 4 to 1.5 and the overall fertility rate from over 60% to about 30%. The residue levels of DDE and DDT in the eggs in the first generation were approximately two-thirds of those in the third generation.

Heinz (1979) demonstrated some changes in reproductive behaviour of mallards exposed to 0.5 ppm of methylmercury in the second and third generations, with the reduction of the number of one-week ducklings produced being lowest in the second generation. Residue levels of mercury in eggs had reached a plateau after the first generation. In all of these studies it should be noted that all of the doses used were quite low.

Species variation The variation in the response of the reproductive cycle to toxic chemicals is illustrated by three examples from the considerable range available. The effect of PCBs on the reproduction of mink has been extensively studied by Aulerich, Ringer and co-workers. Ranch mink fed a diet containing 0.64 ppm PCBs (as Aroclor 1254) experienced complete reproductive failure, and significantly reduced reproductive success was found on a diet of 0.30 ppm (Aulerich and Ringer, 1977). In contrast, rats fed 20 ppm Aroclor 1254 showed no reproductive impairment, even in multi-generational studies, and only slight decreases in survival of young in the second and third generation were noted at 100 ppm (Linder *et al.*, 1974). Mice were similarly insensitive (Sanders and Kirkpatrick, 1977). A diet of 100 ppm causes increased liver weights and evidence of MFO induction, but no effect on reproductive organ weights or the onset of oestrus.

Interesting differences have been found in the ability of specific congeners of PCBs to induce MFO enzymes in mink and rats. A summary of the findings of Gillette *et al.* (1987a) is given in Table 3.3. Detailed pathological studies (Gillette *et al.*, 1987b) revealed that the toxic co-planar congener 3,4,3',4'-TCB caused severe necrotizing enteritis in all mink in this treatment group. Parallel findings have been reported in the relationship between the induction of AHH and immunosuppressive effects in different strains of mice (Vecchi *et al.*, 1983). These workers found that antibody production was strongly inhibited in C57Bl/6 and C3H/HeN strains by low, single doses of TCDD (1.2 μg/kg), whereas other strains (DBA/2 and AKR) were not inhibited. This finding correlated well with the degree of induction of AHH.

DDE-induced eggshell thinning, which caused the decline of several species of raptors, most notably the peregrine falcon, over much of its range in Europe and North America (Hickey, 1969), shows marked inter-species variation. Specifically, 10 ppm caused 22% thinning (which is above the level that causes reproductive failure) in the kestrel, 40 ppm caused 12% in the mallard, and even 300 ppm did not cause significant effects in the chicken (reviewed in Peakall, 1975).

Table 3.3 Comparison of the effects of PCB congeners on the reproduction of mink and rats

PCB congener	Mink	Rat
2,4,2′,4′-TCB	Clinically normal	Clinically normal
	No change in cytochrome P450	No change in cytochrome P450
	No induction of MFO enzymes	Some induction of MFO enzymes
3,3,3′,4′-TCB	Severe anorexia and diarrhoea	Clinically normal
	Increase of cytochrome P450	Increase in cytochrome P450
	No induction of MFO enzymes	Induction of MFO enzymes

After Gillette *et al.* (1987a).

The Rasputin-like qualities of the chicken towards DDT and its major metabolite DDE have given this species and other members of the Galliformes the reputation of being an insensitive group of species for the testing of chemicals. This is not universally true, as the chicken appears to be one of the most sensitive of avian species to PCBs, with 10 ppm causing almost complete reproductive failure (Scott *et al.*, 1971), whereas other species show effects only at much higher doses (reviewed in Peakall, 1987).

There can be little doubt that studies on the effects of PHAHs on reproduction of captive raptor birds were essential to an understanding of the effects that were being seen in the environment. The question is whether or not it is important to maintain colonies of these additional species now that the basic effects of individual PHAHs on avian reproduction have been largely elucidated. I believe that there are a number of areas for which the maintenance of such colonies can be justified. Recent studies on dicofol in the kestrel (Clark *et al.*, 1990) were essential for an understanding of the problems likely to be caused by this pesticide. The question of dioxin equivalents and binding of PHAHs to the Ah receptor (section 5.6) will need to be examined in a range of species in order to validate this concept. The problems of mixtures, both of PHAHs in such contaminated areas as the Baltic and the North American Great Lakes and of heavy metals, caused by acidification, will need to be investigated. Raptorial species are required for testing of the secondary poisoning caused by rodenticides, although these studies are unlikely to involve experiments on the breeding cycle itself. The studies by Reijnders (1986) on harbour seals breeding in captivity were critical in pinpointing the cause of reproductive problems of this species in the Wadden Sea.

3.1.3 Experimental field studies

Field studies suffer from the lack of control over the experiment. There are problems such as predation, vandalism, weather, and also the influence that the observer has on the study being carried out. Additionally, for experimental field studies, there is the difficulty of getting the toxicant to the animal, especially if a chronic dose is to be tested or if the material is highly toxic.

Application of toxicants to free-living individuals Enderson and Berger (1970) exposed nesting prairie falcons to dieldrin by tethering contaminated starlings near their nest site. The results of this study are difficult to interpret because of the high levels of DDE also found in these birds, an example of the type of problem that can occur in field studies. This technique is limited to those species taking live prey or carrion, and 20 years later, the use of live prey would be unlikely to pass an animal care committee. Osborn and Harris (1979) implanted PCBs contained in an open plastic tube below the skin of puffins. The implantation was carried out using a local anaesthetic and no adverse effects were observed. Despite the fact that this technique allows simulation of chronic exposure by slow release – at least for stable, lipophillic compounds such as PHAHs – this interesting approach does not seem to have been used again.

In the case of chemicals with short-lived effects it is necessary to dose frequently. This has been done successfully at US Fish and Wildlife Research Station at Patuxent with starlings. This species, which will nest in artificial nest boxes, can be handled without undue disturbance. This work has already been described in some detail in section 2.1.5.

The effects of oil on both the avian egg (Coon *et al.*, 1979) and the birds themselves (Peakall *et al.*, 1980a) have been studied under field conditions. In the latter case, by using the hole-nesting black guillemot, it was possible by subsequent recapture to measure growth and by taking blood samples to follow changes in biochemical parameters.

A general point about disturbance as far as avian species are concerned emerges from the above discussion. Species that nest in enclosed spaces – holes or nest boxes – have two major advantages. First, the predation rate is considerably lower and, second, the disturbance factor is less. Birds returned to their hole tend to stay there, while open-nesting, precocial birds tend to scatter and become vulnerable to predators. Open-nesting altricial species normally have high predation rates, which makes them difficult to study. This should not be taken to exclude completely the use of the latter, as has been demonstrated by Busby *et al.* (1983) on the white-throated sparrow.

Other manipulative approaches Adverse reproductive effects from toxic chemicals may be caused by direct embryotoxic effects and/or effects on the behaviour of the breeding adults. It is possible to separate these effects by an egg exchange experiment. The basic idea is to move eggs from a highly contaminated (dirty) colony and place them under adults in a relatively uncontaminated (clean) colony, and vice versa. It is also possible to incubate eggs from both 'dirty' and 'clean' colonies artificially, and to examine them for direct embryotoxic effects. While this method is not as realistic as an exchange experiment, it overcomes the logistical difficulties, especially the problem of 'clean' and 'dirty' colonies being asynchronous, and enables detailed studies to be made on embryonic development. Experiments involving egg exchange have been carried out on the osprey (Wiemeyer *et al.*, 1975), herring gull (Peakall *et al.*, 1980b) and Forster's tern (Kubiak *et al.*, 1989).

3.1.4 Observations on untreated individuals

Numerous studies have been made in which pollutant residue levels were correlated with reproductive effects. Only some general considerations are given here. Studies making use of specific biomarkers are considered in sections 3.2 and 3.3.

Wildlife as an early warning system The crash of the population of the peregrine falcon and other raptors throughout most of the Holarctic (Hickey, 1969), and the reproductive failure of trout in New York (Burdick *et al.*, 1964), salmon in New Brunswick (Elson, 1967), seals in the Baltic (Helle *et al.*, 1976) and terns and other fish-eating birds on the Great Lakes (Gilbertson, 1975) are examples where observations on wildlife have identified environmental problems. As the dates reveal, the above is a roll-call of some of the major earlier environmental problems that have been brought to our attention through studies on wildlife. Regrettably this is not all in the past. Seals in the Wadden Sea (Reijnders, 1986), and terns and lake trout in Lake Michigan (Kubiak *et al.*, 1989; Mac, 1988) are current examples of wildlife still experiencing problems.

Observer bias The determination of reproductive success requires observations over a considerable period of time. The dilemma is that accuracy is increased by more observations but so is the disturbance, which may well decrease reproductive success. The detailed studies made by Fetterolf (1983) on a ring-billed gull colony found that while net reproductive output was 89% in the least disturbed section, it decreased to 69% in the moderately disturbed area and decreased further to 45% in the most heavily disturbed. The fate of the chicks in the areas is shown in Figure 3.2. The percentages in this diagram were calculated based on the number of eggs recorded as hatching in each area. The number of chicks found dead in their natal territory does not vary significantly, but there are major increases in the

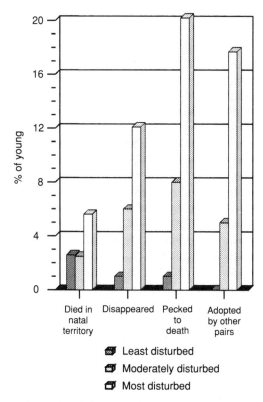

Figure 3.2 Effect of disturbance on the fate of ring-billed gull chicks. After Felterolf (1983).

number of chicks that disappear and that are pecked to death, and those found in other territories. Obviously these types of effects cause major problems to biologists attempting to determine reproductive success of gulls. A protocol to obtain the gross reproductive output of herring gulls with a minimum of effort and disturbance, devised by Mineau and Weseloh (1981), has been used for monitoring this species under the Great Lakes Program.

Detailed studies of nest attentiveness have been made using telemetered eggs that record the core and surface temperatures and the intensity of light falling on the small pole of a wax-filled egg (Varney and Ellis, 1974). This information allows one to determine the amount of time that the adult is incubating the eggs and the extent to which eggs are allowed to cool. A portable detector has been developed which enables the viability of the embryo to be determined by amplifying the heartbeat (Mineau and Pedrosa, 1986). Normal 'candling' procedures cannot be used on the heavily pigmented egg of the herring gull. The detector has been found to have 95%+ accuracy after 16 days of incubation.

3.2 Studies on embryos

Non-mammalian vertebrates have embryos that are easily examined since they develop externally to their parent. Fish, amphibians and reptiles usually produce a large number of eggs, which is advantageous for toxicological studies. The advantages of using embryos of aquatic species to study developmental changes caused by pollutants were reviewed recently by Weis and Weis (1987). Besides the advantages of lower cost and shorter developmental time compared to mammalian studies, some aquatic embryos have the special advantage that non-destructive examination can be made during development so that abnormalities can be observed as they occur.

Avian embryos have been used extensively for toxicity testing over the last 40 years (Ridgway and Karnofsky, 1952; McLaughlin *et al.*, 1963; Khera and Lyon, 1968; Dunachie and Fletcher, 1969). These studies have used mortality as their endpoint. Hoffman and Albers (1984) examined the embryotoxicity and teratogenicity of a range of insecticides, herbicides and petroleum derivatives on mallard eggs. Recently, Hoffman (1990) reviewed the embryotoxicity and teratogenicity of environmental contaminants to avian eggs. The distinction between embryos and eggs is important here as Hoffman is looking at the effects of contaminants that have been transported across the eggshell, whereas most studies, such as those listed at the beginning of this paragraph, look at the effects of compounds injected into the egg.

When considering transportation across the shell, by far the most toxic group of compounds is the petroleum products. Studies on several species have shown that a variety of crude, refined and waste oils are embryotoxic. Toxicity is generally dependent on the concentration of the higher molecular weight PAHs. LD_{50}s are often less than 5 μl of oil per egg. Only a few studies consider biochemical changes in addition to mortality or teratogenic effects. A major problem with biomarkers is the rapid rate of change during embryonic development. For AChE, there is a rapid increase in levels during the latter portion of embryonic development in the precocial mallard (Hoffman and Eastin (1981)), whereas the altricial starling hatches with low levels and the major part of the increase occurs during the fledging period (Grue *et al.*, 1981). Heinrich-Hirsch *et al.* (1990) found that both P450I- and P450II-type enzymes developed as early as day 4 in the chicken embryo although there was considerable variation with time and from enzyme to enzyme. Hamilton *et al.* (1983) found that the basal level of AHH activity in the chicken embryo remained constant throughout incubation, but was capable of stimulation after day 6/7 of incubation by tetrachlorobiphenyl.

Waste crankcase oil placed on the eggs of mallard caused a dose-dependent increase in the number of abnormal young, with a significant increase being found even at the lowest dose, 2 μl (Hoffman *et al.*, 1982). Significant increases of aspartate aminotransferase and decreases of ALAD were seen at the two highest doses (5 and 15 μl) although alkaline phosphatase was not affected. The use of these enzymes as biomarkers is considered in more detail in sections 6.4.2, 6.3.8, and 6.4.5 respectively. In experiments with bobwhite quail, significant changes in the percentage of abnormalities were seen at the two highest doses, which paralleled changes of aspartate aminotransferase.

The induction of four MFO enzymes by microlitre quantities of Prudhoe Bay crude oil (PBCO) placed on the shells of chicken eggs was demonstrated by Lee *et al.* (1986). These workers found a clear dose response for the activity of all four enzymes over the dosage range of 1–5 µl, followed by a plateau at 5–10 µl. The activity of the most sensitive enzyme, 7-ethoxyresorufin 0-deethylase (EROD), was increased 22–24-fold. Exact comparison of induction with mortality is difficult, as the former was determined in embryos older than those of the latter. Crude oil is most toxic when applied to the egg at the early stages of development, whereas enzyme induction does not occur until day 10. The LD_{50} of PBCO at day 8 was determined as 1.3 µl. In a subsequent study (Walters *et al.*, 1987) it was determined that the nitrogen-, oxygen- and sulphur-containing heterocyclic fractions were the most effective inducers, followed by the aromatic fraction. The aliphatic compounds were essentially inactive.

Meiniel (1977) found a good correlation between axial teratogenesis and AChE inhibition in embryos when OPs were placed on eggshells. Malathion caused little effect, but both bidrin and parathion caused both strong inhibition and marked axial deformities. Regrettably no measurements were made on NTE.

The relationship between AChE inhibition and skeletal abnormalities for mallard eggs exposed to three OPs was examined by Hoffman and Eastin (1981). All three compounds – malathion, diazinon and parathion – caused marked inhibition of AChE, but only the highest dose of parathion caused significant elevation of the occurrence of skeletal abnormalities. No relationship between the degree of inhibition of AChE and the occurrence of abnormalities was found. The activity of NTE was not examined in this study. In a subsequent paper (Hoffman and Sileo, 1984) the effects of the OP EPN were examined in mallard eggs. A significant increase in the percentage of abnormal survivors was found at all doses (12, 36 and 108 µg/g egg) although the occurrence rates were not dose dependent. Significant decreases of both AChE and NTE were also found at all doses. Again clear dose dependence was not established, but NTE was almost completely inhibited at the two higher doses.

The amphibian embryo has also been used for many toxicology studies involving mortality and development alterations (Birge *et al.*, 1979); Cooke (1981) and Pough (1976). A review of the acute toxicity of a wide range of pollutants to amphibians has been made by Harfenist *et al.* (1989). Metals are found to be high on the list, whereas many pesticides, especially the organophosphates and carbamates, have high LC_{50}s. This agrees with the

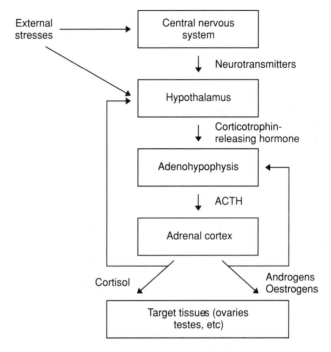

Figure 3.3 Outline of hormonal control of reproduction.

field observations that this class of organisms is little affected by operational use of OPs and carbamates (Pearce and Price, 1975).

3.3 Hormones

The hormonal control of reproduction in vertebrates is a complex, hierarchical system with many feed-back mechanisms. An outline of the control system is given in Figure 3.3. Thus there are diverse points at which pollutants can act on such a complex system, but conversely the system has many built-in mechanisms to combat any perturbation.

The sequence of events that link the brain to behavioural changes through the integration by the nervous system of sensory, neural, and endocrine factors mediating reproductive behaviour is beginning to be understood in some detail. The responses between the brain and the pituitary are mediated by the hypothalamo-hypophysealgonadal axis. Secretion neurons in the brain synthesize gonadotrophic hormone-releasing hormone (GHRH), which is transmitted by a vascular connection to the pituitary portal system. Stimulatory glycoprotein hormones – gonadotrophins – are secreted by the pituitary gland, which in turn initiates hormone secretion by the gonads. The onset of breeding is marked by an increase in the pituitary gonadotrophins (luteinizing hormones, LH, and follicle-stimulating hormone, FSH) and gonadal steroids.

The neural processes involved in avian reproduction have recently been reviewed by Silver and Ball (1989). They describe the sequence of events involved in this process in the following stages: (1) transduction of stimuli from external factors by specialized receptors into neural events; (2) pathways from these receptors to target areas in the brain releasing hormones in the preoptic and hypothalamus; (3) neuromodulatory and enzymatic factors regulating the release of hormones and (4) peripheral steroid hormones acting back on the brain to alter its responses.

The identification of the sensory receptors capable of transducing environmental clues into hormone secretion is by no means complete. The existence of encephalic photoreceptors has been shown in several avian species (Oliver and Baylé, 1980) and studies by Foster and Follett (1985) on quail have measured the effect of light of different wavelengths on the concentration of plasma LH. Photoreceptors in a range of vertebrates are known to contain the protein opsin as part of the mechanism for receiving the photic signal. The mechanism of transduction to the GHRH neurons is less well understood. Interaction of pollutants with this first stage of the reproductive process does not appear to have been examined. As the events that translate external environmental stimuli such as day length and rainfall into the onset of the reproductive process become known, the stage will be set for such investigations. It is possible that this first stage, without the feed-back loops that protect other parts of the system, may be more sensitive to the effects of pollutants.

The effects of pollutants on the endocrine system have been examined in a number of ways. They may damage any of the organs from the pituitary through to the gonads, alter the levels of hormones or act like hormones themselves.

3.3.1 Effects of polyhalogenated aromatic hydrocarbons

Direct oestrogenic effects The oestrogenic effects of PHAHs, as determined by elevation of glycogen in target organs, were examined in both mammalian and avian species by Bitman et al. (1968) and Bitman and Cecil (1970). The main finding, supported by subsequent work reviewed in Kupfer (1975), is that only the o,p-isomers of DDT and related compounds had strong oestrogenic effects. Since the o,p-isomers are very much less stable than the p,p'-isomers, the environmental importance as it relates to DDT is low.

Female–female pairing in western gulls was reported as occurring at a frequency of 8 to 14% in a colony in California (Hunt and Hunt, 1977). The clutch produced was abnormally large (four to six eggs), and intervals between laying of eggs were shorter than those found for normal clutches, indicating that both members of the pair laid eggs. Most of the eggs in supernormal clutches were infertile, indicating the absence of a resident male. Fox and Boersma (1983) reviewed the data available for the ring-billed gull. They concluded that supernormal clutches were more frequent in expanding colonies. Within the Great Lakes the highest values were noted in the expanding colonies in Lake Superior and the lowest values in the saturated colonies in Lake Ontario. The effects of injected DDT, DDE, methoxychlor and oestradiol in the yolk of eggs of the California gull were studied by Fry and Toone (1981). Oestradiol, even at the lowest doses used (0.5 to 2 ppm), caused feminization of the male embryos, indicating the validity of the technique. O,p′-DDT caused feminization of five out of six of the lower-dosed embryos (2 and 5 ppm), no effects were seen with p,p′-DDT, and half (3/6) of the high doses of p,p′-DDE (20 and 100 ppm) showed feminization. Methoxychlor showed feminization of all embryos at all doses (2 to 100 ppm) although the total number of embryos was small (only eight for five dosage studies). The study was bedevilled by poor survival of the embryos; only 108 of the 264 eggs survived to pipping. Being wise after the event, it is evident that the sample size for each dosage group was too small.

A more detailed account was published by Fry *et al.* (1987) although no new experimental toxicological work was included. They concluded that supernormal clutches occurred when two conditions are met; for example, the sex ratio of breeding adults is skewed towards females and nest sites are available for female–female pairs or polygynous trios. This female–female pairing was found to occur in the newly formed or rapidly expanding colonies of ring-billed and California gulls nesting in the Great Lakes and in the Pacific northwest and in populations of herring, western and glaucous-winged gulls breeding in areas polluted with organochlorines. In the final sentence of their paper they state, 'female–female pairing and reduced reproductive success of gulls in southern California and the Great Lakes is a striking example of pollutant effects on entire populations of birds.' Comparison of the levels of PHAHs needed to cause feminization and those found in herring gull eggs on the Great Lakes just do not support this strong statement. The minimum levels found by Fry and co-workers are 2 ppm o,p′DDT, 2 ppm methoxychlor, and 20 ppm DDE. The o,p′-isomers of DDT and DDD have been rarely found in avian environmental samples. Even in the highly contaminated brown pelican eggs from Anacapa Island, California, the highest recorded value is 0.5 ppm (Lamont *et al.*, 1970); similarly, methoxychlor is rarely recorded. Certainly 20 ppm DDE has often been reached, but this value depends on combining the dosage data from the 20 and 100 ppm groups; even this combination gives a group size of only six males.

As far as the use of direct oestrogenic effects as a biomarker is concerned, it would be necessary to examine the specificity of binding to a specific receptor. In view of major advances in our knowledge of both receptors and structure-activity relationships of specific PHAH isomers and congeners, a re-evaluation of such direct action of pollutants would be worthwhile.

Effects mediated by mixed function oxidase systems Following the accidental discovery that chlordane stimulates the hepatic MFO system in rats (Hart *et al.*, 1963), a great deal of work has been done on the induction of this system by PHAHs (Chapter 5). The rapid pace of this work can be demonstrated by the fact that the review by Conney *et al.* (1967) lists 12 PHAHs that stimulated drug metabolism. At this time, concern over DDT was high on the environmental agenda, and this work, coming from the medical community, was eagerly adopted by those studying the environmental effects. At this time also, PCBs were discovered in the environment (Jensen, 1966). There was concern that the efforts to ban DDT would be made more difficult as the chemical companies would be able to blame the effects seen on PCBs. It was for this reason that I undertook to examine the effects of both DDT and PCBs on steroid metabolism. The results (Riseborough *et al.*, 1968) clearly showed that both were strong inducers of the MFO system.

A chronic study (180 days) on the effects of toxaphene on the plasma levels of testosterone and the activity of mixed function oxidases in male rats was carried out by Peakall (1976). The plasma levels of testosterone showed a significant decrease at day 5 but by day 15 had returned to control levels. The MFO activity increased rapidly and after day 5 remained fairly constant although there was some decrease towards the end of the experiment. The results are consistent with the hypothesis that induction of MFO enzymes caused increased metabolism of the steroid, leading to lower levels, but that this decrease was soon compensated for by increased synthesis.

Breeding cycle studies One of the most detailed mammalian studies, from the biochemical viewpoint, was made on rats exposed to polybrominated biphenyls by Johnston *et al.* (1980). These workers found that a dietary level of 100 ppm did not affect plasma concentrations of the luteinizing hormone, prolactin or corticosterone. Nor were the biogenic amines affected (section 2.2). Such a finding does not mean that PHAHs are without effect on mammalian reproduction. The fact that this occurs is clearly demonstrated by studies on mink. Following the finding that pollutants in Great Lakes fish were causing reproductive failure in ranch mink (Hartsough, 1965), toxicological studies have been carried out by Aulerich, Ringer and co-workers. These workers were able to demonstrate that the causative agent was PCBs. Diets containing 0.30 ppm PCBs caused reproductive impairment, and 0.64 ppm caused complete reproductive failure (Aulerich and Ringer, 1977). Subsequent work showed that the adverse effects related to adult and fetotoxicity were associated with PCB congeners that exhibit dioxin-like

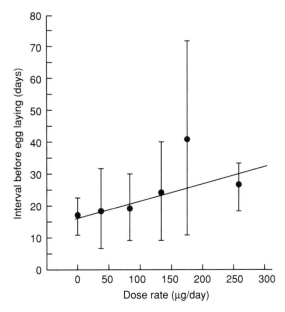

Figure 3.4 Relationship between dose of DDT and delay in onset of breeding. Jefferies (1967).

properties (Aulerich *et al.*, 1987). These workers also found that mink is one of the most sensitive species to poisoning by TCDD (Hochstein *et al.*, 1988). Unfortunately this work on the mink does not include measurements of the levels of hormones.

Delay in the onset of breeding of Bengalese finches caused by p,p′-DDT was discovered by Jefferies (1967). The dose-response that he obtained, Figure 3.4, remains one of the clearest demonstrations of the effect of a PHAH on the breeding cycle. Among birds the effects of PHAHs on reproductive behaviour and biochemistry have been studied in the greatest detail in the ring dove. A summary of the data available on DDE and PCBs is given in Table 3.4. Delay in the onset of breeding is a consistent finding. At the biochemical level this appears to be due to the effect of increased metabolism by the MFO enzymes, but this increased metabolism is soon overcome by the feedback mechanisms and normal hormone levels are achieved. These findings agree with those on the levels of testosterone in both other avian species and a mammalian species. In studies on quail exposed to a diet containing PCBs (Clophen A60) during the second to fourth weeks of life, Biessmann (1982) found that several reproductive parameters – such as delayed breeding and decreased laying capacity – were affected, but that steroid hormone levels were not affected except for an increase in testosterone levels in males at the lowest dosage level. In an

Table 3.4 Effects of PCBs and DDE on reproductive behaviour, success and hormone levels in doves

Chemical	Dose (ppm)	Courtship behaviour	Time to lay	Nest attendance	Reproductive success	LH	Steroid hormones	Reference
DDE	10	Decreased	–	–	–	–	–	(1)
	40	–	?	–	–	No effect	–	(2)
		–	–	–	Decreased	–	–	(3)
		–	Increased	–	–	Fail to show surge	–	(2)
	50	Decreased	–	–	–	–	–	(1)
	100	Decreased	Increased	Decreased	Decreased	–	–	(4)
PCBs	10	–	Increased	Decreased	Decreased	–	–	(5)
		–	Increased	–	Decreased	–	Some changes	(6)
	25	Decreased	Increased	Decreased	Decreased	–	–	(7)
Mixture	Low	No effect	Increased	No effect	Decreased	–	Some changes	(8)
	High	Decreased	Increased	Decreased (brooding only)	Decreased	–	Some changes	(8)

References: (1) Haegele and Hudson (1977); (2) Richie and Peterle (1979); (3) Haegele and Hudson (1973); (4) Keith (1978); (5) Peakall and Peakall (1973); (6) Koval et al. (1987); (7) Farve (1978); (8) McArthur et al. (1983).

experiment on the grey partridge, Abiola *et al.* (1989) found that PCBs (DP5) caused an elevation of hepatic MFO activity but did not alter the testosterone levels. In a study which exposed rats to toxaphene for 180 days, Peakall (1976) found a decrease of testosterone at day 5, but that levels had returned to normal at the next sampling time (15 days) and remained unaltered compared to control values for the remainder of the period.

The effect of PCBs on steroid hormone metabolism has been investigated in cod by Freeman *et al.* (1980), although these studies do not include the reproductive cycle. Altered steroid metabolism in *in vitro* preparation of kidneys was demonstrated in cod treated *in vivo* with Aroclor 1254. However, the fact that the rate of metabolism of the steroid is altered does not mean that the circulating levels are necessarily altered, since increased metabolism can be compensated for by increased synthesis. In this case one is effectively using MFOs as a biomarker rather than the levels of the steroid.

3.3.2 Effects of organophosphates

The effect of parathion on the breeding success and hormone levels in bobwhite quail were examined by Rattner *et al.* (1982a). AChE activity decreased in a dose-dependent manner, being reduced to 30% of control at the highest dose (400 ppm). One major difficulty was decreased food intake and the concomitant loss of body weight. This occurred in a dose-dependent manner and weight loss over a ten-day period was 36% at the highest dose. Ovarian weight decreased markedly, being only 12% of control weight in the two highest groups. This decrease is still marked even if it is expressed on a body weight basis. Egg production fell to 30% of control values. In a second experiment LH and steroid hormone concentrations were determined on quail exposed to a low concentration range (25 and 100 ppm). Because of the problem of weight loss an additional group was used, in which one bird of the pair was fed clean mash on alternate days (pair-fed). A decrease of LH was found at the highest dose, but not in the pair-fed birds. Both corticosterone and progesterone concentrations varied during the course of the day, but the patterns of the experimental birds were not significantly different from controls. In a subsequent experiment Rattner *et al.* (1986) demonstrated that a single dose of parathion caused a marked decrease in LH. At the higher dose (10 mg/kg) the levels were still low after 24 hours, but at the lower dose (5 mg/kg) had returned to control by the end of this period. The changes in LH paralleled those found for AChE.

The effect of organophosphates on the interaction of corticosterone with cold stress has been examined in bobwhite quail (Rattner *et al.*, 1982b) and kestrels (Rattner and Franson, 1984). Cold plus OP caused a marked rise in cortisterone levels. This effect was not found with the pyrethroid, fenvalerate (Rattner and Franson, 1984).

3.3.3 Effects of oil

The effect of chronic exposure to oil on the reproduction of mallard has been extensively studied by Holmes, Cavanaugh and co-workers (Holmes *et al.*, 1978; Cavanaugh *et al.*, 1983; Cavanaugh and Holmes, 1987. They found that the onset of egg-laying was significantly delayed – a parallel with findings with PHAHs – but in contrast with the finding for PHAH, studies found that abnormally low plasma oestrogen concentrations occurred throughout the ovarian cycle and that the normal diurnal changes in steroid hormone concentrations were much less marked. The rise in LH was much slower in birds exposed to oil, but those birds that eventually laid had LH concentrations approaching those of control birds. These workers consider that the primary site is likely to be the steroidogenic cells of the differentiating ovary.

Other effects included a decrease in the incidence of fertilization, abnormal brooding behaviour and an increase in embryonic mortality. A recent study by this group (Holmes and Cavanaugh, 1990) examined the sex-related effects of ingested oil on mallards. They found that when the male, but not the female, was exposed to contaminated food the ovarian development was normal, but the ability of the male to fertilize the eggs was significantly reduced. In the reverse case, exposure of the female only, it was found that ovarian development was delayed and the frequency of fertilization was decreased to a similar extent as when the male only was exposed. These workers suggest that gonadal endocrine function is affected by oil in both the male and female.

The hormone levels of several groups of winter flounder and salmon of different sizes and under different water conditions after exposure to oil were examined by Truscott *et al.* (1983). Changes in the levels of total testosterone were inconsistent, with a few groups showing marked decreases, but most were unaffected. The results for 11-ketotestosterone were more consistent, with decreases found for all but one of the 7- and 14-day groups, although 11-hydroxytestosterone was affected only in a few cases. The most consistent finding was induction of AHH activity, although even here one group, which had shown decreased testosterone activity, did not show induction.

3.3.4 Receptors

The variety of responses of steroid hormones is mediated by binding specifically and with high affinity to intracellular steroid receptor proteins. The steroid hormones themselves are poorly soluble in water and circulate in combination with carriers. These may be either high-capacity, low-affinity carriers, such as serum albumin, or low-capacity, high-affinity carriers such as corticosteroid-binding globulin. At the target cell there are two modes of action. The dominant one is the release of the hormone, which then diffuses through the cell to bind with specific receptors in the cytosol. The second mechanism is the hormone-carrier complex binding with plasma membrane receptors, and this is followed by transfer to intracellular receptors. The outline of these two processes is shown in Figure 3.5 (Wallach, 1987). Specific intracellular receptors exist for each category of steroid hormone, and each consists of a binding site for the hormone and for DNA. When the hormone attaches to its specific binding site, the hormone/receptor complex becomes transformed into an activated DNA-binding unit. The interaction of the DNA-binding site with a specific sequence in the genome of the target cell results in activation.

Complementary DNA clones encoding a number of steroid receptors have been isolated, enabling studies on the structure and function of these proteins to be undertaken. These studies have led to the identification of a family of related genes that bind ligands of remarkable diversity (Evans, 1988). Comparisons of the amino sequence in combination with functional studies have shown a common structure for this superfamily of receptors. This family includes receptors for glucocorticoids, mineralocorticoid, progesterone, oestrogen, vitamin D_3 and the thyroid hormones. It is surprising that there is a common structure for ligands as diverse as the steroid and thyroid hormones, which have neither structural nor biosynthetic similarities (Evans, 1988; Godowski and Picard, 1989; Lefkowitz *et al.*, 1989). The relationship, if any, of this superfamily to the Ah receptor has not been defined. However, a recent paper by Astroff and Safe (1990) has shown that TCDD antagonizes the oestrogen-induced response in the rat uterus and structure-activity data suggest that the Ah receptor is involved in mediating the anti-oestrogenic responses in the target cells.

A number of steroid receptors have been identified. Evans (1988) lists several in his schematic amino acid comparison of the steroid hormone superfamily. These include receptors for glucocorticoid, progesterone, oestrogen and oestrogen-related compounds.

The relationship between AHH induction and the specific binding of the cytosolic oestrogen receptor was studied by Duvivier *et al.* (1981). Plotting the data from four different inducers, including benzo(a)pyrene, a highly significant negative correlation was obtained, indicating that receptor levels and AHH activities are linked by a common biochemical event.

The effects of a variety of PHAHs on the binding of progesterone to its cytoplasmic receptor have been studied by Lundholm (1988) in the shell gland mucosa of the duck and the chicken and in the uterus of the rabbit. He found that the PHAHs studied (DDE, PCB and chlordane) decreased the binding of progesterone. The potency was higher in the duck than in the chicken, while the duck and the rabbit showed similar values.

At present, environmental toxicological studies have been confined largely to work with the Ah receptor (section 5.6), but despite the difficulties of

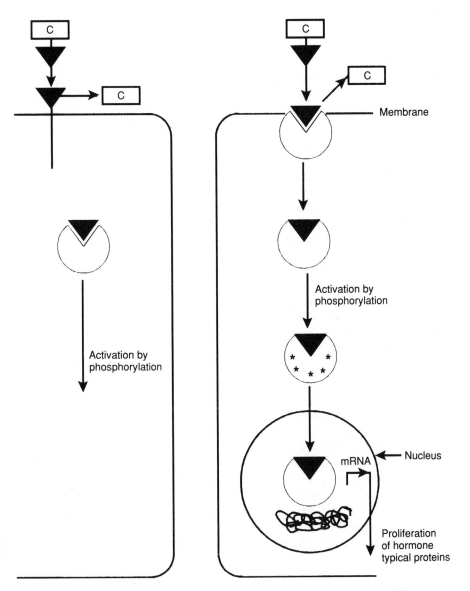

Figure 3.5 Interaction of hormones with carriers and receptors. After Walloch (1987).

working with molecules that are present only in trace amounts (characteristically 10^3 to 10^4 per cell) it is likely that the rapid progress in the molecular biology of receptors will soon lead to their wider use in environmental toxicology.

3.3.5 Summary

Some of the major environmental problems with toxic chemicals have involved reproductive impairment (Table 3.3). These have included the collapse of the population of several avian raptoral species throughout the Holarctic, reproductive failure of fish-eating birds in several areas of North America, and reproductive failure of lake trout and salmon in North America and of seals in the Baltic and Wadden Seas. Furthermore, it is possible, even likely, that chemical stress may have caused animals not to attempt to breed. Such an occurrence would be very difficult to detect. The reproductive process requires a considerable investment by the parent. There are many environmental conditions – poor food supply, adverse weather – that affect reproduction. The phenomenon has been well studied in the Arctic. Large clutch sizes of such predators of voles as buzzards and owls occur when prey is abundant, and under unfavourable conditions breeding may be abandoned altogether (Lack, 1954). This phenomenon is not confined to this harsh region. The breeding of tropical sea birds is readily affected by food shortages caused by the El Niño. For long-lived species it is more advantageous to terminate breeding early than to continue to invest resources and subsequently to fail.

The best evidence for the occurrence of abandonment of breeding caused by pollutants comes from studies on the impact of oil on the reproductive performance of sea birds. Two major studies have been those of Fry *et al.* (1986) on the wedge-tailed shearwater in Hawaii and Butler *et al.* (1988) on Leach's storm-petrel in Newfoundland. Both of these species nest in burrows in large colonies. This enables a large sample size to be checked rapidly. Successful breeding pairs tend to return to the same or adjacent burrow in successive years, making multi-year studies possible. In both studies an increase in nest abandonment and a decrease in incubation attentiveness were noted. A significant difference between the studies was that while Fry and co-workers found a decreased number of birds that had been oiled returning to the colony, no long-term effects were found in the other study.

The impact that biomarkers have made, so far, on reproductive studies is small. Some of the major studies involving the effects of pollutants on reproduction, together with those biomarker studies that were undertaken, are listed in Table 3.2. The studies on the effects of PHAHs on mink have focused on overall reproductive success. Those on the seal have included detailed studies on effects on retinol and thyroid, but not on direct reproductive parameters. Enzyme studies have been part of the studies on DDE-induced eggshell thinning, but this phenomenon, although it has profound population effects, is limited to comparatively few species and to only one major pollutant.

We still lack biomarkers that indicate reproductive impairment. The correlation of reproduction with dioxin equivalent is based on AHH induction (section 5.6) and is one of the most promising advances. Nevertheless it is an indirect biomarker of reproduction. The levels of circulating hormones vary widely during the reproductive cycle, and complex feedback mechanisms exist to overcome perturbations. The possibility of using receptor assays as a direct assay should be investigated.

4 *Studies on genetic material*

4.1 Introduction

The fundamental role of DNA in the reproductive process is so well known that it is unnecessary to define it. Nevertheless, the endpoints used in assessing the damage to DNA by environmental pollutants are specific genotoxic effects, especially the increase of neoplastic disease, rather than effects on the reproductive process. For this reason effects of pollutants on genetic material are considered separately from reproductive effects.

The relationships between DNA changes and harm to the organism are extremely complex. In a research report in *Science*, Maugh (1984) states that 'we have more than 15 years of data correlating DNA damage with carcinogens and mutagens in animals and the tools to study the interactions of chemicals with DNA in humans'. While adduct formation is a good means of assessing the exposure of an individual to chemicals, the exact relationship of adduct formation to harm to the organism is less well understood. For example, although there are good data to show that there is a direct relationship between the extent of cigarette smoking and the number of DNA-BaP adducts, the relationship between DNA-BaP adducts and the occurrence of lung cancer is less well defined (Phillips *et al.*, 1988). In the wildlife toxicology field the establishment of the sequence of events from the initial DNA lesion to harm is even more difficult. Nevertheless, it is reasonable to conclude that reaction of chemicals with DNA results in deleterious conditions such as tumour formation.

There is a continuum of events between the first interaction of a xenobiotic and DNA and mutation, but they may be divided into four broad categories (Shugart, 1990a). The first stage is the formation of adducts. A variety of assays are available and these are discussed in section 4.3. At the next stage, toxic chemicals may cause secondary modifications of DNA such as strand breakage (section 4.4), change in the minor base composition, (section 4.5) or an increase in the rate of DNA repair. The third stage is reached when the structural perturbations to the DNA become fixed. At this stage, affected cells often show altered function. Several cytogenetic assays are available to detect and measure chromosomal aberrations; one of the most widely used is sister chromatid exchange (SCE), which is discussed in section 4.6. Finally damage caused by toxic chemicals may lead to the creation of mutant DNA, which leads to alterations in gene function.

A variety of tests are available to examine for possible damage to, or changes in, genetic material caused by environmental pollutants (Kohn, 1983). The most widely used techniques in studies on wildlife samples are: the ratio of RNA to DNA, formation of adducts, breakage in the individual strands of DNA, the degree of methylation of DNA and the frequency of sister chromatid exchange.

4.2 RNA/DNA ratio

The RNA/DNA ratio has been demonstrated to be a reliable indicator of growth rate in marine phytoplankton (Sutcliffe, 1965) and fish (Bulow, 1970). The basic concept is that the DNA content per viable cell remains more or less constant, whereas RNA concentration in the cell varies and is highest during active protein synthesis. Hence, during periods of growth the RNA/DNA ratio increases. Bulow (1970) demonstrated some decrease in the ratio in golden shiners deprived of food, while resuming feeding resulted in rapid increase which had reached a plateau within four days. In the bluegill, Bulow *et al.* (1981) found marked seasonal variations. Ratios were high in the spring and autumn and low during the summer and winter. They considered that the low summer values were due to high temperatures and low dissolved oxygen.

An early study was that by Keil *et al.* (1971), who examined the effect of a PCB on the growth and nucleic acid concentration in a marine diatom. They noted a decrease in the amount of RNA (halved at a concentration of 0.1 ppm PCB, which gave over 100 ppm in the diatom) but no effect on growth or DNA concentration.

McKee and Knowles (1986) examined the concentration of DNA and RNA in *Daphnia magna* exposed to chlordecone (kepone). Modest, but statistically significant, increases of both DNA and RNA were noted at lower concentration, although the RNA/DNA ratio was not affected. The authors concluded that changes in nucleic acids were less sensitive than either reproduction or survival of the organism.

Wilder and Stanley (1983) found significant correlations between RNA/DNA ratio in salmonid fishes and overall growth, and that restricted food intake caused a decrease of the concentration of RNA. No effect was seen in brook trout from streams exposed to carbaryl via drift from forest spraying. The actual levels of carbaryl are unknown, nor was acetylcholinesterase activity measured, so the exposure of the fish is presumed rather than proved.

The effects of a number of toxicants on the levels of nucleic acids of larval fish were examined by Barron and Adelman (1984). For the toxicants of most environmental relevance, chromium and the hydrocarbons p-cresol and benzophenone, the decrease of RNA was more marked than that of DNA.

The effects of lead, in combination with adequate or deficient levels of iron, were examined in quail by Stone and Fox (1984). They found that an increase of lead levels in the diet caused a slight decrease in RNA in birds on the adequate iron diet and that incorporation of orotic acid (a precursor in RNA synthesis) into RNA was decreased by iron deficiency.

The effect of crude oil on hepatic RNA/DNA in mice was studied by Leighton (1990). Increases on a total liver basis of both RNA and DNA were found, but when the increased liver size was allowed for by expressing the results as mg/g liver, no effect was seen on the concentration of DNA and only a modest (10%) increase in RNA.

The RNA/DNA ratio is a useful, but non-specific, indicator of recent growth and general nutrient condition but is probably not an indicator of DNA damage.

4.3 DNA adducts

The covalent binding of environmental pollutants to DNA – adduct formation – is a clear demonstration of exposure to these agents and an indication of possible adverse effects. The relationship between the degree of adduct formation and environmental levels causing them is complex. Besides the questions of rates of uptake, metabolism and excretion, which are common to any measure, chemical or biological, there is the question of the stability of the adduct itself.

While many methods have been developed, the comparatively few that have been used in environmental studies are based on radioactive post-labelling or identification of specific adducts by fluorescence spectrophotometry and various chromatographic techniques.

The techniques based on mono and polyclonal antibodies are highly sensitive, and enzyme-linked immunosorbent assays (ELISA) are capable of detecting adduct concentrations of 10^{-18}M, which corresponds to about one adduct in 10^8 normal nucleotides (Poirier, 1984). The major limitation of immunological techniques is that specific antibodies have to be developed for each adduct. The advantages are sensitivity and, once the antibodies are available, the rapidity of the assays. So far, these techniques do not seem to have been used in field investigations. However, it may reasonably be assumed that this last statement will soon be out-of-date.

In post-labelling techniques the DNA is enzymatically hydrolysed to the 3'-monophosphates of the normal nucleotides and adducts. The adducts are then concentrated either enzymatically or by chemical extraction. These are labelled with ^{32}P phosphate; then the nucleotides can be separated by thin-layer chromatography and adducts will show up as an unusual spot on the chromatogram. This assay does not require a knowledge of the identity of the chemical adducted to the nucleotide. The sensitivity of the technique can be as high as one adduct in 10^{10} normal nucleotides, which is the equi-

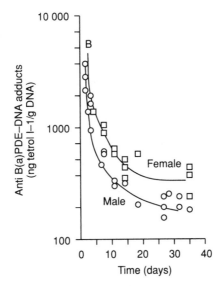

Figure 4.1 Rate of change in BaP adducts in mouse skin with time. Shugart and Kao (1985).

valent of one adduct per mammalian cell (Gupta and Randerath, 1988). The disadvantage of the technique is that it is labour intensive and the adduct is not specifically identified.

The degree of formation of specific adducts can be measured by high-performance liquid chromatography (HPLC) /fluorescence analysis. Essentially the method consists of the acid-induced removal of the adduct from DNA, followed by the separation and quantification of the free tetrols. The method allows fmol quantities of BaP to be detected (Shugart *et al.*, 1983; Shugart, 1986). This technique is also labour intensive, but gives quantitative data on the level of adduct formation.

The two techniques give different information. The former gives an index of the degree of total covalent binding, whereas the latter gives information on the actual degree of binding for a few specific compounds.

The persistence of BaP adducts in mouse skin DNA and haemoglobin was studied by Shugart and Kao (1985). A good dose response of degree of adduct formation and the dose of BaP was found for both haemoglobin and skin DNA. They found that the decrease of the number of adducts in skin was initially very rapid (Figure 4.1) but that a small proportion of adducts persisted after four weeks. The half-life of this more stable population of adducts was about 30 days. A difference in the time course between sexes was noted, with the decay of adducts being more rapid in male mice.

Studies by Varanasi *et al.* (1989a) have demonstrated that benzo [a] pyrene-diolepoxide DNA adducts formed in the liver of English sole are extremely stable. These workers found that the level of the major adduct remained essentially unchanged over a period of 1 to 60 days after exposure to a single dose of BaP. Since BaP itself is rapidly metabolized, this persistence appears to be due to the stability of the adduct, not to a steady state situation in which rates of formation and repair of adducts balance out. The persistence of BaP adducts in the English sole is much higher, by at least an order of magnitude, than that found in mice (Shugart and Kao, 1985).

High incidence of hepatic lesions, including neoplasms and cellular alterations, was found in English sole exposed to PAH-contaminated sediments in Puget Sound in the western United States. The composition of the PAHs was characteristic of creosote, and the stomach contents contained high levels compared to those from a control area which was free from detectable neoplasms. Levels of PAHs in the bile were 20 times higher than those from a control population. In studies on the same species, Varanasi *et al.* (1989a) found the level of adducts was linear with dose over the range 2–100 mgBaP/kg body weight. However, adducts could not be detected at a dosage of 0.1 mg/kg, although significant induction of AHH was observed. The levels of the major adducts remained essentially constant over a period of 60 days following a single injection of BaP, whereas the levels of metabolites in bile decreased by 86% over this period. The detection of the adduct BaP-diolepoxide-DNA, which is considered to be the ultimate carcinogen, gives additional support to the hypothesis that exposure to PAHs is an important factor in the occurrence of neoplasms in fish.

The levels of adduct formation in the liver and haemoglobin in bluegill sunfish exposed to a single dose of benzopyrene were examined by Shugart *et al.* (1987). The degree of adduct formation increased with temperature being 2–3-fold higher in the liver at 20 °C compared to 13 °C. The ratios were even higher in haemoglobin. The number of adducts formed per macromolecule was much higher, roughly two orders of magnitude, for hepatic DNA than haemoglobin.

Use in monitoring The ^{32}P-post-labelling technique has been used by Varanasi *et al.* (1989b) to measure the DNA adducts in English sole and winter flounder. The area of autoradiograms used corresponds to polynucleic aromatic hydrocarbons (PAHs) with four and five benzenoid rings. These workers found markedly elevated levels in both areas of Puget Sound, in the state of Washington, that they examined. Values ranged from 5 to 93 and 3 to 27 nmol/mol nucleotides compared to less than 0.2 nmol/mol nucleotides for a sample collected from a control area. Levels in winter flounder from Boston Harbor, Massachusetts, were in the same range as the Puget Sound samples, for example, 3 to 25, although no control values are given for this species. The mean levels of PAHs in the bile agreed with the mean levels of adduct formation, but the correlation on an individual basis was weak. It is likely that this discrepancy is explained by the difference in the time sequence of the two events since adduct formation represents exposure over weeks (Varanasi *et al.*, 1989a), whereas changes in bile are rapid.

Dunn *et al.* (1987) examined the formation of aromatic DNA adducts in brown bullheads from the Buffalo and Detroit rivers. In both cases chromatograms of the ^{32}P-post-labelling studies showed the presence of PAHs which were absent in aquarium-raised fish. The actual adduct levels were determined on some samples by HPLC. The values for the Buffalo River averaged 70 nmoles adduct/mol nucleotide, while two samples from the Detroit River had values of 52 and 56. The average value for aquarium-raised fish was 15 nmol/mol.

Some preliminary studies have been carried out on beluga whale in the St Lawrence estuary (Shugart *et al.*, 1989). Detectable levels of BaP-DNA adducts have been found in the brains of three whales, ranging from 69 to 206 ng/g DNA. No adducts could be detected in four samples from the Mackenzie Delta.

The use of covalent binding of xenobiotics to haemoglobin rather than to DNA has been reviewed by Neumann (1984). He found that for a given chemical there was a constant ratio between the reaction with tissue DNA and haemoglobin and concluded that the reaction followed first-order kinetics. The use of haemoglobin enables serial studies to be made on the same individual.

The formation of adducts represents a good measure of exposure of organisms to PAHs. The stability of DNA and haemoglobin adducts to this class of compounds means that their presence can be detected after they have otherwise been cleared from the body. The fact that adduct formation to haemoglobin can be studied means that non-destructive testing is possible. A good correlation between the number of adducts found in skin DNA and haemoglobin was found in mice exposed to BaP (Figure 4.2). However, it should be noted that the absolute number of adducts was much lower in haemoglobin than in skin.

Monitoring of adduct formation provides one of the best means to detect exposure to PAHs, which, while not highly persistent, are capable of causing serious harm. The exact relationship between adduct formation and carcinogenesis is under intensive study in relation to human health. In the environmental field the correlation between macromolecular damage and the epidemiological data is in an earlier stage, but has been started for a number of fish studies.

4.4 DNA strand breakage

Breakage in chromosomes can be examined directly under the microscope, using squash preparation or solid tissue preparations (Bloom, 1981), or *in vitro* by the alkaline unwinding assay (Kanter and Schwartz, 1979). The latter technique is based on the fact that DNA strand separation, under

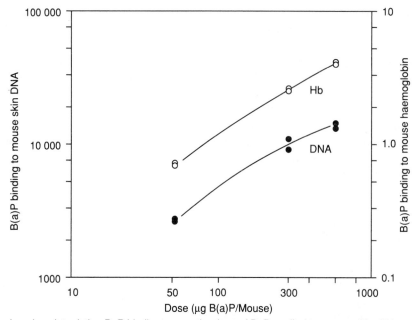

Log–log plot relating BaP binding versus the dose of BaP applied to mouse skin: (○) BaP binding of haemoglobin expressed as pg tetrol I-1/mg haemoglobin; (●) BaP binding to DNA expressed as ng tetrol I-1/g DNA. Each data point represents the average values from two mice.

Figure 4.2 Relationship of adduct formation in mouse skin DNA and haemoglobin after exposure to BaP. Shugart and Kao (1985).

carefully defined conditions of pH and temperature, takes place at the single-stranded breaks within the molecule. The amount of double-stranded DNA remaining after alkaline unwinding is inversely proportional to the number of strand breaks, provided that renaturation of the DNA is prevented.

The time course of BaP-induced changes to DNA in the bluegill sunfish and the fathead minnow over a 40-day exposure period to 1 μg/l followed by a recovery period of 50 days was examined by Shugart (1988). Initially, there was a rapid decrease in the fraction of double-stranded DNA and a concomitant rise in the number of strand breaks (Figure 4.3). The percentage of double-stranded DNA returned to control levels after 30 days and remained essentially constant over the remaining period of the experiment. The eventual disappearance of the damage to DNA indicates the ability of the organism to repair such damage.

Strand breakage and rates of repair of the DNA of hamster cells exposed to various metals were studied by Robinson *et al.* (1984). After incubating cells with soluble metal salts, they found that high concentrations of lead acetate and nickel chloride were required to cause breakage, whereas cal-

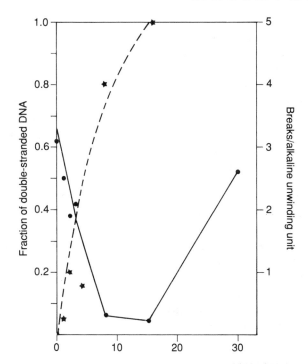

Figure 4.3 Number of breaks and fraction of double-stranded DNA in sunfish exposed to BaP. Shugart (1988).

cium chromate and mercuric chloride caused breakage at much lower concentrations. In preparations of isolated nucleoids, calcium chromate was not effective in causing breaks, but both nickel and mercuric chlorides caused breaks. These results show that chromium requires metabolic processes to produce breaks in DNA, whereas nickel does not require a similar type of activation. Mercuric chloride was by far the strongest inducer of breaks, but is considered to cause cell death rather than a carcinogenic response.

Use in monitoring Increased chromosomal aberrations (breaks and other abnormalities) were found in two rodent species (white-footed mouse and cotton rat) collected from areas close to a petrochemical waste disposal site (McBee *et al.*, 1987). The values of mean number of aberrant cells, chromatid breaks, number of acentric fragments and translocations were similar for both species for both control sites and significantly increased at both waste dump sites, indicating that cytogenetic analysis of small mammals is a feasible test system for environmental mutagenesis. In a subsequent study, McBee and Bickham (1988) demonstrated that flow cytometry could be used to differentiate between DNA of animals from waste dump sites and that from control areas. While giving less detailed information than standard karyology tests, flow cytometry is much less time-consuming.

Chromosome breaks in the gills of the mudminnow were used by Prein *et al.* (1978) to study pollution on the Rhine. The percentage of metaphases with at least one break ranged from 2 to 8% and the number of breaks per metaphase from 0.02 to 0.08 for control fish. After three and seven days the values for mudminnows maintained in water from the Rhine were 13–15% and 0.14–0.16 respectively. After 11 days these values had increased to 28% and 0.31 breaks/metaphase.

The findings of Shugart (1988) that the initial damage caused by BaP to fish DNA disappears, presumably due to repair, after a period of about 16 days indicate the type of problem that needs to be solved before this technique can be used as a reliable biomarker. Nevertheless an increased level of breaks was reported in bluegill for a contaminated stream in Tennessee (Shugart *et al.*, 1989). Similar time sequence data on damage and repair of DNA in other classes of organisms, for example, birds, are lacking.

Examination of liver samples from young double-crested cormorants and herring gulls collected in 1987 from Lake Ontario showed no difference from control material collected from the Atlantic coast (L.R. Shugart and D.B. Peakall, unpublished data).

4.5 Degree of methylation of DNA

In a wide variety of animals, from mammals to crustaceans, DNA is modified after synthesis by enzymatic conversion of cytosine to 5-methylcytosine (5 MeC). It is considered that this methylation plays an important role in the control mechanisms that govern gene function and differentiation (Razin and Riggs, 1980; Razin and Szyf, 1984). In a number of studies reviewed by Razin and co-workers, it has been demonstrated that there is a corrrelation between gene activity and the proportion of 5 MeC in the genetic material.

Shugart (1990b) examined the percentage of 5 MeC in the DNA of bluegill sunfish that had been exposed to BaP. He found a decrease of 38% in the amount of methylated cytidine over the entire period of the experiment, which consisted of 30 days of exposure and a recovery period of 45 days. The experiments were carried out on samples stored from a previous experiment and the sample size was limited. Nevertheless, the data indicate a decrease over the entire period, with the lowest value occurring at the end of the recovery period. No straightforward correlation was found with previous data on either strand breakage or adduct formation.

A decrease in the percentage of 5MeC has been noted in pre-fledgling herring gulls collected in 1987 from Lake Ontario as compared to marine material. No significant difference was noted in double-crested cormorants (L.R. Shugart and D.B. Peakall, unpublished data). These studies of genetic material were carried out after the major effects from toxic chemicals had

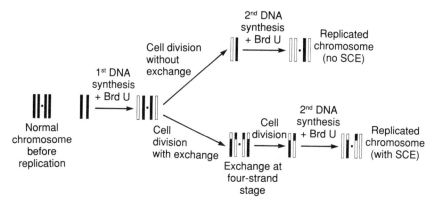

Figure 4.4 Diagram of sister chromatid exchange. After Das (1988).

passed and chemical exposure was greatly reduced. Currently, detailed studies are under way to look for possible changes in DNA in cormorant samples collected from Green Bay, where an elevated prevalence of congenital anomalies still occurs (Environment Canada, 1991).

4.6 Sister chromatid exchange

Sister chromatid exchange (SCE) is the reciprocal interchange of DNA at homologous loci between chromatids at the four-strand stage during the replication of chromosomal DNA. This process is shown diagrammatically in Figure 4.4. Chromosomes of cells that have gone through one DNA replication in the presence of labelled thymidine or nucleic acid analogue 5-bromodeoxyuridine and then replicated again in the absence of the label are generally labelled in only one of the chromatids. The label is observed to be exchanged from one chromatid to the other. The sister chromatid exchanges can be easily visualized using the light microscope in metaphase by auto-radiography or by differential staining using fluorescent dyes in the case of 5-bromodeoxyuridine.

Chromosomes that have undergone SCE should not be regarded as damaged in the conventional sense since they are morphologically intact. Nevertheless, SCE occurs at sites of mutational events including chromatid breakage. In order to produce an SCE, a lesion must pass through the S-phase and then the SCE can be detected during the following mitosis. Thus increased strand breakages during the S-phase lead to increased SCE. In many cases SCE is more readily observed than the damage itself.

Good correlations have been observed between the number of induced SCEs per cell against the dosage of X-ray and the concentration of a number of chemicals known to cause chromosomal aberrations (Perry and Evans, 1975). None of the mutagens studied by Perry and Evans were of environmental importance. Although SCE analysis can be used as a quantitative indicator of mutagenesis, there is a wide range of number of mutations per cell caused by individual chemicals or a class of chemicals (Carrano *et al.*, 1978).

Factors affecting activity The effects of cell culture components on rate of SCE rate observed have been reviewed in some detail by Das (1988). The frequency of SCE varies with different culture media, with the type and amount of serum and antibiotics used, and the anticoagulants used in the collection of blood samples. The concentration of 5-bromodeoxyuridine is also important since it can cause effects on the SCE rate. It is necessary to use the lowest concentration that enables one to visualize the changes. Fortunately, for the use of this technique on field collected samples, there does not seem to be a difference between fresh and stored blood (Lambert *et al.*, 1982). Das (1988) compiled a table summarizing the base level of SCE values in human peripheral blood found in the literature. Based on 26 published studies, SCEs/cell varied from 5 to 24, with the data strongly skewed towards the high end. Eighteen studies fell within the range of 5.1 to 8.1.

The chicken embryo can be readily used for SCE and other chromosomal studies (Bloom, 1981). Topical application of a solution of the chemical onto the inner shell membrane is the simplest and least traumatic technique. Direct application to the embryo can be used where immediate application to a specific area is desired. A window has to be cut into the shell, which is subsequently sealed and the egg incubated in the horizontal position. Direct injection into the yolk-sac can be used, especially for lipid-soluble compounds, but this technique is more often unsuccessful. The spontaneous rates of SCE in several strains of chicken do not show significant differences. In a subsequent review (Bloom, 1984) considered the *in vivo* induction potential of 53 compounds, including both mutagens and nonmutagens. He found that 90% of the mutagens induced SCEs and that all of the nonmutagens failed to induce SCEs above the baseline. The changes in SCEs could be related to mutagenic potency, DNA inhibition and carcinogenic activity. Some clastogens did not induce SCE but caused massive chromosomal damage that was readily seen in the chick assay. In the case of promutagens, which included several PAHs of environmental interest, a parallel between SCE and the induction of the MFO system, which activates such promutagens as benzo[a]pyrene, was found.

A test system involving the detection of direct DNA labelling and SCE induction after placing chemicals on the anal lips of the bursa of Fabricius in the developing chicken has been established (Bloom *et al.*, 1987). The bursa of Fabricius is a primary site for the early development and differentiation of the B-lymphocyte. The uptake of DNA-specific dye (fluorochrome 4′-6-diamidino-2-phenylindole), which labels cell nuclei by binding to DNA and is readily detected by ultraviolet excitation, was studied in addition to the rate of SCE.

Although the data are conflicting, the general impression is that rate of SCE occurrence increases with age in human being (Waksvik *et al.*, 1981). Margolin and Shelby (1985) investigated the statistical aspects of studies on the difference between sexes. They examined the results of 12 human studies, each of which showed small and usually non-significant differences. Using a Wilcoxon ranking test, a significant difference was shown, with the levels of SCE being higher in females than in males. This finding was confirmed in a study with a large sample size undertaken by Soper *et al.* (1984). Margolin (1988) concluded that analysis of a series of small, under-powered studies of a small effect can, in his words, lead to the creation of a scientific myth.

Nayak and Petras (1985) found that wild caught mice and laboratory mice housed in outdoor enclosures had higher SCE values than laboratory housed mice (wild or inbred). Female mice had significantly higher values than males.

Review of experimental studies A dose-dependent response of SCE/cell in mussels to mitomycin C, a compound used to block protein synthesis, was found by Dixon and Clarke (1982), indicating the possibility of using this organism to examine for cytogenetic effects. In a subsequent study a dose-dependent increase in the number of SCEs/cell was found for cyclophosphamide (a widely used promutagen) in mussels that were also exposed to phenobarbital, a classic inducer of the mixed function oxidase system (Dixon *et al.*, 1985). While no mechanistic link was established, the results indicate the MFO system produces metabolites that cause cytogenetic changes.

Studies by Dixon and Prosser (1986) found that tributyl tin oxide (TBTO) did not alter SCE occurrence in mussels or the frequency of occurrence of chromosomal aberrations. However, there was a clear, dose-dependent relationship between TBTO concentration and larval survival.

The frequency of SCE and chromosomal aberrations was examined by Nayak *et al.* (1989a) in maternal bone and marrow and in foetal liver and lung cells in mice exposed to lead on gestational day nine. A moderate, but statistically significant, increase in the frequency of SCEs was found in maternal bone marrow cells. Several specific chromosomal aberrations, mainly deletions, were found in the maternal bone marrow and foetal cells of lead-treated animals. In a parallel study (Nayak *et al.*, 1989b), no significant changes in SCEs in maternal and foetal cells were observed after exposure to cadmium chloride. Foetal tissues showed mitotic inhibition at the two highest doses (8.4 and 11.4 mg/kg); however, significant chromosomal changes were seen only at the highest dose.

Use in monitoring Nayak and Petras (1985) examined the SCE level in wild mice from various locations in southern Ontario. They found significant negative correlations between SCE level and both the distance to the Windsor-Detroit complex and the distance to the nearest industrial complex. The authors consider that SCE analysis in mice could be used as a first-line monitoring procedure for the detection of changes in environmental genotoxicity. While these findings are interesting, it should be noted that the sample sizes used in this study were small.

Chromosomal and chromatid aberrations in cotton rats living close to two hazardous waste dumps were compared to material from a control site by Thompson *et al.* (1988). Considering all types of aberrations together, the occurrence rate was higher from the test sites than from the control area. Individual tests, such as chromatid breaks, showed higher values in rats from the waste dump sites, but these increases were not statistically significant.

The rate of SCE was measured in seven-day herring gull embryos from eggs collected in 1981 from five colonies within the Great Lakes basin and one Atlantic coast colony (Ellenton and McPherson, 1983). No significant differences were found among the colonies, nor was the rate elevated on the Great Lakes as compared to the control colony. Additional tests for genotoxicity of herring gull egg extracts were carried out using the Ames test for induction of point mutations and in Chinese hamster ovary cells for the induction of SCE and chromosome aberrations (Ellenton *et al.*, 1983). None of the extracts were mutagenic in *Salmonella*. In the experiments with mammalian cells, all extracts, including controls, gave positive responses to both SCE and chromosome aberrations. It is considered likely that a lipophilic compound or compounds of biological origin caused the effects.

The frequency of SCE in fish exposed to water from the River Rhine was reported by Van der Gaag *et al.* (1983). They found that for mudminnows the SCE frequency increased from a control value of 0.050 to 0.128 after 3 days and further increased to 0.155 after 11 days. For *Nothobranchius* the values increased from 0.055 to 0.104 after a week.

4.7 Use of genetic material in monitoring

Some surveys, such as those by Nayak and Petras (1985) using the level of SCEs in wild mice in southern Ontario and Dunn *et al.* (1987) on DNA adducts in brown bullheads from the Buffalo and Detroit rivers, indicate that monitoring would be valuable.

Monitoring has been carried out on the incidence of tumours in fish. Despite the fact that these data are pathological rather than biochemical, they are worth reviewing briefly in order to show the extent of the problem. Fish are frequently sampled in considerable numbers and, more importantly, many of their tumours are visible externally. Such data could not be readily collected from other classes of species even when large sample sizes are available, such as muskrat from trappers, or duck from hunters, due to the cost of dissection.

Sampling Locations for Brown Bullheads and White Suckers
(followed by the year when fish were sampled)

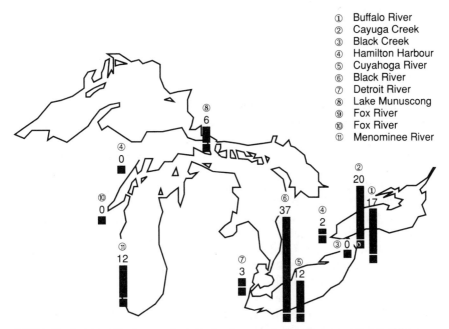

① Buffalo River
② Cayuga Creek
③ Black Creek
④ Hamilton Harbour
⑤ Cuyahoga River
⑥ Black River
⑦ Detroit River
⑧ Lake Munuscong
⑨ Fox River
⑩ Fox River
⑪ Menominee River

NOTE: The height of the bar graphs indicates the prevalence (as a percentage) of liver tumours in populations of brown bullheads and white suckers sampled from Great Lakes tributaries. These data were collected in various years during the 1980s

Figure 4.5 Incidence of tumours in Great Lakes fish. Department of Fisheries and Oceans.

In the North American Great Lakes, surveys to determine the incidence of tumours have been carried out as part of the surveillance programme. The levels of occurrence of tumours in brown bullheads and white suckers sampled from the Great Lakes are shown diagrammatically in Figure 4.5. Given the large number of contaminants in the Great Lakes, it is virtually impossible to link carcinogenesis to a specific chemical, but circumstantial evidence for a chemical origin is strong.

Neoplasms may be caused by viral agents and parasites as well as by chemicals. There is difficulty in proving that chemical exposure is definitely the cause of the tumours observed and even more difficulty in proving which specific chemical caused the tumour. Dawe (1987) provided the criteria used to separate environmental from hereditary cancers in fish. These criteria include an association of diet, geographic, distributional and lifestyle that exposes the population to xenobiotics; a broad range of histotypes; and increasing tumour occurrence in older animals. He concludes that 'a relatively high prevalence (which means as little as 1% or greater) of one or more tumour histotypes in a feral species in a habitat where there is no reason to suspect constrictions on the gene pool, and where known xenobiotic exists, favors an environmental etiology and justifies expenditure of effort towards identifying carcinogens.'

Many neoplasms occur in relatively unpolluted environments and appear to have no obvious association with environmental contaminants. For example, lymphosarcoma, a malignant tumour of pike, is virally induced and occurs in clean as well as polluted sites (Sonstegard, 1977). It is unlikely, given the large number of contaminants in the Great Lakes, that carcinogenesis in feral fish populations can be definitely linked to a specific chemical. The most convincing evidence for chemical carcinogenesis in the Great Lakes has been the occurrence of liver and skin tumours in brown bullheads exposed to PAHs. Baumann *et al.* (1987) and Black (1983) report an increased prevalence of tumours in bullheads from the Black River and the Buffalo River (Figure 4.5). The strength of association increased as laboratory and field studies confirmed the presence of PAHs in the diet of brown bullheads (Maccubbin *et al.*, 1985) and the carcinogenic impact of sediment extracts on tumour induction in bullheads and mice (Black, 1988).

Skin and liver tumours have been reported in white suckers from several sites on Lake Ontario (Figure 4.5). Liver lesions include bile duct hyperplasia and low levels of hepatocellular carcinoma. Hepatocellular tumours were found in white suckers from Hamilton Harbour, Sixteen Mile Creek and the Humber River. They were not found in fish from the other sites. This tumour is often associated with contaminant exposure in laboratory induction studies. The chemicals responsible are unknown, but there are good data showing that Hamilton Harbour sediments are heavily contaminated with PAHs, including benzo[a]pyrene, a potent mammalian carcinogen. Metcalfe *et al.* (1988) successfully induced liver tumours in rainbow trout, following a single exposure to sediment extract from Hamilton Harbour.

Although the presence of tumours in Ontario fish was reported by Gaylord (1910), few systematic tumour surveys were conducted in the Great Lakes prior to 1965. Growing awareness of widespread chemical contamination in the Great Lakes provided incentive for understanding the effects of contaminants on biota. Tumour surveys focused on urban and industrial areas and on species such as the benthic-feeding brown bullhead, which is likely to be exposed to contaminated sediments.

The studies on the Great Lakes have been extensive rather than intensive. The best intensive studies are those that have been carried out on English sole and starry flounder in Puget Sound, Washington. These studies have recently been reviewed in a special issue of *Science to the Total Environment*, devoted to chemical contaminants and fish tumours (Myers *et al.*, 1990; Stein *et al.*, 1990). Much of the work has focused on PAHs rather than PHAHs. While many PAHs are known carcinogens, the potential for some PHAHs, especially TCDD, is also well known. Regression analyses to relate the sediment levels of PAHs, PCBs and chlorinated butadienes to hepatic neoplasms in fish have been carried out (Malins *et al.*, 1988). The prevalence of neoplasms was positively correlated with both PAHs and PCBs, but negatively correlated with the chlorinated butadienes. As would be expected for a condition with a long incubation period, fish size was an important correlate. An analysis based on dioxin equivalents rather than total PHAHs would be interesting. Dioxin equivalents have been related to AHH induction in fish (Casterline *et al.*, 1983). Extension of this approach to other end points, such as has been done for reproduction in fish-eating birds (section 5.6), should be carried out.

Overall, the studies involving DNA have reached a fascinating stage. A great deal of work is being carried out at the molecular level; the linkage of these alterations has led to effects on the organism, and there are strong indications that this information can be used in wildlife toxicology.

5   *Mixed function oxidases*

5.1 Introduction

The haem-containing enzymes known as cytochromes are a major component of the defences of organisms against toxic chemicals in their environment. Originally evolved to handle naturally occurring toxic compounds, they now play an important role in the detoxification of man-made chemicals. Nebert *et al.* (1989) consider that the ancestral cytochrome gene is probably 2000 million years old. The major divergence occurred 800 to 1000 million years ago when animals began using plants as food, and self-defence mechanisms against toxins in plants evolved. Later, additional families of cytochromes evolved in response to the necessity to metabolize combustion products.

Xenobiotic transformation occurs in two stages. In Phase I, polar groups are introduced to the molecule through oxidation, reduction, hydrolysis and transfer of groups. Phase II involves conjugation of the chemical or its metabolites with polar compounds such as glucuronic acid, glutathione, sugars, sulphates and phosphates to form water-soluble compounds which are readily excreted. Phase I and Phase II reactions usually work together in a sequential manner; different Phase I and Phase II enzymes compete for the parent xenobiotics and their metabolites. Xenobiotics generally undergo several different biotransformations simultaneously, which can result in the formation of a large number of metabolites and conjugates. Despite their importance in detoxification, the Phase II enzymes are often ignored. Little is known about the effects of environmental contaminants on their activity. However, the species variation of one of the most important of the Phase II enzymes, UDP-glucuronyltransferase, has been studied by Walker (1980).

In addition to the regulation of cytochrome-dependent enzyme activity by induction, there also exists a more rapid regulation based on phosphorylation. The process is selective, leading to phosphorylation of specific isozymes by defined protein kinases. This can rapidly lead to marked changes in metabolism which are selective for given substrates. Phosphorylation of cytochrome P450II has been shown to influence metabolism by altering cytochrome-dependent enzyme activity as well as cytochrome levels. The field has been recently reviewed by Oesch-Bartlomowicz and Oesch (1990). So far there do not appear to be any studies on the possible interaction of pollutants with this aspect of the mixed function oxidase system.

5.2 Nomenclature

The nomenclature of this important system is confused, although recently there has been some improvement. Initially the system was referred to as 'drug metabolizing enzymes' by the pharmacologists who discovered it. Now there are two names in widespread use: 'the mixed function oxidases', and 'the monooxygenase system'. In my opinion the first name gives a better idea of the physiological role of the system, as it certainly carries out many functions, whereas the second gives a better idea of the underlying biochemistry. Faced with these alternatives, I decided to go with the more physiological name of 'mixed function oxidases'.

The second confusion of nomenclature occurs over individual cytochromes. Initially, division was made into two major classes, the P450s (the number referring to the wavelength at which the cytochrome could be detected) and the P448s, with some falling in between, called mixed inducers. They are sometimes called PB-type and MC-type, referring to the compounds – phenobarbital and 3-methylcholanthrene – which were first used to induce the system, and also by the use of initials, i.e. P450PB and P450MC. Other systems with single initials have also been used, and some workers use P450 to refer to any cytochrome system.

Recent work has shown the complexity of the system. The evolution and classification of the cytochromes have been considered by Nebert and co-workers. In a recent paper in a special issue of *Xenobiotica* devoted to this subject, they state that the P450 superfamily is 'presently known to contain more than 78 members divided into 14 families' (Nebert *et al.*, 1989). A summary of the major families of P450s is given in Table 5.1. In a paper in the same issue, Stegeman (1989) reviews the forms of P450 in fish and discusses their relationship to the mammalian forms. As far as possible, the convention followed in this chapter is based on this recent work. Under this system, P448 becomes P450I and P450 becomes P450II. The term *P450* is used to refer to the system as a whole or when it is not possible to identify the cytochrome specifically.

5.3 General description of the system

The mixed function oxidase (MFO) system is a coupled electron-transport system composed of two enzymes – a cytochrome and a flavoprotein, NADPH-cytochrome reductase. The system occurs in most organs, but the activity is much higher in the liver than elsewhere, and it is the hepatic system that is referred to here unless otherwise stated. Cytochrome P450II dependent oxidations include aromatic hydroxylation, acyclic hydroxylation,

Table 5.1 Classification of P450s

Nomenclature	Induced by/specificity
P450I	Polycyclic aromatic, TCDD
P450II	Phenobarbital-inducible family*
P450IIA	Specific for testosterone hydroxylase
P450IIB	PB inducible
P450IIC	PB inducible
P450IID	Specific for debrisoquine 4-hydroxylase
P450IIE	Ethanol inducible
P450III	Steroid inducible
P450IV	Specific to lauric acid w-hydroxylation
P450XI	Located in mitochondrion
P450XIA	
P450XIB	
P450XVII	Formation of steroid 17-hydroxylases
P450XIX	Involved in synthesis of oestrogens
P450XXI	Formation of steroid 21-hydroxylases
P450LI	Plant/yeast
P450CI	Prokaryote

* PB-inducible genes largely confined to P450IIB and C.
After Nebert and Gonzalez (1987).

dealkylation and deamination. The most widely studied enzymes are benzphetamine N-demethylase and aldrin epoxidase. Cytochrome P450I dependent reactions include N-oxidation and S-oxidation; widely studied enzymes include ethoxyresorufin O-deethylase (EROD), benzo[a]pyrene hydroxylase (BaPH), and aryl hydrocarbon hydroxylase (AHH).

The study of hepatic metabolizing enzymes was pioneered by Conney and co-workers (Conney *et al.*, 1956; see also subsequent reviews, Conney, 1967 and 1982). Initially they were investigating the finding of Richardson *et al.* (1952) that the carcinogenic potential of an aminoazo dye was greatly reduced in rats pretreated with 3-methylcholanthrene.

The first demonstration of the effect of an organochlorine pesticide on hepatic microsomal metabolism was made by Hart *et al.* (1963). This important finding was a fortuitous result of an experiment to examine the effects of food deprivation on drug metabolism. These workers noted an unexpected decrease in the sleeping time of rats given phenobarbital after their cages were sprayed with chlordane. They deduced that this was due to stimulation of hepatic microsomal drug metabolizing enzymes, the correctness of which deduction they were able to demonstrate by controlled experiments with chlordane.

Induction of the MFO system increases the rate of metabolism of a wide variety of substances. Street and Chadwick (1967) demonstrated that rats exposed to 50 ppm DDT showed a tenfold increase in the rate of excretion of the polar metabolites of dieldrin. Organophosphates are also degraded by MFOs. For example, Neal (1967) showed that pretreatment of rats with PB increased the *in vitro* metabolism of parathion. Although the system normally results in detoxification and enhanced excretion, it can also cause the formation of more toxic metabolites such as epoxides, arylamines and oxygen uncoupling (Parke *et al.*, 1985; Guengerich, 1987).

Activity of mixed function oxidases is affected by a wide variety of compounds. Classes of compounds of environmental interest besides the organochlorines include the organophosphates, pyrethroids and polynuclear aromatic hydrocarbons. Studies on the inhibition of MFOs by organophosphate insecticides date back to 1959 (Welch *et al.*, 1959). Induction by pyrethrum was demonstrated by Springfield *et al.* (1973). Induction by polynuclear aromatic hydrocarbons was shown by Arcos *et al.* (1961). These workers examined 57 PAHs; the most potent inducers were 20-methyl cholanthrene, naphthacene and anthanthrene, while other PAHs such as coronene and 3,4-benzophenanthrene had no effect. The use of MFO systems to examine for pollution by this class of compounds is considered in section 5.7. Protection by MFO induction against the lethal effects of warfarin (Ikeda *et al.*, 1968) and strychnine (Kato, 1961) have been demonstrated.

The regulation of the P450 gene by the Ah receptor is shown diagrammatically in Figure 5.1 (after Nebert and Gonzalez, 1987). Initially the TCDD (or other environmental contaminant that is capable of binding with the Ah receptor) enters the cell and binds to a cytosolic receptor. This inducer-receptor complex is then transported into the nucleus of the cell. After interaction with a site in the nucleus, which is not well understood, transcription of specific messenger RNA occurs. This leads, in turn, to the induction of specific P450 proteins which are incorporated into the endoplasmic reticulum. Metabolism of TCDD and other xenobiotics can then occur. The metabolites produced may be innocuous products that can be excreted, or reactive intermediates that can react with critical receptors. In the latter case direct toxic effects or the initiation of cancer may occur.

5.4 Factors influencing activity

The ability of an organism to metabolize xenobiotics via the MFO system is the summation of enzyme activity, the concentration of the enzyme per unit of microsomal protein, the concentration of microsomal protein in the liver, and the ratio of liver to body weight. This summation is referred to as the total hepatic induction index (THI) (Boersma *et al.*, 1984) and can be expressed as follows:

$$\text{THI} = \frac{\text{Enzyme activity}}{\text{Cytochrome P450}} \times \frac{\text{Cytochrome P450}}{\text{Microsomal protein}}$$

$$\frac{\text{Microsomal protein}}{\text{Liver weight}} \times \frac{\text{Liver weight}}{\text{Body weight}}$$

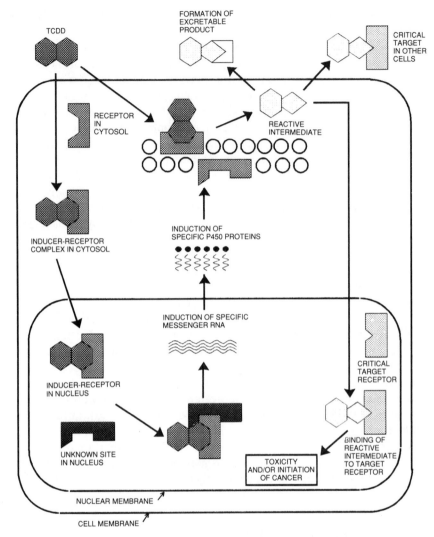

Figure 5.1 Diagram of MFO system. Nebert and Gonzalez (1987).

This equation can be cancelled down to:

$$\text{THI} \; = \; \frac{\text{Enzyme activity}}{\text{Body weight}}$$

but this summary equation does not give the insight into the processes involved. Boersma and co-workers examined the components of THI by exposing rats to benzanthracene. The results are shown diagrammatically in

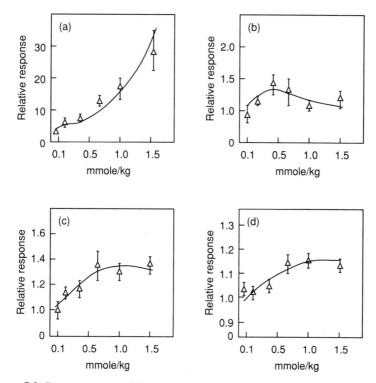

Figure 5.2 Dose response of factors comprising the total hepatic induction index relative to control. (a) Specific activity of the enzyme normalized to cytochrome P450 concentration, $F = 86.8$ ($p < 0.001$); (b) change in cytochrome P450 concentration normalized to microsomal protein concentration, $F = 1.15$ ($p > 0.05$); (c) change in microsomal protein concentration per liver mass, $F_{2,38} = 6.64$ ($p < 0.01$); (d) change in liver mass per body mass, $F = 8.54$ ($p < 0.005$). Boersma *et al.* (1984).

Figure 5.2. While the largest single factor is the increase of specific activity of the enzyme per unit of cytochrome, the other factors also make significant contributions to the overall ability of the animal to metabolize this xenobiotic.

Species variation The species variation of MFO activity covering several classes of organisms has been reviewed by Walker (1978 and 1980). The enzymatic activity has been measured by many different workers using many different substrates. Walker has used a relative activity formula, based on the activity of the male rat, taking into account the liver:body weight ratio. This formula is:

$$\frac{\text{Relative}}{\text{activity}} = \frac{\text{Specific activity (species A)}}{\text{Specific activity (male rat)}} \times \frac{\text{liver wt/body wt (species A)}}{\text{liver wt/body wt (male rat)}}$$

In general terms the relative activity of epoxide hydrase follows phylogenetic lines with mammals > birds and amphibia > fish, with little overlap between the three groupings. Glucuronyltransferase activities were much higher in mammals than in fish with virtually no overlap. The aldrin epoxidase activity follows the same trend, but in this case there is considerable overlap. A linear log relationship was found between relative activity and body weight for mammals (Figure 5.3.) with man being the outlier, having activity lower than would be expected by body weight. No data on the largest mammals are available. The data on birds have been updated in a recent review by Ronis and Walker (1989). Fish-eating birds show a high correlation coefficient for the regression of activity against body weight, with the values for relative activity being approximately an order of magnitude lower than for mammals of comparable weight. Other species of birds tend to have values intermediate between fish-eating birds and mammals, but the correlation of activity against body weight is considerably weaker. Low values for fish-eating birds are considered to be due to the fact that until xenobiotics contaminated aquatic food chains they had less need of these defence mechanisms. Regrettably no data appear to be available on fish-eating mammals. It would be most interesting to see if animals such as otters more closely resemble other mammals or other fish-eaters.

An exception to the lower activity of fish-eating birds is the puffin. Walker and Knight (1981) examined the enzyme activity of several species of sea birds. They found that the epoxide hydrolase activities fell within the general range for birds, but that the monooxygenase activities were low in all but this one species. There do not appear to be dietary or other ecological factors that would explain these differences. The authors consider that a possible explanation is that MFO activity is part of a resistance mechanism, but there is no firm evidence.

The kinetics of species variation have been studied in the rat, pigeon, frog and trout by Ronis and Walker (1985). By examining a range of concentration, these workers were able to examine conditions that were environmentally realistic. At saturation values they found that the ratio for activity of aldrin epoxidation was 1:0.067:0.014:0.008, whereas at lower concentrations the values changed to 1:0.057:0.1:0.03, indicating a change in the relative effectiveness of the frog to metabolize aldrin at low concentrations.

The low level of activity in fish has been considered to be due to the fact that until historical times this group of organisms has found the excretory route across the gills effective for removing most xenobiotics. It is clear that this mechanism is quite inadequate to deal with highly liposoluble organochlorine molecules that have been released into the environment by man in the last few decades. A general trend for terrestrial species to have higher enzyme activity was found by Walker (1980), who reviewed the data on several enzyme systems. The demarcation was most clear-cut with epoxide hydratase.

The relationship between feeding habits and MFO activity has also been

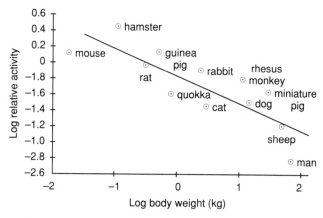

Figure 5.3 Relationship of body weight to MFO activity in mammals. Walker (1978 and 1980).

demonstrated in insects. A detailed review of the data for aldrin epoxidase has been made by Brattsten (1979). He found that aldrin epoxidase activity is lowest in monophagous species, considerably higher in oligophagous species, and some tenfold higher in the most polyphagous species.

The comparison of Phase I and Phase II hepatic transformation in quail and trout to that of a number of mammalian species commonly used in toxicity testing has been made by Gregus *et al.* (1983). They found that the overall metabolism of xenobiotics could vary several hundredfold between species.

Intra-species variation The variation in MFO activity under highly controlled conditions caused by using a specific strain of mouse obtained from three different suppliers was examined by Litterst (1978). Several substrates were used and, in a number of cases, significant differences were found. The degree of variation was quite small, the largest difference being 25%.

Hardly surprisingly, much wider variations have been found in the wild. Pedersen *et al.* (1976) examined the variation of several hepatic enzyme systems in six strains of rainbow trout. These workers found considerable differences from one enzyme system to another; AHH induction varied 20-fold, whereas other enzymes varied little.

Within-group variation can be considerable. The coefficient of variation for even the cleanest group of gulls examined, those collected on the east coast of Newfoundland, was 0.35 to 0.45 (Peakall *et al.*, 1986), and those from industrial areas were higher. The coefficient of variation in the trout for various sites ranged up to 1.0 (Pedersen *et al.*, 1976).

Sex, age and seasonal variation The variation of activity between the sexes has often been found in mammals and birds both in laboratory-maintained species and in those collected from the field. The direction of these variations is not usually consistent and is frequently different for different substrates. A review of MFO activities in fish states that activities are higher and show more seasonal variation in the female (Dixon *et al.*, 1985a). The complexities of the sex differences were demonstrated by Stegeman and Chevion (1980) in rainbow and brook trout. They found that the levels of cytochrome P450 and aminopyrine demethylase activity were markedly higher in the males, but that when the enzyme activity was normalized to the cytochrome content it was greater in the females.

The activities of aldrin epoxidase (AE), EROD, and 7-ethoxycoumarin-O-deethylase (ECOD) were measured in whole embryos, liver and yolk-sac tissue of the chick embryo from day 4 to 15 (Heinrich-Hirsch *et al.*, 1990). In yolk-sac tissue substantial activity of AE and ECOD was found on day 4. Using the whole embryo, AE could be detected on day 4, but EROD and ECOD were not detected until day 6 or 7. In liver tissue the activities of all three enzymes were present from day 9 and had reached maximum values by days 11 to 13. Hamilton and Bloom (1986) found that the chick embryo had substantial basal activity of AHH and that this enzyme was highly inducible by xenobiotics after day 6. The induction of enzymes in embryos is considered further in section 3.2.

The changes in a number of MFO systems were examined in the developing nestling herring gull (Peakall *et al.*, 1986). The activity of aminopyrine N-demethylase decreased significantly with age, but no consistent changes were found with EROD, BaPH, or cytochrome P450.

Marked seasonal variation of MFO activity was found in wild gulls and pigeons (Fossi *et al.*, 1990) with values being substantially higher in the breeding season. However, studies at various stages of the breeding cycle of the herring gull showed only small variations (Peakall *et al.*, 1986). Detailed studies of seasonal variation of MFO activity have been made in fish. Luxon *et al.* (1987) found that the AHH activity of lake trout decreased sharply in the autumn at the time of spawning and increased thereafter. Maximal seasonal variation was about 10-fold, but most measurements over the two-year period were within much narrower limits. No correlations with sex, age or size were found. Edwards *et al.* (1988) measured the MFO activity of winter flounder monthly for a two-year period. They found that the seasonal variation of activity was up to sixfold, being highest shortly after spawning. Variation was larger in females than in males, which in the latter did not exceed twofold. While there were a few differences in detail between the two studies, they form a good basis for the annual differences that can be expected in fish. Payne (1984) has suggested that kidney MFO activity, although lower than that found in the liver, might be better for monitoring in order to avoid the fluctuations due to the reproductive cycle.

Temperature and feeding regimes Jimenez *et al.* (1988) found that changes in temperature and feeding regimes had marked effects on EROD activity in bluegill sunfish. Animals that were not fed and were allowed to acclimatize for two weeks at 4, 13 or 26 °C showed low and almost uniform activity. In contrast fed fish had more variable activity and at the higher temperatures much higher activity (sevenfold at 26 °C).

Sensitivity and duration of effect Considering the number of possible inducers and the number of species that have been studied, only a very limited amount of the literature can be considered. As regards the duration of effects, a clear distinction must be made between the inducers that are themselves highly persistent, the PHAHs, and those that are readily metabolized, the PAHs. In the case of the PHAHs, it can be assumed that the duration of the effect is dependent on the clearance rate of the inducer, which is typically slow. In the case of the PAHs which are rapidly eliminated, we are looking at the stability of the MFO system itself. The similarity between those two sets of initials makes me regret the abandonment of the somewhat less specific term *organochlorines*, or *OCs*, but one can only swim against the tide so much.

Using a wide range of dietary levels of DDT, Gillett (1968) established that the 'no-effect' level for the induction of aldrin epoxidase in rats was 2 ppm. A dietary level of 1 ppm of DDT was found to increase the rate of metabolism of estradiol in ring doves (Peakall, 1970). In brook trout, Addison *et al.* (1981) demonstrated that MFO activity was induced by a single dose of PCBs (Aroclor 1254) and remained elevated for 20 days. They found a good correlation between body burden of PCBs and enzyme activity and calculated that an increase over background (twofold increase) occurred with a body burden of 4 $\mu g/g$.

Single doses of 0.1 ml Prudhoe Bay crude oil caused induction of aminopyrine N-demethylase and EROD in herring gulls (Peakall *et al.*, 1989). Levels had decreased significantly after 24 hours and were back to control within 72 hours.

The induction by several PAHs in five species of marine fish was studied by James and Bend (1980). In time course studies with sheepshead, the enzyme activity was maximal after three days and had returned to control values after 14 days in the summer (water temperature 26 °C), whereas in winter (water temperature 14 °C) maximal values were not reached until 8 days and still remained elevated after 28 days. Payne and Fancey (1982) give values of 0.15 to 0.60 ppm of oil in the water column to cause threefold induction in cod and sculpin. Kurelec *et al.* (1977) found that 0.17 ppm diesel oil caused 10-fold induction.

5.5 Interactions with other biomarkers

In view of the fact that MFOs are induced by such a wide variety of compounds, it is hardly surprising that such induction can be correlated with almost any other effect. Cause and effect relationships are more difficult to establish. The role of the induction of mixed function oxidases is closely involved with effects on retinol levels and PHAH-induced porphyria. These interactions are discussed in sections 6.2 and 6.3 respectively. The role of MFOs in the metabolism of indigenous steroid hormones has been considered in section 3.3 and with the immune system in section 8.2.

A strong empirical relationship between tumour promotion and induction of the hepatic P450 system was found by Lubet *et al.* (1989). The authors note that the potent inducers ($> \times 40$) – phenobarbital, barbital, ethyl-phenylhydantoin and DDT – are all potent liver tumour promoters, whereas structural analogues that are not P450 inducers all fail to display significant liver-tumour-promoting activity. The dose of DDT used over the 72-week period of the experiment was high (500 ppm). While it is stated in the abstract that DDT is a potent tumour promoter, this compound – the only one of environmental interest – is not included in either the figure relating enzyme activity to relative promotion index or in the tables. Although it has been debated, the consensus of scientific opinion is that DDT is not a carcinogen (Coulston, 1985).

5.6 Mechanism of action, receptors and dioxin equivalents

The studies into the mechanism of action, the interaction of PHAHs with receptors and the expressing of results in terms of dioxin equivalents have only been made possible by congener-specific work. Differences between the actions of different congeners of PCBs have been known for some time. Hutzinger *et al.* (1972) show differences in the metabolism of four different PCBs in pigeons, rats and trout. Johnstone *et al.* (1974) found marked differences in the ability of different isometrically pure PCBs to induce MFO activity in the rat, and similar findings were reported by Bunyan and Page (1978) in quail. However, work using individual PCB congeners has only really taken off during the last decade.

The mechanism of action of PCBs, PCDFs and PCDDs is considered to proceed via initial binding to a high-affinity, low-capacity cytosolic receptor protein. The earlier work in this field was summarized by Poland and Knutson (1982), who found a good correlation between potency of PCDDs to induce AHH activity and their toxic potency. They concluded that the cytosol-binding protein, by a variety of criteria, had the *in vitro* properties expected of the receptor for the induction of cytochrome P450I. The correlation of several lines of investigation strongly suggest that, in mice, the Ah locus is the structural gene for the cytosol receptor. Examining the toxicological and receptor-binding data, they concluded that it is likely that PHAHs exert their toxicity through the cytosol receptor.

The identification of the Ah receptor (Poland *et al.*, 1976) with stereo-specific, high affinity binding to 2,3,7,8-TCDD was a key finding in bringing molecular biology into the realm of toxicology. They found that the binding affinity of 23 PCDFs and PCDDs closely correlated with the ability of these compounds to induce AHH activity.

The ability of specific PCBs, PCDFs and PCDDs to induce the P450I system is greatly influenced by the degree of chlorination and the chlorine substitution pattern. The most toxic PCBs are those that are unsubstituted in the ortho positions, i.e. 3,3',4,4'-tetrachlorobiphenyl (TCB), 3,3',4,4',5-pentachloro-biphenyl, and 3,3',4,4',5,5'-hexachlorobiphenyl, which allows the molecule to assume a co-planar configuration. The structure-activity relationships of 50 PCB congeners for which MFO induction data were available has been carried out by Clarke (1986). She divided compounds into five groups: weak P450 or inactive, primary P448, mixed type inducers, primary P450 and entirely P450. It is the second group that contains the most toxic congeners, including those listed above, and is characterized by having both the para positions substituted and the ortho positions unsubstituted. The most toxic PCDFs and PCDDs are those substituted in the 2, 3, 7 and 8 positions (Mason *et al.*, 1986).

Sawyer and Safe (1982) showed that 15 PCB isomers and congeners had – with one exception – similar potencies to induce AHH, EROD and BaPH. Subsequently these workers showed the AHH and EROD activity of mixtures of PCBs – Arochlors, Kanechlor and environmentally important mixtures – were essentially a summation of the individual PCBs and PCDFs (Sawyer *et al.*, 1984; Sawyer and Safe 1985). In all these studies it is clear that there is a very close relationship between the induction of the various P450I enzyme systems.

The close relationship between AHH induction and body weight loss, AHH induction and thymic atrophy, and *in vivo* and *in vitro* AHH activity for 22 PCBs, PCDDs and PCDFs was demonstrated by Safe (1987). While data on receptor binding are tabulated with AHH and EROD activity values, they are not discussed nor are r values given. In his summary Safe states that 'the precise role of the Ah receptor in this process has not been delineated.' Mason *et al.* (1986) examined 14 PCDDs for receptor binding, AHH, EROD induction and toxic response (body weight loss and thymic atrophy). The relationship between receptor binding and AHH binding was not linear, but there were good correlations between AHH and EROD induction and toxic responses.

For the PCDFs the receptor-binding avidity for 2,3,4,6,7,8-PCDF was four orders of magnitude greater than most mono- and dichlorodibenzofurans and was approximately half that of 2,3,7,8-TCDD (Bandiera *et al.*, 1984). The induction potencies of AHH and EROD varied over five to six orders of magnitude. The correlation between AHH and EROD was very high ($r = 0.99$), suggesting that both oxidation processes are catalysed by the same cytochrome isozyme. The correlation between receptor binding and AHH or EROD was poor ($r = 0.55$) in contrast to earlier findings for PCBs and PCDD.

The studies on PHAHs were extended to the brominated compounds by Mason *et al.* (1987), who examined the receptor-binding affinities, AHH, EROD, and toxic response of the polybrominated dioxins in rats. Good correlations were found for AHH against thymic atrophy, which parallels earlier work by these workers on the chlorinated compounds. Similarly high correlations were found between AHH and body weight loss.

In the *Annual Review of Pharmacology and Toxicology*, Safe (1986) makes it quite clear that the interactions of enzyme induction, receptor binding and toxicological manifestations are very complex and that our knowledge is far from complete. This careful review should be required reading for anyone interested in the subject and is quoted at some length below. Safe states that 'the criteria for receptor response specificity are supported by numerous studies with genetically inbred responsive and nonresponsive strains of mice and with some mammalian cells in culture', but that 'these data that support the receptor-mediated response specifically are in contrast to data in several other studies with animals and cell cultures. Hepatic 2,3,7,8-TCDD receptor levels in guinea pigs, rats, mice, hamsters and non-human primates vary less than tenfold and these levels show no correlation between their maximal AHH inductibility and their susceptibility to the toxic effects of 2,3,7,8-TCDD and related PHAHs.' He concludes this section by stating that 'it is apparent that response specificity to Ah receptor ligands is a highly complex process that depends not only on receptor levels but also on many other factors.'

Regarding the structure-activity studies, he states that 'the SARs for PCDDs and PCDFs clearly support an Ah receptor-mediated mechanism of action of these compounds; comparable studies have been reported for other classes of PHAHs. It is assumed that the persistent effects elicited by these toxins are related to a sustained receptor-ligand occupancy of nuclear binding sites, but this has not yet been demonstrated experimentally.'

Problems that have to be faced before one can use this approach for wildlife toxicological investigations in the field are extrapolations from species to species and extrapolations from cell culture to the intact animal. Brunström and co-workers have carried out studies on avian embryos. Marked differences were found in the sensitivity of the chicken, pheasant, turkey, duck and gull (Brunström and Reutergardh, 1986; Brunström and Lund, 1988; Brunström, 1988). They found that the pheasant was 50 times less sensitive than the chicken, and other species even less sensitive. This emphasizes the difficulties of inter-species comparison since the chicken and pheasant both belong to the order Phasianidae. Gasiewicz and Rucci (1984) present evidence that the Ah receptor is homologous among several mammalian species. However, there is no relationship between sensitivity to TCDD toxicity and the hepatic concentrations of the Ah receptor. Brunström and Andersson (1988) found that the ranking of the toxicity of a small series of co-planar PCBs to the chicken embryo and the degree of induction of EROD were in the same order. Brunström (1990) examined the toxicity and potential to induce EROD of six mono-ortho-chlorinated PCBs in the chicken embryo. He found that those compounds having a chlorine adjacent to the ortho-chlorine were ten times more potent than those with a hydrogen in this position. Nevertheless, the activity was some three orders of magnitude lower than for 3,3',4,4',5-pentachlorobiphenyl. These data sets expand the work on cell cultures from mice to the avian embryo. Conversely, Bursain *et al.* (1983) found no correlation between MFO induction and effects on egg production in quail fed polybrominated biphenyls.

Elliott and co-workers (Elliott *et al.*, 1990) examined the effects of individual PCB congeners on induction of P450I and P450II enzymes in quail and kestrels and compared them to the response which had been found in rats. An overview of the results is given in Figure 5.4. The congeners were selected to cover a range of substitution patterns (non-ortho, mono-ortho, and di-ortho), and the dosages used were based on their relative presence in environmental samples and their efficacy in causing induction of mixed function oxidase enzymes in rats. The dose of the non-ortho (3,3',4,4',5) and mono-ortho (2,3,3',4,4') was based on the ED_{50} value for *in vivo* AHH induction in the rat. However, in the quail, the mono-ortho was a much more potent inducer of both P450I and P450II enzymes. The pattern of enzyme induction found in the kestrel is quite different from both the quail and the rat.

The application of this complex biochemistry into field investigations has been carried out by means of expressing the complex mixtures of PHAHs as 'dioxin equivalents'. This concept is based on the linear correlation which has been found between the *in vitro* $-\log EC50$ of AHH induction and the *in vivo* $-\log ED50$ for toxic effects (weight loss and thymic atrophy) for a large number of PCBs, PCDFs and PCDDs (Safe, 1987). This bioassay approach is rapid and inexpensive compared to the conventional chemical analysis by gas chromatography and mass spectrometry. However, the approach is based on data obtained from mice. It is quite clear that the response of other organisms, in studies cited above, are quite different and that 'dioxin equivalents' calculated on quail – if the database was large enough – would be quite different from those based on mammalian data. Even within classes of organisms there are marked differences.

In summary, the studies on the structure-activity relationship of a large number of PHAHs show that the toxic and biochemical responses are interrelated. However, it has not been established whether induction of cytochrome-mediated enzyme activities leads either directly or indirectly to toxicity or whether enzyme induction and toxicity are independent aspects of the response.

A valuable tool in the identification and characterization of cytochrome P450 isozymes is the use of monoclonal antibodies. Monoclonal antibodies can be used as highly specific probes to define cytochrome P450 epitopes.

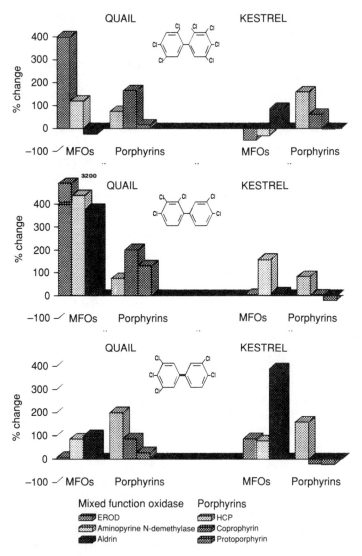

Figure 5.4 MFO and porphyrin responses of quail and kestrels to specific PCB congeners. After Elliott *et al.* (1990).

Initially developed for laboratory studies of rodents (Park *et al.*, 1982), they have subsequently been used in environmental studies in fish (Park *et al.*, 1986; Stegeman *et al.*, 1986, 1987) and birds (Ronis *et al.*, 1989a). A comparison of activities of EROD and AHH and the Western immunoblot assay based on the use of monoclonal antibodies (MAB) was made on winter flounder from coastal Massachusetts by Stegeman *et al.* (1987). These workers found that enzyme activity data unequivocally substantiated the immunoblot analysis. Highly significant correlations were found between both EROD and AHH activity and the P450E equivalents calculated from the amount of immunoreactive protein detected by MAB 1–12–3. Ronis *et al.* (1989a) examined the reactivity of liver microsomes of five species of sea birds against MABs P450IA1 and P450IIB1 from rats and P450E from scup. The latter is the fish orthologue of P450IA1. All of the avian species tested showed cross-reactivity to both rat P450IA1 and the fish orthologue P450E. The authors claim that this coincides with the appearance of PCB residues and suggests some degree of environmental induction. However, the correlation coefficients between expression of P450IA1 and PCB content of tissues are poor. In contrast only weak cross-reactivity was seen against rat P450IIB1, suggesting that there was little phenobarbital-type of induction in these species. One of the difficulties with environmental work using this technique is that the MABs used have, largely, been developed in laboratory rodents and their specificity to other orders is open to question.

The cross-reactivity against various rat cytochromes has been used to study the evolution of P450 gene families in non-mammalian vertebrates (Ronis *et al.*, 1989b). These workers examined the cross-reactivity, using polyclonal antibodies for several vertebrate groups (primitive fish, teleost, reptiles and several species of sea birds). It was found that the P450I family is well represented in all vertebrate groups. In contrast the P450II group, which is complex in mammals, is represented by only a few isozymes in other vertebrates. These findings would explain the ready inducibility in lower vertebrate groups by PAHs and other P450I inducers and the relative lack of effect of the phenobarbital, non-coplanar PCBs type of inducers.

5.7 Use of mixed function oxidases as biomarkers

The concept of using induction of MFO activity in fish as a monitor of pollution of the marine environment by oil was put forward by several workers in the mid-1970s (Ahokos *et al.*, 1976; Burns, 1976; Payne, 1976). The subject has been reviewed by Payne (1984) and Payne *et al.* (1987). The first of these reviews gives a number of case studies, a few of which can briefly be mentioned. Following an oil spill in the Adriatic, Kurelec *et al.* (1977) found an eight-to tenfold increase in MFO activity in fish, which, after a month, decreased to two- to threefold control levels. The activity remained at this elevated level for another five months. An extensive two-year study was carried out on flatfish along the coast of California (Spies *et al.*, 1982). These workers were able to demonstrate elevated levels of activity associated with sewage outfalls and an off-shore petroleum seep. Another, more recent, extensive study is that of Luxon *et al.* (1987), who measured AHH activity in lake trout from several sites around the Great Lakes and used material from an interior provincial park as a control. The activity in material collected from the western end of Lake Ontario averaged ten times higher than control material from Lake Superior and was similar to controls, while that from Lake Huron was substantially lower. While residue level data of a number of PHAHs and PAHs are given for sediment and for the total PCB values for lake trout muscle, it is not possible to make any specific correlations. Congener-specific PHAH data are not available so that it is not possible to examine the degree of induction against dioxin equivalents. This approach has been used by Casterline *et al.* (1983). They examined a small series of fish (eight from five locations) to test the hypothesis that AHH induction could be used as a biological screen to select samples for chemical analysis of PCBs, PCDDs and PCDFs. The TCDD equivalency was calculated for each sample; seven fell on a curve relating AHH activity to TCDD equivalents; the last sample had much higher AHH activity than would be predicted. The results of this experiment suggest that this preliminary biological screen could be used to select samples for much more expensive chemical analysis, but the sample size of this particular experiment is too small for the results to be considered conclusive.

In their more recent review, Payne and co-workers (Payne *et al.*, 1987) give excellent diagrams summarizing the data available on MFO induction in fish and invertebrates associated with hydrocarbon pollution and also that in fish associated with mixed organic contaminants (Figure 5.5).

The relationship of induction of EROD and AHH in the deep-sea fish *Corphaenoides armatus* to PCBs on a congener-specific basis has been examined by Stegeman *et al.* (1986). Although the sample size was small a good correlation was established. The presence of induced forms of P450I was confirmed by the immunoblot technique. This technique was also used by this group (Stegeman *et al.*, 1987) to study induction in winter flounder in coastal Massachusetts. They found a strong correlation between EROD and AHH activity and between the activities of both of these enzymes and P450I equivalents calculated from the immunoblot technique. A reasonable, but less strong, correlation between EROD and AHH was also found by Foureman *et al.* (1983) in the same species. These workers examined a large number of fish, over 400, during a three-year period. The large sample size is important to establish frequency distributions as there were large variations (over 500-fold in the case of EROD) in activity. Although the authors conclude that elevated AHH and EROD activities in many of the winter flounder are due to exposure to PAHs, they caution that the causative agent cannot be limited to PAHs. While agreeing with these conclusions,

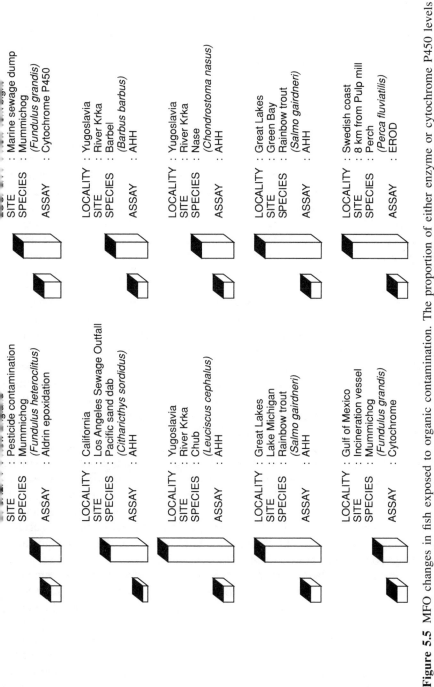

Figure 5.5 MFO changes in fish exposed to organic contamination. The proportion of either enzyme or cytochrome P450 levels detected at reference (short towers) and experimental sites (long towers) is presented in schematic form. All differences between reference and experimental sites were statistically significant ($P < 0.05$ or better). Payne *et al.* (1987).

we should mention that they are based on a broad knowledge of the distribution of PAHs rather than on specific analysis. Based on a smaller sample size, but using specific analysis, Stegeman *et al.* (1987) found that the high activity in fish from Boston Harbor correlated well with high PAH concentrations. But specimens from other areas did not show close correlations with either PAHs or PCBs.

The relationship of industrial pollution to the activity of benzo[a]pyrene in non-migratory fish in Yugoslavia has been examined by Kezić *et al.* (1983). These workers found much higher activity in fish from the contaminated River Sava than in those from two other rivers with lower industrial activity and population. Since detailed analytical data are not available, these domestic and industrial outputs are expressed as population equivalents. However, examination of the relationship of these population equivalents to enzyme activity along the River Sava does not bear out the summary statement that it is highly correlated to the recent pollution history of the river. Ahokos *et al.* (1976), examining the effect of industrial effluent on pike, found a decrease of benzo[a]pyrene hydroxylase and an increase of AHH activity. Gallagher and Di Giulio (1989) found that the EROD activity in fish in a stream exposed to a wide variety of industrial pollutants and discharge from waste-water treatment plants was not elevated over control values. However, the liver to body weight ratio and the occurrence of lip and lower jaw lesions were increased in the polluted site. The complexity of responses to a wide and variable range of pollutants is hardly surprising.

The relationships of cytochrome P450E levels and the degree of induction of EROD in the intestine and liver of spot, a marine fish, to the degree of PAH contamination of sediments were examined by van Veld *et al.* (1990). While the degree of induction of EROD activity in the intestine was more marked than that in the liver, the degree of correlation with sediment concentration was much poorer (Figure 5.6). Similarly, the level of cytochrome P450E increased markedly in the intestine, but again the correlation with sediment levels was not as strong as for the liver. The fact that increased cytochrome levels and EROD activity are readily induced in the intestine indicates their importance in the absorption and metabolism of PAHs, but since the changes are similar at all contaminated sites they are not useful as indicators of exposure. In contrast, the correlation with these changes in the liver corresponds well to sediment concentrations.

Significantly elevated hepatic AHH activity was found in embryonic herring gulls collected from Lake Ontario and Saginaw Bay in Lake Huron in 1981 when compared to control material collected from marine colonies (Ellenton *et al.*, 1985). No significant difference was found for material from Lake Superior or Georgian Bay in Lake Huron. A good correlation was found between enzyme activity and the concentration of 2,3,7,8-TCDD. Subsequent studies conducted in 1982, using different enzyme systems, showed no significant differences between the Great Lakes and control material

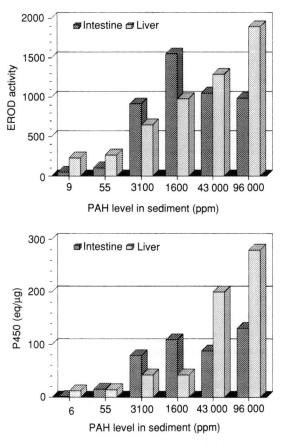

Figure 5.6 Relationship of sediment concentration of PAHs to EROD activity in liver and intestine of spot. After Van Veld *et al.* (1990).

(Boersma *et al.*, 1986). Hepatic tissue from embryonic Forster's terns collected in 1983 from Green Bay, Lake Michigan, had AHH activities that were threefold higher than in material collected away from the Great Lakes (Hoffman *et al.*, 1987).

Tillitt *et al.* (1988; 1989) reported on EROD-induction found in the rat hepatoma cell line, using 217 egg samples from fish-eating birds collected from 41 colonies in Michigan and Ontario. They found significant differences in the ability of egg composites from various regions around the Great Lakes to cause induction. The relative ranking of colonies correlates well with known areas of contamination, with material from Green Bay and Saginaw Bay giving the highest and those on Lake Superior the lowest values. Detailed chemical analysis is not available at this time for these samples, so that the dioxin equivalents calculated from the MFO activity

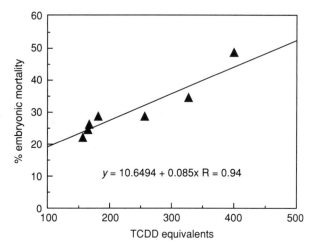

Figure 5.7 Relationship of embryonic mortality of caspian tern and dioxin equivalents. Tillitt *et al.* (1988).

cannot be checked against that obtained from the chemical composition. In studies in Green Bay, where congener-specific analytical data are available (Kubiak *et al.*, 1989), it was found that over 90% of the estimated TCDD equivalents could be accounted for by two co-planar PCB isomers (2,3,3',4,4' and 3,3',4,4'5-pentachlorobiphenyl). The relationship between the degree of embryonic mortality and the TCDD equivalents in the eggs of caspian terns is shown in Figure 5.7. The high degree of correlation is one of most exciting correlations of field studies and laboratory bioeffects studies that has been found.

A variety of physiological and biochemical changes in fish at different distances from a pulp mill was measured by Andersson *et al.* (1988). They found that EROD was strongly induced, 8-fold at 2 kms and 2-fold at 10 km from the point of discharge. Red blood cell and total white cell count (opposite directions) and blood glucose also altered in a distance-dependent manner. Changes in ascorbic acid levels, and liver and gonad somatic indices were also seen, but these were not clearly related to the distance from the source. Regrettably the statistical significance of these changes are not given. The central role of MFOs in mediating the responses of fish to pulp-mill effluent has been put forward by Lehtinen (1990). A simplified version of his diagram is given in Figure 5.8. In this diagram I have omitted the short-term effects via the formation of free radicals and concentrated on the longer-term effects. The effects on growth are complex; they may be stimulated or depressed. In the cases where it is stimulated, the body composition, especially the percentage of fat, is often altered. He points out that we must take into account the time-dependent variability of the MFO system.

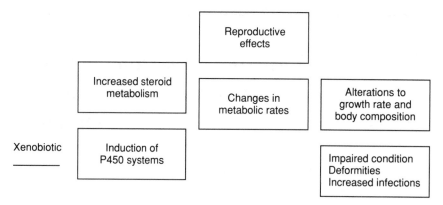

Figure 5.8 Model for the interactive process in fish exposed to pulp-mill effluent. Lehtinen (1990).

The usefulness of mixed function oxidases as a biological monitor has been clearly demonstrated for hydrocarbon pollution in fish and aquatic invertebrates and for PHAH contamination in a wide range of organisms. Since the response is caused by a very wide variety of chemicals, it means that the system is capable of showing exposure sufficient to cause a biological response to many xenobiotics. Conversely, it tells little about the causative agent(s), but can be used to delimit an area for which it is worth the time and expense of more detailed investigations. Nevertheless the system should not be pushed too far. Payne *et al.* (1987), in discussing MFOs as a practical tool for environmental managers, state that 'concerns about oil pollution are quite justified but in the case of the North Sea, scientific evidence is available to allay such fears.' The fact that induction of MFOs can only be demonstrated within a few kilometres of rig sites is one small piece of evidence of the effects of oil and does not in itself make it unnecessary for environmental managers to have concerns over the effects of oil development on fish stocks.

From a practical, biochemical viewpoint the considerable variation within a specific population means that the sample size has to be fairly large. The fact that the system is induced by a large variety of natural compounds as well as xenobiotics and affected by a wide variety of other parameters – temperature, diet, etc. – means that great care must be taken to ensure that controls really are controls. As with any other measurements, quality assurance should always be undertaken.

6   *Thyroid function, retinols, haem and regulatory enzymes*

6.1 Thyroid function

6.1.1 Introduction

The specific function of the thyroid gland is to produce the thyroid hormones, thyroxine (T_4) and triiodothyronine (T_3), which convert plasma iodide to iodine. The overall regulatory mechanisms involved in thyroid hormone control are shown in Figure 6.1, and an outline of the metabolic pathway of the thyroid hormones is shown in Figure 6.2. In addition to providing T_3 and T_4, the thyroid gland is also the source of calcitonin, a polypeptide hormone involved in calcium homeostasis.

The thyroid gland consists of many spherical follicles surrounded by cuboidal epithelial cells. The follicles contain the thyroid hormones bound to thyroglobulin, which appear histologically as colloidal substance. A hypothyroid state (clinically referred to as goitre) occurs when the diet is deficient in iodine. The follicles are enlarged and the epithelial cells flattened. In a hyperthyroid state, the follicles are small, contain little colloidal, and the epithelial cells are more columnar.

The thyroid has an important role in metabolic processes, particularly those related to metabolism, development and growth. A lower metabolic rate is associated with hypothyroidism, in which the effect is similar but not identical to the calorigenesis caused by catecholamines.

The two most significant environmental studies on the thyroid are the experimental studies of Jefferies and co-workers and those of Leatherland, Sonstegard, Moccia and others on the occurrence of hyperplasia in salmon in the North American Great Lakes. These studies, together with a brief review of the effect of PHAHs on thyroid function in mammals, which now include a study on seals, are reviewed in the next section.

6.1.2 Summary of experimental data

Jefferies and co-workers clearly established that DDE, DDT, dieldrin and PCBs cause alterations to the thyroid in several avian species: the pigeon

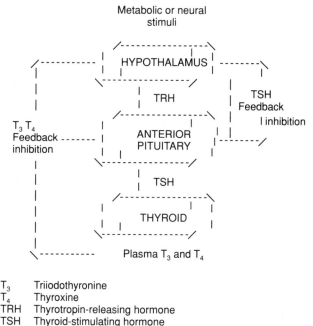

Metabolic or neural
stimuli

T$_3$ Triiodothyronine	
T$_4$ Thyroxine	
TRH Thyrotropin-releasing hormone	
TSH Thyroid-stimulating hormone	

Figure 6.1 Overall regulation of thyroid hormones.

(Jefferies and French, 1971, 1972); the lesser black-backed gull (Jefferies and Parslow, 1972); and the common guillemot or murre (Jefferies and Parslow, 1976). A comprehensive review of these studies was made by Jefferies (1975). The basic finding was that PHAHs cause hypothyroidism – increased thyroid weight, follicle size and colloid area – at lower concentrations and hyperthyroidism at higher dosages. However, the residue levels associated with hyperthyroidism indicate that this occurs only at close to lethal dosages. The experiments of Jefferies and co-workers did not include determination of thyroid hormones. In ring doves exposed to 3,4,3′,4′,-TCB, Spear and Moon (1986) found that T$_4$ was reduced, but that T$_3$ and thyroid weight were unaffected. The complexity of interactions between dietary iodine and PHAHs was demonstrated by the finding that an injection of TCB reversed the thyroid hyperplasia in ring doves caused by a low iodine diet (Spear and Moon, 1986). These workers considered that this effect was due to opposing influences on the thyroid-stimulating hormone.

Feeding studies of PCBs and mirex, both separately and in combination, to coho salmon on serum T$_3$ and T$_4$ levels were carried out by Leatherland and Sonstegard (1978). Both T$_3$ and T$_4$ levels were significantly reduced in fish exposed to the high-mirex diet (50 μg/g in dry feed) for three months, but only T$_3$ was reduced by PCBs (500 μg/g). Serum levels of both T$_3$ and

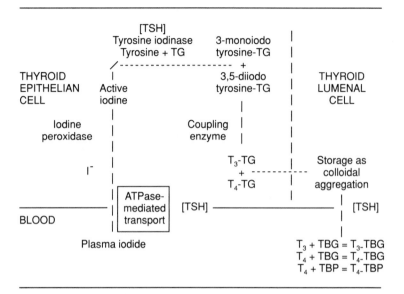

Figure 6.2 Metabolic pathways of the thyroid hormones.

T_4 were reduced by a combination diet (5 µg/g mirex + 50 µg/g PCBs). There was no histological evidence of thyroid hyperplasia or goitre formation. Similar studies on rainbow trout also did not indicate that thyroid hyperplasma was induced even at the highest doses used (500 µg/g PCBs or mirex at 50 µg/g in the diet) (Leatherland and Sonstegard, 1979, 1980). The effects of the thyroid hormones were less marked in rainbow trout than in coho salmon and were confined to reduction of both T_3 and T_4 at the highest dose of mirex only. It seems likely that some of the results of these experiments were due to poor uptake of the PHAHs. The ratios of levels in carcasses were very different from those in the diets, and absolute values of the carcass levels seem low considering the high dietary level.

These difficulties were overcome in a subsequent experiment (Leatherland and Sonstegard, 1982a) in which yearling coho salmon were fed diets containing flesh of mature salmon from three different sites in the Great Lakes. A group fed a diet containing the flesh of Pacific salmon served as the control. No direct comparison can be made between the two experiments, as both species used and the duration of the experiment (27 weeks versus

2 weeks) are different. The highest PHAH load of the food was that using Lake Michigan salmon (total PHAH 8.8 ppm with PCBs at 6.3 ppm and DDE at 1.8 ppm). The figures for the lowest (Lake Erie) were total PHAH 2.2 ppm, PCBs 1.7 ppm and DDE 0.2 ppm. Mirex was found in the Lake Ontario salmon (0.26 ppm). The PHAH levels in the Pacific salmon were almost an order of magnitude lower than those from Lake Erie. The effects on thyroid hormone levels were not marked. No effects were seen on T_4, and for T_3, only the values of fish exposed to Lake Erie (50% of control) and Lake Ontario (61%) salmon flesh were significantly lower. In contrast, a reduction in thyroid epithelial cell height was seen only in the Lake Michigan salmon. The thyroid colloid vesicular index was decreased in the Lake Michigan and Lake Ontario groups.

Marked hypothyroidism was seen in rats fed diets of salmon from the Great Lakes (Sonstegard and Leatherland, 1979). After a month on a diet of salmon from Lake Ontario, the serum T_4 levels had fallen to only 20% of control values, while T_3 values were not significantly altered. The ratio T_4/T_3 was reduced to 18% of the control group fed Pacific salmon by one month and down to 10% by two months. Significant, but less marked, changes were seen in rats fed on salmon from Lake Michigan, while those fed on fish from Lake Erie were only affected at the longer time period. Some changes in the cell height of the thyroid were also noted. These effects parallel the amount of total PHAHs measured in the flesh of the salmon, in contrast to the lack of correlation with goitre frequency (Moccia *et al.*, 1978). Villeneuve *et al.* (1981) also noted mild, but statistically significant, histological changes in the thyroids of rats fed Lake Ontario salmon, but these workers did not measure serum T_3 and T_4 levels.

Studies along these lines, using the common seal, have been carried out by Brouwer *et al.* (1989a). These workers fed two groups of seals (12 in each) diets containing different levels of contaminants for nearly two years. The seals were allowed to breed during this period and at the end of the experiment the high group were fed for six months on the 'clean' diet before they were released into the environment. The average daily diet of the highly contaminated group contained 1.5 mg PCBs and 0.4 mg DDE (fish from the Wadden Sea) and the lower group 0.22 mg PCBs and 0.13 mg DDE (fish from NW Atlantic). Exposure of seals to fish from the Wadden Sea resulted in a marked reduction of plasma retinol, but levels returned to normal after six months on the clean diet. The effects on T_3 and T_4 were less clear-cut. At the first time point a small, but significant, reduction of T_4 and larger reduction of T_3 were seen, but these changes were not significant at the second sampling point, perhaps as a consequence of compensation by the thyroid.

The changes found by Sonstegard and Leatherland (1979) in rats fed Great Lakes salmon are paralleled by the experiments of Collins and Capen (1980) using PCBs. These workers studied the effects of 5, 50 and 500 ppm

Aroclor 1254 on serum T_3 and T_4 and thyroid structure. The ratio of T_4/T_3 decreased steadily as concentration increased (33.6, 26.1 and 10.8 compared to control value of 42.7). At the two higher concentrations this was due to a decrease of T_3, but at 5 ppm there was a significant increase of T_3, while T_4 was unchanged from control values. Exposure to PCBs caused a dose-dependent hypertrophy and hyperplasia of follicular cells with an abnormal accumulation of large colloid droplets.

The changes to the thyroid hormones caused by HCB were the reverse of those caused by PCBs. The effects of a high dose of HCB (1 g/kg, daily for one or eight weeks) on thyroid function and metabolism in rats was studied by Pisarev *et al.* (1990). Serum levels of T_4 were reduced to about a quarter of control values in both time periods, whereas the levels of T_3 were not significantly affected. A reduction of 50% in the protein-bound iodide was observed and studies on liver slices demonstrated an increase in T_4 dehalogenation.

Some congener-specific work has been carried out using polybrominated biphenyls. Akoso *et al.* (1982) found that 100 ppm in the diet of the mixture Firemaster FM BP-6 or both the 2,2',4,4,'5,5'-hexabromo biphenyl (HBB) and 2,3',4,4',5,5'-HBB caused an increase of T_3 and a marked decrease of T_4, causing a marked decrease in the T_4/T_3 ratio. A similar effect was found with a dietary level of 10 ppm 3,3'4,4'5,5'-HBB. Thyroid weights were significantly reduced by 3,3',4,4',5,5'-HBB, at both dietary levels of 1 and 10 ppm. These effects parallel the ability of the congeners to induce both AHH activity and UDP-glucuronyl transferase. Spear *et al.* (1990) found that 3,3',4,4',5,5'-HBB at 40 ppm caused a decrease in the levels of T_4 but no significant effect on T_3. The decrease of T_4 was caused by a doubling of peripheral metabolism of T_4 and a maintenance of normal serum T_3. The increased T_4 metabolism is explained by the marked increase in UDP-glucuronyltransferase activity in the liver. The mechanism for maintaining the T_3 levels is less certain, as no effect on 5'-deiodinase activity was found. It is possible that T_4 binding in the liver is increased and thus maintains the substrate concentration available for the conversion of T_4 to T_3. Most of the studies of environmental pollutants on thyroid function have focussed on the PHAHs. However, ingested lead shot was found by Goldman *et al.* (1977) to cause a modest (15%) but significant increase of the thyroid weight of ducklings. ALAD measurements were not made, but at the dosage used it would be expected that marked inhibition would have occurred. The effect of PAHs appears to be small. Holmes *et al.* (1978) found that South Louisiana crude oil caused a doubling of thyroid weight in sea-water adapted mallard, whereas no effect was seen with Kuwait crude or No. 2 fuel oil. Under conditions of cold stress, effects were seen with South Louisiana and No. 2 fuel oil, being most marked for the latter oil. Rattner and Eastin (1981) found that chronic ingestion of oil did not affect thyroxine levels in young mallards. No histological studies were undertaken.

6.1.3 Factors influencing activity

Species differences Jefferies and co-workers (Jefferies and French, 1971; 1972; Jefferies and Parslow, 1972, 1976) examined the effects of DDT, DDE and dieldrin on the thyroid of the pigeon, and of PCBs on the lesser black-backed gull and the common guillemot (or murre). They found that the guillemot was considerably more sensitive to PCBs than the lesser black-backed gull in that the guillemot exhibited the same effects more than twice as rapidly. Hurst *et al.* (1974) found no clear-cut effects of PCBs on the thyroid of quail, even at dosages and duration times in excess of those used by Jefferies. Differences between species inhabiting the iodide-rich marine and iodine-poor terrestrial areas would certainly be expected.

Spear and Moon (1985) made a comparison of the effects of PHAHs on thyroid histology between different classes of animals. Studies of a variety of PHAHs on rats consistently showed hyperplasia, although data on other mammalian species were not available. The avian data available are a complex mixture of species and of PHAHs, which makes comparisons difficult, but they certainly show a variety of responses. The experimental data on rainbow trout, with mirex and PCBs, showed no effects on thyroid histology. However, experiments involving the feeding of diets containing flesh of mature salmon from the Great Lakes to yearling coho salmon did show some histological effects (section 6.1.2).

Other factors In the gull, Jefferies and Parslow (1972) noted small differences between the sexes and suggested that the females might be more affected by PCBs. In general few significant differences between the sexes have been noted. Natural physiological factors – other than iodine concentration – can affect the levels of the thyroid hormones. Oishi and Konishi (1978) showed that temperature and photoperiod affected the levels of T_4 but not those of T_3 in quail. John and George (1977) found that heat stress and dehydration affect T_3 but not T_4 levels. The duration of the effects of high dietary levels of DDT, DDE and DDA was studied by Richert and Prahlad (1972). Quail were dosed for 120 days and then examined after 85 days on clean diets. Histological alterations were found in all experimental groups and a doubling of thyroid weight was found in the case of DDE. The pharmacodynamics of these compounds are such that considerable body burden would still be expected, especially of DDE, and this cannot be separated from the duration of the thyroid effects *per se*. The numerous environmental factors that can affect the thyroid gland in fish – temperature, salinity, diet – have been reviewed by Leatherland (1982).

6.1.4 Interactions with (or relationship to) other biomarkers

The interrelationship of metabolic pathways and hence of biomarkers makes it difficult to treat specific biomarkers separately. Conversely, treating them together creates a very complex picture.

Recent studies (Brouwer and van den Berg, 1986; Brouwer *et al.*, 1986; Spear and Moon, 1986) have shown the interrelationship of effects of PHAHs on the thyroid and retinol pathways. The outline of these pathways is shown in Figures 6.2 and 6.4. Basically, the pathways are dissimilar, but have a commonality in their transport in the circulatory system. This commonality is discussed in the section on retinol (6.2).

McKinney *et al.* (1985) have examined the likelihood that TCDD acts as a potent and persistent thyroxine agonist by binding thyroxine-binding prealbumin (TBPA). The large amount of literature available on the binding of TCDDs and other PHAHs to the cytosolic Ah receptor is reviewed in the section on mixed function oxidase enzymes (5.6). There is no *a priori* reason to consider the two hypotheses as mutually exclusive, since PHAHs could bind with both receptors.

6.1.5 Use of thyroid function as a biomarker

There are serious limitations in the use of changes in thyroid structure as a biomarker. The frequency of goitre in coho salmon does not correlate with the reported iodine content of the water (Moccia *et al.*, 1977). Subsequent work showed no relationship between tissue burden of PHAHs and goitre frequency; indeed the highest goiter frequency was found in salmon with the lowest total PHAHs levels (Moccia *et al.*, 1978). Furthermore, as discussion in the previous section showed, there is no clear relationship between PHAHs and histological changes in the thyroid. Leatherland and Sonstegard (1984) tabulated the data on lake water iodide, goiter frequency, thyroid goitre index and tissue iodide context for mature coho salmon in Lakes Ontario, Erie and Michigan. Only the ratio between tissue iodide and lake water iodide context was reasonably constant (0.72–1.00). There was no relationship between iodide context and frequency of goitre. Interlake (Ontario, Erie and Michigan) differences were observed in body weight, gonadosomatic index, serum osmotic pressure and concentrations of ions were found in prespawning coho salmon (Leatherland *et al.*, 1981). Highest body weights and GSI were found in fish from Lakes Ontario and Michigan; serum calcium levels were lowest in those from Lake Erie. No correlations were found between these parameters and thyroid dysfunction. Noltie *et al.* (1988) found that all Lake Superior and Lake Erie pink salmon exhibited thyroid hyperplasia, whereas salmon from British Columbia were unaffected. Lesion size was positively correlated with body size in fish from Lake Superior, whereas in Lake Erie nearly all fish showed extreme hyperplasia regardless of body size. Thyroid hormone levels were negatively correlated with lesion size, providing evidence of hypothyroidism. Residue analysis showed that the Lake Superior and Lake Erie fish had significantly different contaminant profiles. The authors consider that these findings warn of potential water quality problems even in the most pristine of the Great Lakes, but the problem of variable iodine does not appear to have been taken into account.

In a companion paper Noltie (1988) examined the relationship between thyroid hyperplasia and a number of parameters associated with their reproductive physiology in Lake Superior pink salmon. Females with overt thyroid lesions had enlarged livers and showed reduced secondary sexual characteristics and water content of eggs, although ovarian weight was not decreased.

The use of thyroid hormones shows more promise. Blood and thyroids were collected from adult nesting herring gull in the Great Lakes over the period 1980–1983; material was also collected from a marine colony (Kent Island, New Brunswick), 1977–1982. Some material collected from Lake Ontario in 1974 was also examined (Moccia *et al.*, 1986). These workers measured the relative mass of the thyroid gland (weight of gland divided by body weight), the follicular diameter, the epithelial area, the epithelial cell height, and the prevalence of epithelial hyperplasia, colloid-containing follicles and colloidal vacuolation. No clear-cut changes emerge for epithelial cell height or prevalence of colloid-containing follicles, but all other parameters showed differences between the marine and Great Lakes material. The results of these studies are shown diagrammatically in Figure 6.3. The relative mass was lower and the other measurements higher in the material from the Bay of Fundy than the material from the Great Lakes. The prevalence of epithelial hyperplasia was higher and that of colloidal vacuolation lower. Furthermore, in all cases except colloidal vacuolation, the effect was most marked in the birds collected from Lake Erie. Beyond that there is no clear-cut pattern within the Great Lakes material and thus the changes reported do not follow the pattern of PHAH contamination. However, the pattern in the herring gull parallels findings that the highest goitre frequency was recorded in salmon in the western end of Lake Erie.

Comparison of material collected in 1974 with that collected in 1980–1982 enables temporal differences to be examined for Lake Ontario. For almost all measurements there has been improvement; that is, the values have moved closer to those made on the thyroids of gulls from the Bay of Fundy.

Although a number of gulls from the Atlantic coast had plasma-free iodine values higher than those from the Great Lakes, the ranges of all the Great Lakes collections overlapped those observed in the Atlantic coast. Thyroxine levels showed no clear-cut difference between material from the Great Lakes and the Atlantic coast material. Variation within any one collection varied from 2 to 25× with the greatest variation occurring in Lake Superior. The Free Thyroid Hormone Index (FTHI) is a measure of the free hormone in the plasma and is normally correlated with the metabolic status of the individual. The FTHI values for specimens collected at Middle Sister Island (Lake Erie), Lake Superior and Green Bay are very low and suggest a hypothyroid state, whereas those of gulls collected in Lake Ontario and Lake Huron are elevated and suggest hyperthyroidism (G.A. Fox and R.D. Moccia, unpublished data).

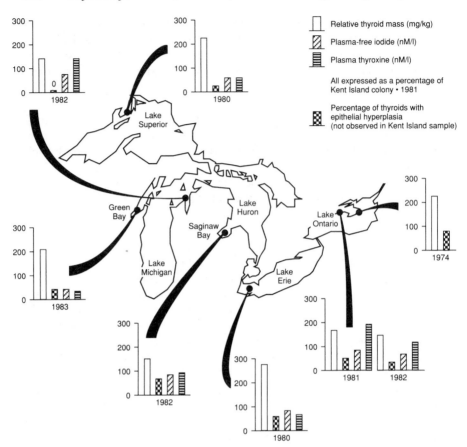

Figure 6.3 Variation of thyroid-related parameters in herring gulls on the Great Lakes. G. Fox, Canadian Wildlife Service, Environment Canada (1991).

It appears that no field studies have been made on possible effects in the thyroids of fish-eating mammals. In view of the sensitivity of reproductive parameters of mink to low levels of PHAHs (section 3.1) and effects found in rats fed diets of salmon from the Great Lakes (Sonstegard and Leatherland, 1979; Villeneuve *et al.*, 1981) and of common seals fed fish from the Wadden Sea (Brouwer *et al.*, 1989), such investigations would be worthwhile. It is difficult to extrapolate the findings on the rat, which is not naturally a fish eater, to the target species such as mink and otter. Indeed, Leatherland and Sonstegard (1982b), using a formulated diet in which Great Lakes salmon was added to a diet of rat chow fortified with vitamins, minerals and lipids, found much less marked effects than those in rats fed a total fish diet. Studies on the common seal indicate that for this target species the effects are modest.

Most studies were carried out without analysis on a congener-specific basis. In view of our rapidly advancing knowledge of the differing effects of specific compounds, work on the thyroid at this level would be interesting. At the present time the T_4/T_3 ratio seems to be the most sensitive indicator. While it is clear that factors other than iodine intake are involved in the changes found in thyroid chemistry and histopathology, it is not clear what these xenobiotic factors are. The degree of variation within sample groups from the marine site is similar to those from the Great Lakes, but in the case of the thyroid the iodine-rich marine site is not a true control. Certainly the separation of chemical factors from others causing iodine deficiency is a difficult problem.

Hetzel (1989) estimates that one-fifth of the world's human population is iodine deficient. The degree to which iodine deficiency occurs in wildlife is unknown, but is a question that should not be ignored.

6.2 Vitamin A (retinoids)

6.2.1 Introduction

In animals, imbalance of vitamin A, a lipid-soluble vitamin, is associated with a diversity of anomalies, including reproductive impairment, embryonic mortality, growth retardation and bone deformities (Thompson, 1976). Animals are unable to synthesize retinoids; the source is the carotenoids which are photosynthesized by certain microorganisms and plants. In the view of IUPAC-IUB (1982) the term *vitamin A* should be used as 'the generic description for retinoids exhibiting qualitatively the biological activity of retinol'. *Retinoid* is a general term for a group of closely related compounds, both natural and synthetic analogues. They consist of four isoprenoid units joined in a head-to-tail manner. The most important of these are retinol, retinyl palmitate and retinoic acid. An outline of the main metabolic pathways involved is given in Figure 6.4.

Recent studies have shown that retinoic acid and its immediate precursors have specific roles in development and regeneration. Thaller and Eichele (1990) have demonstrated that 3,4-didehydroretinoic acid (ddRA) is present in significant quantities in the chick embryo and that both this compound and retinoic acid are equipotent in evoking digit duplications in the chick limb bud assay. This assay has been used to analyse other regions of the chick embryo for polarizing activity. Wagner *et al.* (1990) have examined the role of retinoids in the polarizing activity and development of the floor plate of the neural tube. They find that both retinoic acid and 3,4-didehydroretinol (the precursor of ddRA) are synthesized *in vitro* in the floor plate. These roles of retinoids in controlling development are not confined to the chicken. In a recent review Brockes (1990) cites work that shows the effects of retinoids on diverse cell types and influence on cell interactions.

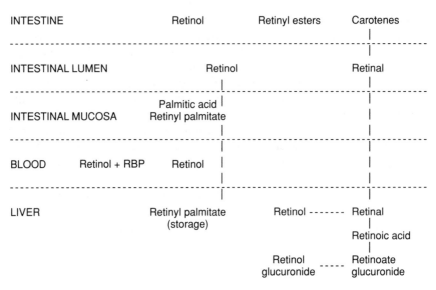

Figure 6.4 Metabolic pathways of retinols.

These include the cranial neural crest cells of the mouse and the urodele blastema. At the present moment the exact role of retinoic acid is open to considerable discussion. Two papers, appearing in the same issue of *Nature*, provide a significantly different interpretation for the role of retinoic acid than that put forward previously. Studies reported by Noji *et al.* (1991) and Wanek *et al.* (1991) consider that retinoic acid changes anterior cells into a zone of polarizing activity, rather than altering anterior cells so that they produce retinoic acid to provide a graded signal. These results suggest that retinoic acid has a secondary role rather than acting as a primary vertebrate morthogen.

The role of retinoids in cell differentiation appears to be that of a switch which starts the onset of developmental mechanisms. The diversity of the effects may be due to the diversity of retinoid receptors that are now being uncovered. While the exact roles of retinoids in the reproductive process remain to be elucidated, they are obviously compounds of importance. As their role is more closely defined, their value as biomarkers will become better defined.

6.2.2 Summary of experimental data

The symptoms of dioxin and furan exposure are similar to those associated with vitamin A deficiency (Kimbrough, 1972). While TCDD reduces the hepatic storage of retinol, there is no evidence that vitamin A deficiency is responsible for any toxic responses of TCDD (Poland and Knutson, 1982).

Increasing documentation of the interdependence of toxic responses and our increasing knowledge of the interrelationship of receptors may well provide this evidence within a few years.

The effects of DDT on vitamin A utilization in the rat were studied by Phillips (1963). He found that both the total amount and the concentration of vitamin A were reduced at dietary levels of 100 ppm, but not at 10 ppm after 73 days of exposure. These effects could be overcome if additional carotene was added to the diet. In another, shorter-term experiment, he found that these changes occurred at 20 ppm, but not at 2 ppm.

Cecil *et al.* (1973) found that 100 ppm of DDT or Aroclor 1242 caused the hepatic level of vitamin A in rats to decrease to half after exposure for two months. The initial levels, expressed on either a µg/g or mg/liver basis, were somewhat higher in males, but the percentage reduction was similar between the sexes. In quail, the sex difference was marked, with the levels in the males being four times higher. In the male, DDT and Aroclor caused a reduction to 80% and 46% of controls respectively, whereas the much lower levels in the female were not affected except by Aroclor on those maintained in the dark. The lower levels in females were considered to be due to egg production. Some decrease in egg production was seen in the second month on PCBs.

The effects of two PCBs of differing degrees of chlorination on retinol levels in pregnant rabbits and their fetuses were studied by Villeneuve *et al.* (1971). The lower chlorinated PCB (Aroclor 1221) produced no consistent changes, but Aroclor 1254 caused a reduction to about 50% of control in the storage levels of retinol in the does at both 1 and 10 mg/kg dietary levels. The effect on the foetus was much less marked. At the higher level induction of MFOs enzymes was found, but no effects on overall reproduction were found.

The interrelationships of dietary retinol levels and effects of TCDD were studied in rats by Thunberg *et al.* (1980). TCDD produced a dose-dependent reduction of hepatic retinol in all three dietary groups. The group with the highest level of retinol in its diet (6 mg/kg compared to the normal level of 3 mg/kg) had approximately twice the hepatic retinol level. Reduction of dietary level to 1.5 mg/kg produced a much less marked effect. The two highest doses of TCDD (1 and 10 µg/kg) produced a marked increase in the activity of UDP-glucuronosyltransferase (UDPGT) (10-fold at highest dose) with little variation caused by retinol levels in diet.

The effects of 3,4,3′,4′-tetrachlorobiphenyl (TCB) and TCDD on retinoid levels in various peripheral organs of rats were studied by Brouwer *et al.* (1988), (1989b). They found that a single dose of 15 mg/kg TCB caused significant reduction of both retinol and retinyl palmitate in the liver (to 25% of control), heart (35%), and lung (44%). No significant alterations were found in the kidney and skin. A dose of 10 µg/kg TCDD almost completely depleted retinol and retinyl palmitate in the liver to only 3–5%

of control values. Reductions in lung, intestine and adrenals were to 40–50% of controls. In contrast, the levels in kidney were four times and serum one and a half times control values.

The effect of several PHAHs, known to have different inducing capacity on the MFO system, on hepatic retinol levels in rats was studied by Azais *et al.* (1987). TCB, known to be a strong inducer of the P448 system, showed marked effects on the hepatic vitamin A level (retinol plus retinyl palmitate). Nevertheless no strong correlation between retinoid depletion and AHH induction was found. DDT and 2,2′,4,4′,5,5′-hexachlorobiphenyl, known inducers of the P450 system, and the weak inducer 2,2′,5,5′-tetrachlorobiphenyl did not cause significant changes to vitamin A levels.

Spear *et al.* (1986) gave ring doves a single injection of the dioxin analogue 3,4,3′,4′-TCB. Hepatic retinol levels were halved, and the levels correlated well with AHH activity. Retinyl palmitate levels were not significantly affected. In a subsequent experiment, the effect of this compound on the retinoid dynamics of nesting ring doves was studied. It was found that egg-laying was delayed and that nearly half of embryos in the treated group died between day four and seven of incubation. The levels of retinol and retinyl palmitate decreased during this period in the exposed group, but there was no corresponding decrease in the controls. The ratio of retinol to retinyl palmitate in the viable eggs of the exposed group was higher than either the controls or the nonviable eggs of the exposed group. No changes were seen in the uroporphyrin levels.

6.2.3 Factors influencing activity

Species differences A difference in the response to TCB was found by Brouwer and van der Berg (1984) between two strains of mice. In one (C57BL/Rij), a dose-dependent decrease of both retinol and retinyl palmitate was found in the liver and retinol in the serum. But in DBA/2 mice no significant alterations were found in the liver, but serum levels were affected similarly to the other strain.

Broadly, retinoid concentrations tend to reflect vitamin availability in the diet and thus are higher in marine species – such as seals and gulls – which feed on marine fish with high retinoid and carotenoid concentrations than in such species as rats and quail. Additionally, there is high individual variation. In a sample of 20 herring gull livers, Spear *et al.* (1986) found that the retinol levels varied from 43 to 2012 μg/g and those of retinyl palmitate from 141 to 5029 μg/g.

Sex differences In rats, Cecil *et al.* (1973) found that the initial level was higher in males, but that the effect on the hepatic level of vitamin A of 100 ppm DDT caused a similar reduction in both sexes. In quail, the same initial difference was found, but in this species DDT caused a marked effect only in the male.

6.2.4 Interactions with other biomarkers

In an experiment involving ring doves injected with 3,3′,4,4′-TCB, liver retinol showed a significant inverse relationship with cytochrome P450-dependent AHH activity in liver microsomes (Spear *et al.*, 1986). More recent data show the same correlation for vitamin A depletion in herring gulls collected from the Great Lakes (P.A. Spear, unpublished data).

A direct link between induction of the P450 enzyme system and specific steps in the vitamin A metabolic pathway was demonstrated by Spear *et al.* (1988). They found increased retinoic acid metabolism in rats following a single injection of 3,3′,4,4′,5,5′-polybrominated biphenyl. Serum retinol was not affected, although the hepatic retinol and retinyl palmitate were decreased by 20%. The rate of retinoic acid and its subsequent oxidation were significantly elevated. The rate of conjugation by UDPGT was also increased.

Thyroid–vitamin A interactions have been studied by Spear and Moon (1986). They found that chickens that were on a low vitamin A diet and exposed to 3,4,3′,4′-TCB became hypothyroid (section 6.1.3). However, growth rate was not affected by this diet. Effects of growth rate were seen only with diets that were low in both vitamin A and iodine. The authors concluded that growth rate may have been altered by circulating levels of retinol and that vitamin A insufficiency may predispose birds to the hypothyroid effects of PCBs.

Brouwer and van den Berg (1984) examined the effect of the dose-dependent response of TCB on mice responsive (C57BL/Rij) and nonresponsive (DBA/2) to inhibition of the Ah receptor. They found that, as would be expected, the DBA/2 mice did not show AHH induction, over the entire range of dosages used (1–100 mg/kg). In the sensitive mice, AHH induction was found, but only at the highest dose. Changes in retinol were much more sensitive. In DBA/2 mice, significant decreases of serum retinol were found over the range 15–100 mg/kg, although liver retinol and retinyl palmitate were not affected. In the C578L/Rij mice all these parameters were affected, the liver retinyl palmitate at dosages as low as 5 mg/kg.

The time responses of a single dose of 15 mg/kg TCB were examined in a subsequent study (Brouwer *et al.*, 1985). In responsive mice hepatic retinol and retinyl palmitate were decreased by 30–40% within two days and remained low until the end of the two-week period. No effects on hepatic retinoid levels were seen in nonresponsive mice. The pattern of AHH activity was a rapid rise in responsive mice one day after exposure, followed by a sharp decrease to control values within four days. No induction of AHH was found in the nonresponsive mice. These workers suggest the difference in the duration of responses of AHH activity and changes in retinol concentrations argues against direct involvement of the MFO system in the alterations of retinoid levels. In view of the fact that AHH induction occurs more rapidly than the onset of retinoid changes, this does not seem a strong argument. What is clear from these studies is that changes in retinoids can be a sensitive indicator of exposure to TCB.

The interrelationship between retinol storage and the activity of UDP-glucuronosyltransferase (UDPGT) has been studied by Thunberg and co-workers. In an early paper, Thunberg *et al.* (1980) demonstrated a dose-dependent effect of TCDD on the hepatic storage of retinol in rats. The increase in activity of UDPGT was in inverse proportion to the effect on retinol storage. These workers hypothesized that UDPGT increased the rate of conjugation and thus of excretion of retinol. In subsequent studies these workers (Thunberg *et al.*, 1984) compared the effect of TCDD to several other compounds (brominated dioxin (TBrDD), two chlorinated predoxins, toxaphene, and the widely used MFO inducers, 3-methylcholanthrene (3-MC) and phenobarbital (PB)). TCDD, 3-MC, PB and TBrDD all reduced the levels of retinol, increased the activity of UDPGT and induced AHH activity. Toxaphene affected UDPGT and AHH, but did not decrease the hepatic retinol levels. The predioxins did not cause significant effects. Despite the fact that the strength of the effects lies in the same sequence, the authors conclude there was no correlation between hepatic retinol storage, AHH induction and UDPGT activity.

A mechanism of action that involves P450 induction followed by binding of a metabolite of TCB with the retinol-binding protein, leading to an inhibition of the formation of the serum transport protein complex for both retinol and thyroxine, has been put forward by Brouwer and van den Berg (1986). These workers analysed retinol-labelled proteins by polyacrylamide gel electrophoresis and showed an association of retinol with retinol-binding protein and a prealbumin, transthyretin. The amount of labelled retinol, both in the serum and associated with binding proteins, was markedly reduced by TCB. In a subsequent paper, Brouwer *et al.* (1986) showed that the direct interaction of the metabolite of TCB with the carrier protein led to reduced levels of plasma retinol and thyroxine in both rats and mice.

A family of retinoic acid receptor-related genes have now been identified (Giguere *et al.*, 1987). This implies the existence of several closely related proteins that could mediate the different effects of retinol on cellular function. The finding of the similarity of the retinoid receptor to the thyroid hormone receptor was surprising in view of the structural dissimilarity of the thyroid hormones and the retinols. It is proposed that the interaction of retinoids with their intracellular receptors induces a cascade of regulatory events. The demonstration that the retinoic-acid receptor is part of the steroid receptor superfamily indicates that mechanisms controlling morphogenesis and homoeostasis are closely linked. Petkovich *et al.* (1987) cloned a cDNA corresponding to a protein that binds specifically to retinoic acid with high affinity. This protein is homologous to the receptors for steroid and thyroid hormones. It appears to be a retinoic acid-inducible factor, suggesting that the molecular mechanisms are similar to those found for other members of the nuclear receptor family.

6.2.5 *Use of retinoids as biomarkers*

Hepatic retinoid concentrations were measured in herring gulls collected from three Great Lakes sites and an Atlantic coast colony (Kent Island, New Brunswick) in 1982 (Spear *et al.*, 1986). The levels in all Great Lakes material were significantly lower than those from the coastal colony. Within the Great Lakes the levels were lowest in Lake Ontario and highest in Lake Superior. Liver retinol showed a statistically significant inverse relationship to AHH activity. While chemical analysis was not carried out on these specific eggs, the changes in retinoids found paralleled the levels of TCDD found in herring gull eggs collected in these areas in 1980 and 1983.

A more intensive study has since been conducted (G.A. Fox and S. Trudeau, in preparation; data cited in Environment Canada, 1991). They found that liver retinol and retinyl palmitate levels varied widely in gulls from the Atlantic coast colonies, both between locations and between years, but that the levels were generally higher than in those gulls collected in Great Lakes colonies. Due to the high degree of variation, a clearer picture is obtained if the data are expressed as distributions (Figure 6.5). Retinol levels in livers of gulls from the Atlantic coast and Lake Superior did not differ significantly from each other, but both were significantly higher than those from the other Great Lakes. The concentrations of retinyl palmitate in livers of gulls from the Atlantic coast were significantly higher ($p < 0.05$) than in gulls from any of the Great Lakes. Within the Great Lakes, the retinyl palmitate stores of gulls from the Detroit River and western Lake Erie were significantly lower than for the other locations.

The most marked retinoid depletion was observed in the collections from the western basin of Lake Erie. This is the same area that had the greatest change in thyroid pathology (Moccia *et al.*, 1986). In contrast, the correspondence between elevation of porphyrin levels (section 6.3) and retinoid depletion is not strong, suggesting that a different agent is responsible for these two biochemical manifestations of toxicity.

In a subsequent study retinoids were measured in the yolk of the herring gull egg (Spear *et al.*, 1990). This investigation was supported with congener-specific analysis of PCDDs and PCDFs. These workers found that there were marked differences in the levels of retinol and retinyl palmitate at different stages of incubation. The levels of retinol dropped sharply and those of retinyl palmitate rose. The net change in the molar ratio of retinol to retinyl palmitate decreased 15-fold in one colony and over 7-fold in another, emphasizing the importance of using material at the same degree of incubation.

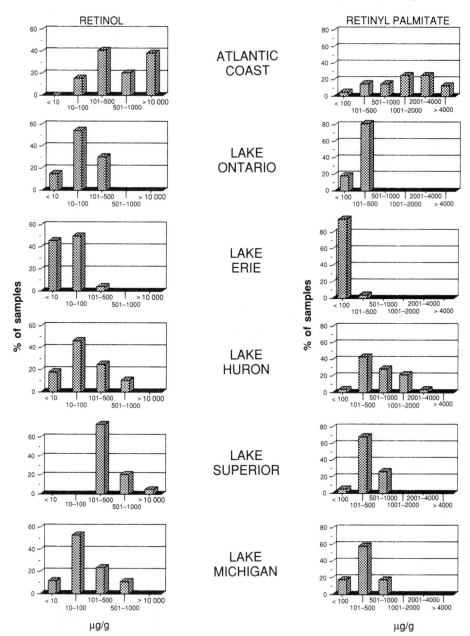

Figure 6.5 Retinol and retinyl palmitate in livers of herring gulls. After Environment Canada (1991).

Eggs in the 2–12-day period of incubation were collected from six colonies in three of the Great Lakes (Ontario, Erie, and Huron). One important difference between this study and those on hepatic levels in adult gulls was the much smaller coefficient of variance found for the eggs. The levels of retinol were significantly lower in eggs from Lake Erie, and those from the north channel of Lake Huron were significantly higher than those from other colonies. The levels of retinyl palmitate in eggs from the north channel of Lake Huron were higher than elsewhere. The molar ratio of retinol to retinyl palmitate was much lower in the north channel of Lake Huron than in any other colony, and those from Lake Erie were significantly lower than in all the remaining colonies.

Five PCDDs and three PCDFs were found above detection limits at one or more sites. In all but one site the major contaminant was 2,3,7,8-TCDD. A highly significant correlation was found between the molar ratio of retinol and retinyl palmitate and the dioxin equivalents (Figure 6.6), but not between either retinol or retinyl palmitate individually. The subject of dioxin equivalents is discussed in section 5.6.

Alterations of vitamin A are known to be caused by a variety of pollutants, including both polynuclear aromatics and PHAHs, and to be correlated with MFO activity. It could be a useful probe for bioeffect studies, but additional work is needed to establish the natural variation.

6.3 Haem, porphyrins and inhibition of aminolevulinic acid dehydratase (ALAD) by lead

6.3.1 Porphyrins: an introduction

Porphyrins consist of four pyrrole rings linked by methylene bridges to form a cyclic tetraphyrrole ring. The source is the haem biosynthetic pathway, which plays a crucial role in all living systems. A major review on the structure, extraction, separation, and identification of the various porphyrins has been produced by Rossi and Curnow (1986). For a detailed review of the use of high-performance liquid chromatography to identify the various porphyrins and their metabolites, the reader is referred to Lim *et al.* (1988).

Two major disruptions of haem biosynthesis by environmentally important agents have been studied. These are the formation of excess amounts of porphyrins by some PHAHs and the inhibition by lead on the enzyme ALAD. Although both of these effects occur in the haem metabolic pathway (Figure 6.7), the studies are quite distinct and are treated separately here.

Haem biosynthesis is normally carefully regulated and levels of porphyrins and their precursors are ordinarily very low. Hepatic porphyria is characterized by massive liver accumulation and urinary excretion of uroporphyrin and heptacarboxylic acid porphyrin. While the mechanism of PHAH-induced porphyria has not been completely elucidated, it is considered by

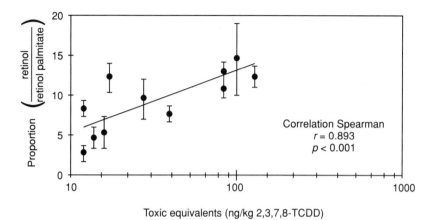

Figure 6.6 Relationship of retinol and retinyl palmitate to dioxin equivalent. Spear (in press).

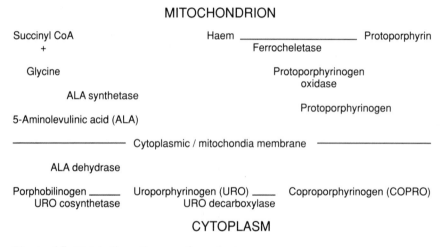

Figure 6.7 Metabolic pathways of porphyrins.

several workers that inhibition of uroporphyrinogen decarboxylase (UROD) is the proximal cause (reviewed in Marks, 1985). However, Sinclair *et al.* (1986) have found that piperonyl butoxide immediately reverses the accumulation of porphyrinogens and concluded that the action of PHAHs could not be accounted for by covalent bonding to UROD. These workers suggest that PHAHs bind to P450I isoenzymes and generate a form of activated oxygen which can oxidize porphyrinogens. The role of piperonyl butoxide would be to prevent the binding of PHAHs to P450I enzymes. The results of Sinclair and co-workers could be explained more easily if the binding of PHAHs to UROD is considered as reversible. The two hypotheses – that porphyria is due to a decrease in activity of UROD or that it is due to oxidation of porphyrinogens to porphyrins – were considered at some length by Bonkovsky *et al.* (1987). These workers considered that a primary problem with the data supporting the first hypothesis was that changes in homogenates did not reflect changes in intact cells. This group came out in favour of oxidation and considered that P450MC (or P450I, section 5.2) has a critical role. Recent experiments, (S. Kennedy, unpublished data) have found porphyria without induction of P450s. This finding and the recent observations of Smith *et al.* (1989) that iron overload can induce porphyria suggest that much is still to be unravelled on the mechanism of induction of porphyria.

Haem is incorporated into haemoglobin and several critical enzymes. The rate of haem production is dependent upon the requirements for these enzymes. Many toxic chemicals enhance the rate of haem bio-synthesis several-fold to meet the requirement for mixed function oxidases (Chapter 5).

6.3.2 Summary of experimental studies

Studies on hexachlorobenzene Of the PHAHs, porphyria caused by HCB has been studied in most detail. Diets containing 200–500 mg/kg HCB cause severe porphyria in rats. The response can be divided into three main periods. Initially there is little increase in porphyrin excretion, followed after a few weeks by a rapid increase, and after about 10 weeks the increase reaches a plateau (Elder, 1978). Neurological lesions are also seen, but these do not seem to be related to porphyria as they may occur before porphyria develops. In mice and guinea pigs, neurotoxic effects usually cause death before the onset of porphyria (De Matteis *et al.*, 1961). Kennedy and Wigfield (1986) found that rats on an HCB-corn oil diet showed severe neurological symptoms, whereas those on a diet that contained HCB but no corn oil showed none or only slight symptoms. Nevertheless, the levels of porphyrins in these animals were ten times higher than in those showing severe neurological symptoms.

Grant *et al.* (1974) examined porphyrin levels and MFO activity in rats exposed chronically to HCB. For female rats no significant differences were found for MFO induction (AHH, N-demethylase) over the entire dose range (10–160 ppm; nine months); porphyrin levels were significantly elevated at 40 ppm and above. Male rats showed a modest (1.5–4) but significant increase in AHH activity over the dose range 10–160 ppm and a similar effect on N-demethylase over the range 20–160 ppm. In contrast there was virtually no effect on the porphyrin levels.

A marked sex-linked difference in the reaction of rats to both HCB and PCBs was found by Smith *et al.* (1990). The levels of hepatic porphyrins in male rats exposed to HCB were 1000-fold higher than in females. There was a 30-fold difference between the sexes at the lower dose of Aroclor 1254 (0.005% in diet) but no differences at the higher dose (0.02%). At this higher dose individuals became jaundiced and severe liver damage, as revealed by increased levels of alanine aminotransferase (section 6.4.2) were seen. Increased levels of liver porphyrins were correlated with a marked decrease in the activity of UROD.

Using HPLC analysis, Kennedy *et al.* (1986) have shown that while there is a latent period in the build-up of total porphyrins in rats exposed to HCB, there is little or no delay in the accumulation of the highly carboxylated porphyrins (HCPs). This point is of considerable importance to the use of porphyrins as biomarkers. It points out the critical need to determine the levels of individual compounds rather than total porphyrins. Otherwise the amount of time that has passed since exposure started could confound the results.

The time course of administering HCB to quail at a dosage of 500 mg/ kg was followed by Carpenter *et al.* (1985). Only the total porphyrin content was measured. Porphyrin levels were significantly increased by day one (2.5-fold) and the increase continued throughout the period, being nearly 30-fold by day 10. Cytochrome P450 was also significantly increased by day one (nearly fourfold), but subsequent increases were modest. Increases of AHH and EROD were also marked by day one (four- and eightfold respectively). Even without measurements of HCPs, it is clear that the response in quail is quite different from that in the rat. These workers noted marked variation in the response of individual quail, with some individuals being non-porphyric. However, small sample size (four) makes it difficult to evaluate the importance of three out of four being porphyric. This marked individual variation was also noted by Elliott *et al.* (1990). By environmental standards, the dietary levels of HCB used in the studies discussed above, typically in the hundreds of ppm range, are high. Although HCB is widely found in the environment, levels are usually no higher than a few parts per million. Among the highest values found were those in cormorants with whole body values ranging from 0.4 to 25 ppm (wet weight basis with an average of 6.9) in the Netherlands (Koeman *et al.*, 1973) and 1.3 to 14.7 ppm, dry weight basis (equivalent to 0.3 to 3.5 ppm, wet weight) in common tern eggs from Lake Ontario (Gilbertson and Reynolds, 1972). Typically, the values in herring gull eggs from the Great Lakes Monitoring Program have been less than 0.5 ppm.

Studies with PCBs The other compound, or rather group of compounds, that has been studied for its porphyrinogenic effects is the PCBs. As is typical of studies with PCBs, the earlier studies were carried out with commercial mixtures, and only more recently has congener-specific data become available.

The comparative toxicity of different commercial PCB mixtures with the same degree of chlorination was studied by Vos and Koeman (1970), using dietary levels of 400 ppm. While porphyria, oedema and mortality were found for all three preparations, there were marked differences. The mortality caused by Aroclor was 15%, whereas mortality of the entire group occurred with both Phenoclor and Clophen. The resulting residue levels were high by environmental standards (120 to 2900 ppm in liver; 40–700 ppm in brain) and also highly variable.

The effect of Aroclor 1254 on rats at 100 ppm in their diet was studied by Goldstein *et al.* (1974). Levels of cytochrome P450 had increased significantly by day two and reached maximal levels within a week. In contrast there was no increase of urinary uroporphyrin until two months after the start of the experiment; then levels rose rapidly and by seven months were 500 times control values. These increases then reached a plateau. No mortality or neurotoxicity effects were found throughout the 13-month period of the experiment.

A large single dose of Aroclor 1254 (500 mg/kg) caused a 1700-fold increase of porphyrins in the kidney of quail within 48 hours (Miranda *et al.*, 1986). The increase in the liver was large (76-fold) but less marked than in the kidney. Increases in the activity of ALA synthetase and uroporphyrinogen synthetase were found in both organs (the former more marked in the liver and the latter in the kidney); decreased activity of UROD was found in the kidney only.

Elliott and co-workers (Elliott *et al.*, 1990) examined the effects of individual PCB congeners on hepatic porphyrin levels and induction of MFOs in quail and kestrels and compared them to the response which had been found in rats. This work has already been discussed under mixed function oxidases (section 5.6) and an overview of the results is given in Figure 5.4. The congeners were selected to cover a range of substitution patterns (non-ortho, mono-ortho, and di-ortho). In the quail modest, but statistically significant, increases in HCPs and coproporphyrin were seen with mono-ortho (2,3,3′,4,4′) and some with the di-ortho (2,2′,4,4′,5,5′). No statistically significant differences were seen in the kestrel. It is likely that increases are masked by the high individual variability.

6.3.3 Interrelationship with other biomarkers

The dependence of the porphyrogenic effect of TCDD on responsiveness of MFO induction was studied by Jones and Sweeney (1980). These workers studied the induction of porphyria in mice that were responsive (strain C57BL/6J) and non-responsive (DBA/2J) to induction of AHH. They found that a weekly injection of 25 µg/kg of TCDD for six weeks caused a sevenfold increase in urine porphyrin excretion and a halving of UROD activity in responsive mice; no significant effects were seen in the non-responsive mice. Studies with backcrosses showed that susceptibility to porphyria was inherited together with responsiveness to AHH induction.

These findings have not been borne out by subsequent studies involving a wider series of strains of mice (Greig *et al.*, 1984). These workers found that the inheritance of increased porphyrin levels and of increased levels of plasma enzymes which indicate hepatic necrosis (alanine aminotransferase and sorbitol dehydrogenase, both discussed in more detail in section 6.4.2) is complex. They suggest that the lack of correlation in the F_2 generation is due to segregation of alleles at more than one locus and that genes other than the Ah influence hepatotoxicity.

The key role that P450I plays in PHAH-induced uroporphyria was shown by Bonkovsky *et al.* (1987). They demonstrated, in cultured chick embryo liver cells, that P450I inhibitors, but not P450II inhibitors, cause rapid termination of uroporphyrin accumulation. The first stage is considered to be the binding of the PHAH to the cytosolic Ah receptor followed by induction of the P450I system. The interactions of PHAHs with the Ah receptor are considered in more detail in section 5.6. The second stage, the formation of an active oxidant, has already been discussed in section 6.3.1.

The effect of Aroclor 1242 and two specific PCB isomers (2,4,2′,4′ and 3,4,3′,4′-tetrachlorobiphenyl) on porphyrin and cytochrome levels and aminolevulinic acid synthesase (ALAS), ethoxycourmarin-o-deethylase (ECOD) and EROD activity in the small intestine and liver of quail was studied by Miranda *et al.* (1987). Both porphyrin and cytochrome levels were increased by all compounds in both the intestine and liver. Although ALAS activity was increased in both organs by all compounds, the increase was much more marked in the liver. ECOD activity was unchanged or reduced, whereas EROD activity was increased in both organs, but again was much more marked in the liver. The authors comment that 'tissue differences in porphyrin levels are not readily explained on the basis of induction of ALAS activity'.

Lambrecht *et al.* (1988) examined the effects of TCDD and 3,4,3,4′-tetrachlorobiphenyl (TCB) on uroporphyrin (URO) accumulation in cultured chick embryo hepatocytes. They found that TCDD alone caused only a slight increase in URO, whereas TCB caused a considerable increase associated with increased 5-ALAS activity. However, in the presence of exogenous 5-ALA, TCDD was more potent than TCB in causing URO accumulation. The concentrations of TCDD and TCB that cause maximal induction of EROD were lower than those required for maximal accumulation of URO. Pretreatment with 3-MC enhanced URO accumulation, whereas inhibitors of cytochrome P450I decreased it. URO accumulation occurred without a decrease in URO-decarboxylase activity. The authors conclude that URO accumulation requires two separate actions: induction of cytochrome P450 at low concentrations of PHAHs and increased uroporphyrinogen oxidation which is catalysed by the induced cytochrome at higher concentrations of PHAHs.

The effect of piperonyl butoxide, an inhibitor of cytochrome P450I, on URO accumulation in cultures of chick embryo hepatocytes by TCB, 2,4,5,3′,4′-pentabromobiphenyl and 3,4,5,3′,4′,5′-hexachlorobiphenyl was examined by Sinclair *et al.* (1986). They demonstrated that inhibition of cytochrome P450I restored haem synthesis as shown by the incorporation of labelled ALA and a decrease in the induced ALA synthetase activity. It was considered that the effect of piperonyl butoxide was not due to inhibition of metabolism of PHAHs, as the hexachlorobiphenyl remained virtually unmetabolized. The authors consider that the evidence 'seems to exclude irreversible covalent binding to UROD by an inhibitory metabolite of biphenyl as the mechanism for URO accumulation'. They suggest that the role of P450I is to bind the biphenyls and to generate a form of activated oxygen. They point out the need to confirm the changes of activity of UROD in intact cells. Direct inhibition, however, is not excluded. Sinclair *et al.* (1987) suggest two roles for PHAHs in the development of porphyria: induction of P450I and some interaction of the PHAH with cytochrome P450I that leads to urogen oxidation.

The effects of a series of di-, tetra- and hexachlorobiphenyls on porphyrin biosynthesis in chicken embryo liver cells were examined by Kawanishi *et al.* (1978). They found that the structural requirements for potent porphyrin-inducing activity were para and meta substituted structures which caused a nearly co-planar conformation. The isomers 3,4,3′,4′,-tetra and 3,4,5,3′,4′,5′-hexa were the most active inducers. Sassa *et al.* (1986) discussed the effects of a similar series of experiments in terms of the three-dimensional structure of these congeners, as determined by molecular orbital calculations. They calculated the conformational energy as a function of the angle between the diphenyl rings. The conformation of the active PCBs was relatively flexible, which enabled them to assume a co-planar formation, whereas the inactive species had a rigid angular structure.

The effects of injected PHAHs (TCB, 2,3,6,2′,3′,6′ and 3,4,5,3′,4′,5′-hexachlorobiphenyls and TCDD) on survival of chick embryos, MFO induction, and UROD were studied by Rifkind *et al.* (1985). ALA synthetase was increased 10- to 20-fold after both one and nine days of exposure except by 2,3,6,2′,3′,6′-hexachlorobiphenyl. Hepatic porphyrins and UROD activities were virtually unaffected. Induction of AHH was increased 10- to 12-fold and EROD 28- to 55-fold (again except for 2,3,6,2′,3′,6′). The PHAHs decreased survival at nine days and were accompanied by decreased thymus weights and increased oedema. The effects on survival were greatest for TCB and least for TCDD. These findings do not correlate with MFO induction. For AHH, induction was the same for all three compounds, and TCDD was the strongest inducer of EROD. The authors conclude that dose-response relationships for lethality and MFO induction are dissociated, and furthermore that PHAH lethality occurs independently of effects on UROD.

Kennedy and James (unpublished data) examined the effects of three co-planar PCBs (3,4,3',4'; 3,4,5,3',4' and 3,4,5,3',4',5') and three non-planar PCBs (2,4,2',4'; 2,4,5,2',5' and 2,4,6,2',4',6') on the accumulation of URO in chicken embryo liver cell cultures. These workers found that the co-planar PCBs were potent in producing URO.

The interrelationship of thyroid function and HCB-induced porphyria was examined in rats by Pisarev *et al.* (1990). The dosage used (1000 mg/kg/day) was high; both at one and eight weeks a significant (40%) increase in liver/body weight ratio was seen and the thyroid/body weight ratio was not affected. There was a marked decrease in serum T_4 to a quarter of control values at both time periods, whereas T_3 levels were not affected. Based on total porphyrins, no effect was seen at one week, but a large (70-fold) increase was seen after eight weeks. An increase in the activity of ALAS at eight weeks and a decrease of porphyrinogen carboxylase at both times were reported. The authors conclude that serum T_4 is the most sensitive indicator, occurring sooner and at low doses, but this statement is based on measuring only total porphyrins. It would be interesting to make a comparison with the levels of specific HCPs.

6.3.4 Use of porphyrins as biomarkers

Koss *et al.* (1986) found that the patterns of hepatic porphyrins were markedly different in pike collected from the River Rhine from those from the River Lahn. The levels of PHAHs were up to 40-fold higher in the fish from the Rhine than those from the Lahn; however, the concentrations of heavy metals differed by less than twofold. Quantitative data on the levels of the highly carboxylated porphyrins are not given, but activity of hepatic uroporphyrinogen decarboxylase in pike from the Rhine was only 30% of that from the Lahn.

The levels of hepatic HCPs were markedly elevated in herring gulls collected from the Great Lakes when compared to those from the Atlantic coast (Fox *et al.*, 1988). These results are shown diagrammatically in Figure 6.8. The highest levels were those from eastern Lake Ontario in 1974 (×38 coastal), Green Bay, Lake Michigan in 1983 and 1985 (×28) and Saginaw Bay, Lake Huron in 1982 and 1985 (×25). Several porphyrinogenic compounds, of widely different potencies, are present in the Great Lakes. Although possible correlations between the concentration of these compounds and the HCP levels were examined, it was not possible to determine which of the potentially porphyrinogenic compounds was responsible for the observed porphyria.

The argument that elevated levels of HCPs in Great Lakes herring gulls are caused by exposure to OCs rests on the fact that there are very few other

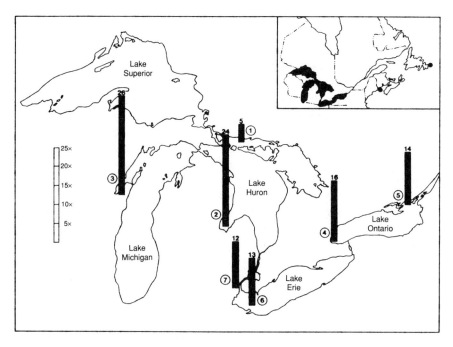

1. DOUBLE ISLAND, LAKE HURON
2. SAGINAW BAY, LAKE HURON
3. GREEN BAY, LAKE MICHIGAN
4. HAMILTON HARBOUR, LAKE HURON
5. SNAKE ISLAND, LAKE HURON
6. MIDDLE ISLAND, LAKE ERIE
7. DETROIT RIVER
ATLANTIC COAST SITES SHOWN IN MAP INSERT

Figure 6.8 Levels of porphyrins in herring gulls in the Great Lakes. Median levels of highly carboxylated porphyrin levels in livers collected from seven Great Lake colonies are expressed as multiples of the median levels in herring gulls from the Atlantic coast. Location of Great Lakes sites are identified with circled numbers. After Fox *et al.* (1988).

known causes of elevated liver HCPs. In humans, the only other reported causes are rare genetic forms of porphyria, steroid therapy in individuals genetically predisposed to the disorder, and, in very rare cases, alcoholism. In experimental mammals, the only reported cause of elevated liver HCPs other than OCs appears to be aflatoxin B_1. OCs are the only known cause of HCP elevation in birds.

While there is no evidence that the elevated porphyrin levels observed in herring gulls on the Great Lakes are harmful to their health (HCP elevations of 10 000-fold can be induced in experimental animals and have been observed in humans poisoned by PCBs, TCDD and/or HCB), HCPs appear to be a sensitive indicator of PHAH-induced changes.

The variation in the mean of the hepatic levels of HCPs in seven species of birds (covering five orders) was only twofold (Fox *et al*., 1988) and the total range was 4–22 pmol/g. No comparable study appears to have been made on any other class of organism. These baseline data, collected from areas of low contamination, show only small variation, but in view of variability of response to PHAHs in experimental studies, variability in areas of high contamination is a problem. Marked differences in the response of male and female rats to HCB and PCBs (section 6.3.2) indicate another potential problem in the use of porphyrins as a biomarker. At present, there is not a large body of information on the inter-species variation on the induction of porphyria. The interaction with MFO induction, which is likely to be more sensitive but less specific than porphyrins, needs to be examined.

6.3.5 Inhibition of ALAD by lead

Inhibition of ALAD was first studied by Hernberg *et al*. (1970) as a means of detecting environmental lead exposure in humans and has since become the standard bioassay for this purpose. The position of ALAD in the biosynthesis of haem is shown in Figure 6.8. The enzyme ferrochelatase is also inhibited but does not appear to be rate limiting. ALAD inhibition has been widely used in measuring exposure to lead and has been comprehensively reviewed quite recently (Scheuhammer, 1987a,b).

6.3.6 Factors affecting ALAD activity

Specificity The effects of cadmium, copper, mercury and zinc *in vivo* on ALAD activity in trout were examined by Hodson *et al*. (1977). They found no inhibition until the levels of these metals approached lethality. Scheuhammer (1987a) has examined the effects of cadmium, copper and mercury (both inorganic and methyl) on the inhibition of ALAD in an *in vitro* avian system. He found that these elements were 10 000 times less potent as inhibitors than lead. The IC_{50} for lead was found to be 0.03–0.04 μmol/ml blood. Similar specificity has been found in fish. Hodson *et al*. (1977) found a dose-related response for lead, whereas even near-lethal doses of cadmium, copper, mercury and zinc did not cause significant inhibition of ALAD.

Dose-response Hodson *et al*. (1977) established relationships between ALAD activity and the log of both the concentration of lead in the blood and in the water. Scheuhammer (1987a) gives plots of ALAD activity against the concentration of metals added *in vitro*. The curves all have the same shape, with ALAD activity falling rapidly over a small concentration range. The curves for cadmium, copper, and mercury are all displaced to much lower concentrations than that for lead.

In *in vivo* studies there was a lack of any dose-response in the red blood cells of kestrels exposed to 25, 125 or 625 mg/kg. All groups showed 45–60% decrease over controls (Hoffman *et al.*, 1985). However, there was a clear dose-response relationship in kidney, with activity reduced to 58% at the lowest dose and down to 20% at the highest dose. There was some indication of a dose response in brain; at the higher doses 43 and 53% decreases of activity were recorded. The group of kestrels exposed to the highest dose had 40% mortality over a 10-day period.

A detailed table relating confidence limits to group size for various degrees of ALAD inhibition was compiled by Hernberg *et al.* (1970). The calculations are based on blood samples from 159 persons with lead levels ranging from 5 to 95 µg/100 ml. The regression is linear over the entire range.

Age/sex A slight increase in liver ALAD activity and marked increase in brain ALAD in nestlings of the American kestrel during the first ten days of their lives were found by Hoffman *et al.* (1985). The activity of ALAD in nestling barn swallows was approximately double that of adults (Grue *et al.*, 1984), but no differences were observed between sexes in adults. No sex-associated differences in plasma levels of ALAD were found in mallards (Finley *et al.*, 1976) or canvasbacks (Dieter *et al.*, 1976). No differences were found between immatures and adults for either sex (Dieter *et al.*, 1976).

Species variation There appears to be no detailed survey of inter-species variation of the activity of this enzyme comparable to those for AChE (section 2.1.3). Comparing different studies, with all the difficulties of different techniques, there appears to be about a sixfold difference among avian species. Values in plasma ranged from 50 units in the pigeon (Hutton, 1980) to 300 in the kestrel (Hoffman *et al.*, 1985).

6.3.7 Interaction with other biomarkers

A range of biochemical parameters associated with the nervous system in young rats exposed to lead was examined by Sobotka *et al.* (1975). They found no consistent changes of any of the biogenic amines, nor were there any changes in the levels of brain protein or DNA. Some inconsistent changes in AChE and butyrylcholinesterase activities were found, whereas ALAD was consistently inhibited even at the lowest dose of lead.

In the kestrel, Hoffman *et al.* (1985) also found no effects of lead on the biogenic amines, although some alterations to the RNA/DNA ratio were found. ALAD was inhibited at all lead concentrations used, although no clear dose-response was demonstrated. Dieter and Finley (1979) found no change in AChE activity in the brains of mallards after dosage (*c*. 200 mg) of lead shot, but there was a significant increase of butyrylcholinesterase in both cerebellum (28%) and cerebral hemisphere (22%). The degree of ALAD inhibition was greater (50%) in the cerebellum than the cerebral hemisphere (35%).

6.3.8 Use of ALAD as a biomarker

ALAD has been used as an indicator of lead exposure both for general problems, such as in urban areas and along highways and also specifically to study the 'lead-shot problem' in waterfowl.

A threefold difference in the blood ALAD activity and a significant, but less marked, difference in the kidney between rats from a rural and an urban site in Michigan were found by Mouw *et al.* (1975). The main physiological indications of lead toxicity in the urban rats were an increase in kidney weight and the incidence of intranuclear inclusions. Both effects could be correlated with lead levels. Similarly, Hutton (1980) was able to demonstrate marked differences in the ALAD activity between rural, outer urban, suburban and central feral pigeons in London. Differences were found in blood, liver, and kidney, but were most marked in blood, where the figures were 50.0, 10.5, 3.7 and 2.2 units, respectively.

The lead levels, ALAD activity and reproductive success of barn swallows and starlings along highways with different traffic densities were examined by Grue and co-workers. For the swallow, they found a significant increase of the lead levels in the feathers and carcasses of both adults and nestlings and a 30–34% decrease in plasma ALAD activity (Grue *et al.*, 1984). However, the number of eggs laid, number of young fledged, and pre-fledgling body weights were not affected. The findings on starlings were similar (Grue *et al.*, 1986). A 3- to 13-fold increase in lead levels and a 43–60% decrease of ALAD activity were found in starlings between the rural and the heaviest travelled route. As with the swallows, no change in reproductive success was found, indicating that lead from automotive emissions does not pose a serious hazard to birds nesting close to motorways.

The inhibition of ALAD has been shown to be a reliable indicator of exposure to lead in studies on several species of fish (Hodson *et al.*, 1977). These workers found a linear regression in plotting ALAD activity against both concentration of lead in the blood (10–540 µg/100 ml) and the concentration of lead in the water (2–100 µg/l).

Andersson *et al.* (1988) found some changes of ALAD in fish at different distances from a Kraft pulp mill. In one series of measurements there was a clear decrease in ALAD activity from source to control areas, but these differences were not consistently found when the measurements were repeated.

Mortality of waterfowl due to the ingestion of lead shot has been of serious concern for many years, the issue first being raised over 70 years ago (Wetmore, 1919). The problem is caused by ducks and geese ingesting spent lead shot during the course of their feeding. Bellrose (1959) found that 12% of the gizzard samples examined contained at least one lead shot and considered that 2–3% of all waterfowl in North America died from lead poisoning. Ingestion of a single lead shot (*c.* 200 mg lead) can kill a mallard maintained on a corn diet within two weeks (Finley and Dieter, 1979), although Dieter and Finley (1979), maintaining mallards on a half corn, half commercial pellet diet, had no mortality from a single ingested pellet. Sub-lethal effects, as manifested by decreased body weight, were found in canvasbacks that had lead shot in their gizzards (Hohman *et al.*, 1990).

Secondary poisoning of bald eagles feeding on waterfowl has also been of concern. Hoffman *et al.* (1981) demonstrated that the ingestion of ten pellets could cause 80% inhibition of ALAD and death within ten days. National surveys of eagles found dead showed that *c.* 5% had died from lead poisoning (Pattee and Hennes, 1983).

One of the best-documented cases of lead poisoning in a wildlife population has been the case of mute swans on the River Thames in England. This case has received additional prominence since these swans have, since the Middle Ages, been considered as the property of the Crown. In 1983 and 1984, 50–60% of the 200–240 swans examined were diagnosed as lead poisoned. The introduction of a voluntary ban of the use of lead weights by anglers reduced this percentage to 36–40% in 1985 and 1986. Following legislation banning the sale and use of lead weights in angling, lead poisoning as the cause of death dropped to 24% of the 213 swans examined in 1987 and 16% of the 241 in 1988 (Sears, 1989).

ALAD inhibition is rapidly induced. The effect is only slowly reversed, with ALAD values returning to normal values only after about four months (Dieter and Finley, 1978) and is sufficiently sensitive to detect the effect of a single pellet. A strong negative correlation between blood lead concentration and log ALAD activity has been found by many workers. This correlation may be improved still further if ALAD activity is expressed as an activity ratio (Scheuhammer, 1987b). In this method, one aliquot is assayed normally; the other is treated with an agent that removes the effects of inhibitors other than lead. Using this approach, Scheuhammer (1989) concluded that 'in the absence of elevated lead exposure, birds had comparable ALAD activity ratios regardless of species, geographical location, or time of year sample,' and that 'underestimation of lead exposure did not occur using the ALAD activity ratio method.' The relation of ALAD activity ratio to blood lead concentration is shown in Figure 6.9, which clearly demonstrates that it is an accurate method of determining the exposure of wild birds to lead. The ALAD assay is a simple one, which can be carried out without expensive equipment or lengthy training. The use of

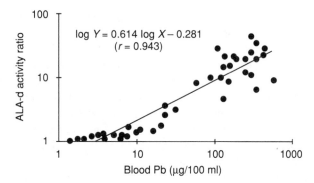

Figure 6.9 Relationship of ALA-d to blood lead. Concentration of blood in free-living mallards from Lalu St Clair plotted against ALAD activity ratio. Lines and equations describe the best-fit linear regressions. Scheuhammer (1989).
Status: Copyright needed.

protoporphyrin as an indicator of lead exposure has been proposed (Passer *et al.*, 1989), but it does not correlate as well and, in addition, is a more difficult assay to perform.

ALAD inhibition represents one end of the biomarker spectrum. It is a sensitive, dose-dependent measurement that is specific for a single environmental pollutant, lead.

6.4 Enzyme activity

6.4.1 Introduction

A large number of enzymes are involved in the regulation of metabolism in vertebrates. The effects of environmental contaminants on three enzyme systems, the induction of mixed function oxidases, and the inhibition of esterases and aminolevulinic dehydratase have been studied in great detail, and these studies have been considered in separate sections (4; 2.1; and 6.3 respectively).

The best data sets available on other enzymes are those on the transaminases (glutamic oxaloacetic transaminase (GOT) and glutamic pyruvate transaminase (GPT), lactate dehydrogenase (LDH), alkaline phosphatases (ALP) and the adenosine triphosphatases (ATPases)). These enzymes are considered in this section. For other enzymes the data are fragmentary and no attempt has been made to catalogue them.

Ideally one would like to cover all the major metabolic pathways. This has been attempted by Christensen and co-workers, who have published a series of papers on the *in vitro* inhibition of enzymes by a range of environmentally important compounds (reviewed in Christensen *et al.*, 1982). The enzymes considered were GOT, LDH, carbonic anhydrase, ATPase, AChE, RNase, lipase and urease. While this approach allows rapid screening, it has not been possible to relate *in vitro* inhibition with *in vivo* changes. Enzymes such as GOT and LDH are inhibited by PHAHs *in vitro*, but in the intact animal an increase is seen since the physiological mechanism is the disruption of membranes, which allows the enzyme to reach the plasma. *In vitro* inhibition of rabbit muscle LDH has been shown by Hendrickson and Bowden (1976) for a range of PHAHs – mirex, DDT and dieldrin. Meany and Pocker (1979) considered that the inactivation of LDH by PHAHs is due to coprecipitation rather than inhibition. Jackim *et al.* (1970) pointed out that changes of enzyme activity in fish exposed to heavy metals were not always the same in either magnitude or direction. In some cases, for example, the work on ATPase by Fattah and Crowder (1980), good correlations between *in vitro* and *in vivo* have been found. Nevertheless, *in vitro* enzyme studies will not be considered further.

In all cases it is necessary to establish a firm baseline from which to judge any changes of enzyme activity caused by pollutants. Such factors as the effects of age, sex, season and inter-species variations have to be borne in mind, but are not specifically considered here. Westlake *et al.* (1983) have produced a major review of inter-species variation of a range of enzymes for a wide variety of wild birds and mammals collected in the United Kingdom. Hill and Murray (1987) have made a critical evaluation of the inter-species, inter-sex and seasonal variations of a wide range of biochemical parameters in four avian species. More recently, Fairbrother *et al.* (1990) have examined the effects of age, sex and reproductive condition of the mallard on many serum chemistry parameters including the activities of several enzymes.

6.4.2 Amino acid metabolism

Two transaminases of particular clinical importance are glutamic oxaloacetic transaminase (GOT) and glutamic pyruvate transaminase (GPT). These enzymes control the conversion of aspartate to glutamate and of alanine to glutamate, respectively, and are also referred to as alanine and aspartate aminotransferases, respectively. The enzyme glutamate dehydrogenase (GDH) controls the conversion of ketoglutarate to glutamate, and defects in GDH can affect the level of glutamate and subsequently of purines. The information available on the effects of pollutants on enzymes involved in amino acid metabolism is summarized in Table 6.1.

These are intracellular enzymes, the serum levels of which are normally low. Significant tissue damage can lead to markedly increased levels of both of these enzymes in serum. These enzymes are compartmentalized, and more are localized in the mitochondria than in the cytosol. Full release would occur only if both the mitochondrial and plasma membranes were

Table 6.1 Effect of pollutants on enzymes involved in amino acid metabolism

	GOT	GPT	GHD	Reference
PHAHs				
DDE		+ Quail = Starling		Dieter (1974) Dieter (1975)
Dieldrin	+ Sailfin			Lane and Scura (1970)
PCBs		+ Quail + Starling		Dieter (1974) Dieter (1975) Ito (1973)
Photomirex	+ Carp + Rat			Chu et al. (1981)
OPs				
Carbophenthion	+ Quail + Geese = Chicken = Pigeon		+ Quail	Westlake et al. (1978)
Methyl pirimiphos	+ Quail + Quail		+ Quail – Quail	Westlake et al. (1981a)
Malathion		+ Quail = Starling		Dieter (1974) Dieter (1975)
Methyl malathion	= Kestrel			Rattner and Franson (1984)
Methylparathion Phosmethylan Carbendazim		+ Chicken + Chicken + Chicken		Somlyay and Várnagy (1989)
Dichlorvos	+ Catfish	+ Catfish		Verma et al. (1981)
Fenthion	+ Chicken	+ Chicken		Singh et al. (1989)

Carbamates				
Aldicarb	+ Quail		+ Quail	Westlake et al. (1981b)
Methiocarb	+ Quail		= Quail	
Oxyamyl	+ Quail		= Quail	
Primicarb	+ Quail		= Quail	
Thiofanox	+ Quail		= Quail	
Methidathion	+ Carp			Asztalos et al. (1990)
Carbofuran	+ Catfish	+ Catfish		Verma et al. (1981)
Metals				
Cadmium	= Trout			Christensen et al. (1977)
Copper	= Mullet			Helmy et al. (1979)
	+ Carp			Asztalos et al. (1990)
Lead	= Calf			Logner et al. (1984)
Lead	? Carp			Helmy et al. (1979)
	+ Mullet			Hoffman et al. (1981)
	+ Eagle		= Eagle	Sastry and Sharma (1980)
Mercury	+ Fish	+ Fish		Hilmy et al. (1981)
	+ Fish			Helmy et al. (1979)
	= Carp			Dieter (1974)
Methyl-mercury	= Fish	= Quail		Christensen et al. (1977)
		+ Starling		Dieter (1975)

Table 6.1 (cont.)

	GOT	GPT	GHD	Reference
Others				
Acidity	+ Fish			Adams *et al.* (1985)
Sewage effluent	+ Fish			Wieser and Hinterleitner (1980)
Water quality	+ Carp			Renqing (1990)
Fenvalerate	+ Kestrel			Rattner and Franson (1984)
	+ Chicken	+ Chicken		Mohamed and El-Sheamy (1989)
Cypermethrin	+ Chicken	+ Chicken		
Oil	+ Duck			Szaro *et al.* (1978)
	+ Mallard	+ Mallard		Hoffman *et al.* (1982)
	= Quail	– Quail		
		= Fish		Chambers *et al.* (1979)
Paraquat	+ Carp			Asztalos *et al.* (1990)
2, 4-D		+ Chicken		Somlyay and Varnagy (1989)
Krenite	– Mallard	– Mallard		Hoffman (1988)

= No effect – Decreased + Increased.

damaged. Purely cytosolic enzymes, such as LDH, correlate better with cell variability as measured by trypan blue exclusion than does GOT (Story *et al.*, 1983). The comparison of the changes in levels of transaminases with LDH is given later (section 6.4.3). Acosta *et al.* (1985) found that argininosuccinate lyase (ASAL) is more sensitive than GOT, GPT or LDH as a measure of leakage of cytosolic enzymes in primary cultured hepatocytes exposed to several drugs. No studies on ASAL appear to have been made using compounds of environmental interest.

By far the best database is that on GOT. The release of this enzyme is considered to be an index to liver cell damage rather than being directly related to any changes in liver function. Increased levels have been reported to be caused by a wide variety of substances. The only notable exceptions are the heavy metals, for which a variety of responses have been recorded. A small decrease of GOT was observed in brook trout exposed to lead (Christensen *et al.*, 1977). The difference was considered to be statistically significant, but if we look at the standard deviations this seems unlikely to be the case. No effect was seen with cadmium up to 6 µg/l or methylmercury (3 µg/l).

Some marked inter-species variations were found by Westlake *et al.* (1978) in studies with carbophenothion. Quail exposed to an LD_{50} dose had a two-fold increase, and those birds surviving the dose had levels raised to tenfold after 24 hours, whereas no effect was seen on chickens. These workers found a marked species difference within geese, with three Anser showing a much more marked response than did Branta. A comparison of the time course of the changes to plasma GOT activity in the Canada goose and the white-fronted goose is given in Figure 6.10 as a cautionary illustration of the dangers of even limited inter-species extrapolation.

The effects of five organophosphates on a number of enzymes, including GOT and GDH in quail, were examined by Westlake *et al.* (1981a). Plasma GOT was markedly elevated after 24 hours in all quail surviving carbo-phenothion and methyl pirimiphos. No clear-cut effects were seen with the other three OPs, but no quail survived the 24-hour period. Clear increases in GDH were found with carbophenothion at both 2 and 24 hours, but not with any of the other OPs.

A parallel study on the effects of five carbamates was also carried out (Westlake *et al.*, 1981b). A marked increase in plasma GOT levels was found in all quail that survived an LD_{50} dosage of any of five carbamates. In contrast, GDH was elevated only by aldicarb. The dosages used by these workers, both in this experiment and those reported in the previous paragraph, were at the lethal level or close to it. The lowest doses used were 0.33–0.5 of the LD_{50}. As would be expected, severe inhibition of AChE was observed in all cases.

Some interesting differences between the effects of representatives of classes of compounds of major interest (PHAHs, heavy metals and OPs) on

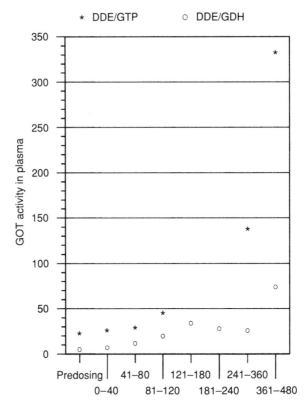

Figure 6.10 Plasma activity of GOT in Canada and white-fronted geese following exposure to carbophenothion. After Westlake *et al.* (1978).

the quail and starling were found by Dieter (1974, 1975) on GPT. Methylmercury (8 ppm in diet) and Aroclor 1254 (100 ppm in diet) increased activity levels 2.4- and 3.4-fold, respectively, in starlings, whereas DDE (100 ppm) and malathion (160 pm) had no significant effect. Quite different effects were found with quail where DDE was the strongest inducer (2.5-fold), followed by Aroclor 1254 and malathion. Mercury (this time inorganic) did not cause an effect. Hoffman *et al.* (1981) found a 50% increase in the activity of GOT in bald eagles exposed to ingested lead shot, whereas GPT was not affected.

In a survey Dieter *et al.* (1976) found that 11% of the blood samples collected from canvasback ducks captured in Chesapeake Bay in the early 1970s had abnormally high GPT levels. Wiser and Hinterleitner (1980) found that GOT and GPT activity in rainbow trout could be related to the distance from a sewage outflow. No difference in the appearance of the gills or liver was noted. These workers considered that the increase in the activity of GOT was correlated with the nitrogen content of the water.

Table 6.2 Effects of pollutants on LDH

PHAHs		
DDE	+ Quail	Dieter (1974)
	+ Starling	Dieter (1975)
DDT	= Redstart	Karlsson *et al.* (1974)
PCBs	= Redstart	
	+ Quail	Dieter (1974)
	+ Starling	Dieter (1975)
Endrin	− Fish	Sharma *et al.* (1979)
	(*Ophiocephalus*)	
Photomirex	+ Rat	Chu *et al.* (1981)
OPs		
Malathion	+ Rat	Dragomirescu *et al.* (1975)
	+ Quail	Dieter (1974)
	+ Starling	Dieter (1975)
	− Carp	Dragomirescu *et al.* (1975)
Methylparathion	+ Chicken	Somlyay *et al.* (1989)
Phosmethylan	+ Chicken	
Methidathion	+ Carp	Asztalos *et al.* (1990)
Metals		
Cadmium chloride	= Brook trout	Christensen *et al.* (1977)
Copper sulphate	+ Carp	Dragomirescu *et al.* (1975)
Lead nitrate	= Brook trout	Christensen *et al.* (1977)
Mercuric chloride	+ Quail	Dieter (1974)
	= Brook trout	Christensen *et al.* (1977)
	+ Fish	Verma and Chand (1986)
	(*Notopterus*)	
Methylmercury	+ Starling	Dieter (1975)
Others		
Oil	= Striped mullet	Chambers *et al.* (1979)
Paraquat	+ Carp	Asztalos *et al.* (1990)

6.4.3 Carbohydrate metabolism: glycolysis

The best-studied enzyme is lactate dehydrogenase (LDH), which controls the conversion of pyruvate to lactic acid and is a vital part of the Cori Cycle. This cycle is central to glycolysis, involving lactic acid, pyruvate, glucose and glycogen. The available data are summarized in Table 6.2.

Increases are consistently seen with the PHAHs, the only exception being that of Karlsson *et al.* (1974), who found no significant increase of LDH activity in redstarts, a small passerine species, treated with a low dose of DDT or PCB (10–11 µg/bird, roughly 0.5 ppm) for 10 days. Dieter (1974) found that activity of plasma LDH in quail was directly related to the log of the concentration of DDE and Aroclor 1254) over the range 5–100 ppm

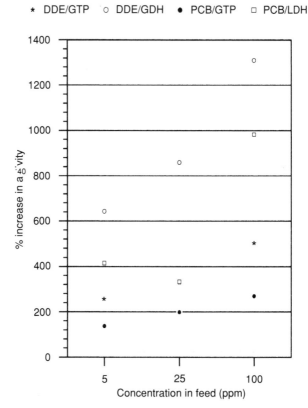

Figure 6.11 Relative increase of GPT and LDH in the plasma of quail caused by DDE and Aroclor 1254. After Dieter (1974).

in diet. Overall, there was a ninefold increase in activity. Increases were also found in starlings (Dieter, 1975), but these were much less marked (×1.5 DDE, ×3 PCB). Effects of OPs and metals were less consistent.

As mentioned in the previous section, some differentiation of which membranes are affected may be possible by examining the relative increase of transaminases and LDH. The data from Dieter (1974) on the effect of DDE and Aroclor 1254 on these enzymes in quail are redrawn as a percentage change in Figure 6.11. While the actual numbers should not be taken too seriously (since it was difficult to read off the control values of LDH accurately enough), the data make two points clearly; first, that LDH is more sensitive than GPT, and, second, that DDE causes a more marked effect than Aroclor 1254. The findings of the same worker (Dieter, 1975) on the starling confirm neither of these findings. The data (Figure 6.12) show that the responses of GPT and LDH are essentially identical for both compounds, but that the response to Aroclor 1254 was more marked than that to DDE.

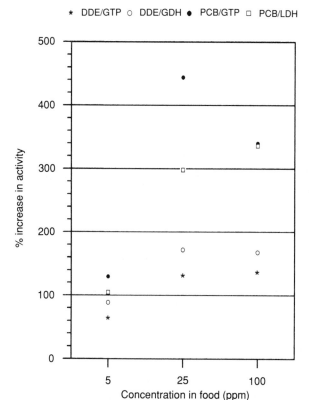

Figure 6.12 Relative increase of GPT and LDH in the plasma of starlings caused by DDE and Aroclor 1254. After Dieter (1975).

The effect of OPs on both GPT and LDH is less marked than that caused by PHAHs. The studies of Somlyay and Varnagy (1989) for three OPs over a wide dosage range do not show a differential effect between the two enzymes.

In field studies Dieter *et al.* (1976) found that 11% of the canvasback ducks examined in the Chesapeake Bay had abnormal LDH activity. Tinsley (1965) studied the effect of DDT in rats on the activity of two enzymes of the hexose monophosphate shunt. He found that dietary levels as low as glucose-6-phosphate dehydrogenase (G6PD) caused a reduction of activity, whereas the activity of 6-phosphogluconate dehydrogenase was not affected. He considered that the ratio could be used as an index of DDT stress. In contrast, Buhler and Benville (1969) found that DDT had no effect on G6PD in coho salmon or rainbow trout. The organophosphate malathion decreased the activity of glucose-6-phosphatase in both the rat and the fish (Dragomirescu *et al.*, 1975).

Kacew and Singhal (1973) found that an acute oral dose of several PHAHs (DDT, chlordane, heptachlor and endrin) to rats caused a 50% increase in serum glucose and a lowering of hepatic glycogen to 60% of control values. Increases were found in four enzymes that play a key role in gluconeogenesis – pyruvate carboxylase, phosphoenolpyruvate carboxykinase, fructose 1,6-diphosphatase and G6P phosphatase – and the effects of the various PHAHs were similar (1.5- to 2.4-fold increases) for all four enzymes.

6.4.4 Oxidative phosphorylation

Oxidative phosphorylation is the process whereby electrical energy is converted to chemical by the formation of the high-energy ATP by the phosphorylation of ADP. ATP can then be used by the tissue for other metabolic processes.

One approach that has been used to study the effects of contaminants on oxidative phosphorylation is the Adenylate Energy Charge (AEC). The concept was put forward by Atkinson and Walton (1967) and is defined as:

$$AEC = \frac{(ATP) + (1/2\ ADP)}{(ATP) + (ADP) + (AMP)}$$

Most of these studies have involved invertebrates, a group largely ignored in this monograph. Din and Brooks (1986) examined the effects of two industrial wastes to see if AEC could be used as an index of sub-lethal stress on phytoplankton. Marked changes were seen only at the highest dosages used (4000 ppm), which seem high but are hard to evaluate in view of the lack of any chemical analysis of the wastes. In the case of one waste, this level was close to the lethal level for the phytoplankton. The ratio of ADP/ATP was found to be one of the least sensitive of a number of tests carried out on the effects of chlordecone on *Daphnia magna* by McKee and Knowles (1986). The data on vertebrates seem limited. Reinert and Hohreiter (1984) state that the only published information on AEC in fish relates to environmental factors such as temperature, dissolved oxygen, and pH.

The interaction of environmental pollutants with ATPases has, however, evoked a good deal of interest. One major area of interest is studies geared towards examining the mechanism of action of DDT and other PHAHs. Despite all the work, from that of Koch *et al.* (1969) onwards on the inhibition of ATPases, general agreement on the exact mechanism of the best-studied of all insecticides has not yet been achieved.

Detailed studies have been carried out on the effect of inhibition of ATPases in osmoregulation in a variety of organisms and the involvement in DDT-induced eggshell thinning. Desaiah *et al.* (1975) carried out one of the few chronic studies on the effects of an environmental contaminant on an enzyme system. These workers studied the effect of DDT on the levels of ATPases in the gills and brain of fathead minnows. At the end of the experiment (seven months) they found that there was a 50% reduction of activity of oligomycin-sensitive MgATPase, but a 40% increase of oligomycin-insensitive in the brain. There was only a small effect on $Na^+K^+ATPase$.

The effects of chlordecone and mirex on the activity of cardiac ATPases of the rat were studied by Desaiah (1980). He found that the potency of chlordecone to inhibit the ATPase system paralleled the decrease of dopamine and norepinephrine binding. It was suggested that chlordecone altered the sodium pump by inhibiting both ATP hydrolysis and ATP synthesis, and thus reducing catecholamine uptake. In contrast the closely related PHAH mirex did not cause these effects.

Janicki and Kinter (1971) found that DDT impaired fluid absorption in intestinal sacs from eels adapted to salt water, and this correlated (log log relationship) with inhibition of $Na^+K^+ATPase$. The effect on $Mg^{2+}ATPase$ was less marked. Waggoner and Zeeman (1975) found that DDT caused increases in plasma osmotic concentration in black surfperch only at levels approaching lethality, and considered that osmoregulatory failure was not the cause of mortality in fish poisoned by DDT. Weisbart and Feiner (1974) found significant, but inconsistent, changes in concentration of osmotic regulation in goldfish exposed to DDT. They also found a lack of correlation of *in vivo* and *in vitro* results. The concentrations used in Janicki and Kinter's experiments cannot be directly related to LD_{50} for the intact organism, but the levels were high by environmental standards. Lock *et al.* (1981) did not find a good correlation between water inflow and ATPase inhibition in trout exposed to mercury. The latter decreased only at levels approaching lethality.

Another ATPase-dependent organ is the avian salt gland. This gland enables birds to maintain their water balance in the marine environment by ingesting salt water and excreting a highly concentrated salt solution (Schmidt-Nielsen, 1960). In pelagic sea birds this gland is always operational, but in species such as freshwater ducks it can be induced by exposure to salt. The effect of DDE on the salt excretion of the mallard was examined by Friend *et al.* (1973). A significant reduction was found in birds not previously exposed to salt. A maximum effect was seen at 10 ppm, with no further increase even up to dietary levels of 1000 ppm. No effect was seen in mallards already exposed to salt water. Miller *et al.* (1976b) examined the effect of DDE on ducks and two species of pelagic sea birds – the black guillemot and the Atlantic puffin – and found that DDE at environmental levels did not affect osmoregulation or salt gland $Na^+K^+ATPase$ activity. Eastin *et al.* (1982) examined the effect of an organophosphate, fenthion, on AChE and ATPase in salt glands. A significant decrease in the activity of $Na^+K^+ATPase$ was found in birds exposed to salt water (to 76% of control), but no effect was found on those maintained on salt water. There were no significant changes to $Mg^{2+}ATPase$. Overall, there seems to be little evidence that environmental pollutants have significant effects on osmoregulation.

DDE-induced eggshell thinning is one of the best environmental case studies. While this is not the place to review the story in detail since much of it involves population dynamics rather than biochemistry, I will outline it (the chronology of the phenomenon is given in Table 10.1). Eggshell thinning of two species of raptoral birds was demonstrated by Ratcliffe (1967) to have started, quite abruptly, in 1946. It was soon found to have occurred for many raptoral and fish-eating birds in many parts of the world (Anderson and Hickey, 1972). Correlations with the residue levels of DDE in eggs (Cade *et al.*, 1971) and cause and effect studies on kestrel (Lincer, 1975) established DDE as the causative agent. For a fuller account the reader is referred to the review of Risebrough (1986), and those requiring even more to the proceedings of a conference on the peregrine falcon edited by Cade *et al.* (1988). A wide range of possible mechanisms of eggshell thinning was reviewed by Cooke (1973). These included effects on absorption of calcium in the gut, laying down and mobilization of calcium in the medullary bone and changes in the shell gland. The finding that blood calcium levels were unaffected, even when severe eggshell thinning occurred (Peakall *et al.*, 1975b), focused the investigation on the shell gland rather than on general calcium balance. Effects of DDE on the activity of Ca^{2+}ATPase in the shell gland of ducks (Miller *et al.*, 1976) and kestrels (Bird *et al.*, 1983) were demonstrated. More recently, detailed studies have been carried out by Lundholm (reviewed in Lundholm, 1987). Studying two varieties of ducks with different sensitivities to eggshell thinning, he found that the effect of DDE was localized to the translocation of calcium from the mucosa cells of the gland to the shell gland cavity. This effect was related to sensitivity and was absent in the chicken; this species is resistant to eggshell thinning. In *in vivo* experiments the reduction in the rate of ATP-dependent calcium binding was related to sensitivity, but *in vitro* no such dependence was found.

6.4.5 Phosphatases

There are a number of enzymes present in serum that can hydrolyse phosphate mono-esters. These enzymes have low substrate specificity and are classified into acid and alkaline phosphatases on the basis of the pH of the optimal activity. In humans elevated alkaline phosphatase (ALP) levels are associated with a variety of disorders involving calcium balance including vitamin D deficiency and hyperparathyroidism.

Jackim *et al.* (1970) found small, but significant, changes in activity of both acid and alkaline phosphatase in killifish surviving after being exposed to 96-hr TL_m of a number of metals. Increases in alkaline phosphatase were noted in the case of mercury and lead, and decreases with beryllium and silver. No changes were reported for cadmium. In contrast, acid phosphatase decreased for all metals, except silver, for which no change was found.

Histochemical localization of acid and alkaline phosphatase activity was used by Swartz (1984) in his studies of the gonadal development of chicken embryos exposed to DDT. Acid phosphatase was used as an indicator of cell degeneration, whereas ALP was considered to be involved in transport across the cell surface. Increased alkaline phosphatase was limited to the stromal cells of the female gonads. He related this change to the oestrogenic effect of DDT and considered that it could be involved in feminization of embryos (section 3.3.1). Acid phosphatase activity was much reduced or absent in the sex cords of 12-day-old ovaries of embryos exposed to DDT.

Alkaline phosphatase activity was measured by Hoffman and co-workers (Hoffman *et al.*, 1984; Hoffman and Sileo, 1984) as part of their studies of neurotoxic effects of an OP. Inhibition of plasma ALP activity and increase of GOT activity were noted. These studies are considered in more detail under neurotoxic esterases (Chapter 2).

6.4.6 Use of enzyme activity as a biomarker

In an attempt to evaluate where we stand on the use of enzymes as biomarkers of the effects of environmental contaminants, I leafed through a book on clinical chemistry. My first feeling was that it is much more sophisticated than anything that could be written on wildlife. But, as I examined the book in more detail, there were certainly a considerable number of diagnostic tests that had been applied in the wildlife toxicology field.

Also, there is a fundamental difference in approach between clinical chemistry in humans and the use of this approach in wildlife. In human medicine, clinical chemistry is usually undertaken in response to an adverse condition of an individual. Even when clinical tests in humans are undertaken on a screening basis, the objective is more to uncover disease than exposure to environmental agents. In wildlife toxicology, the reverse is the case; here one is using biochemical parameters to examine for the effect of environmental agents and disease is a confounding factor. Regrettably it is a confounding factor that is all too often not fully taken into account. There is a need to have veterinary scientists work more closely with toxicologists.

An important evaluation of the use of enzymes and other biochemical parameters as diagnostic tools in wildlife has been made by Hill and Murray (1987). These workers measured a variety of biochemicals periodically in bobwhite quail, starlings, red-winged blackbirds and common grackles throughout the year. All the species were maintained outdoors, but only the bobwhite came into breeding condition. A summary of their findings that apply to the enzymes that have been considered here in some detail is given in Table 6.3. A quick look at this table shows that inter-species and inter-seasonal differences are important, whereas differences between the sexes are few. This impression is confirmed in an examination of the more extensive table presented by Hill and Murray. The matrix that they present shows that 42 out of 51 of the inter-species variations are statistically

Table 6.3 Summary of statistical comparison of differences in enzyme activity among species, sex and season

Enzyme	Species variation		Sex variation				Seasonal variation			
	Quail v Passerines	Blackbird v Grackle	Quail	Starling	Blackbird	Grackle	Quail	Starling	Blackbird	Grackle
BChE	*		*				*			
GOT	*	*						*	*	*
GPT	*	*						*	*	*
LDH	*	*					*	*	*	*
ALP	*	*		*			*		*	*

Quail = Bobwhite quail   Blackbird = red-winged blackbird
BChE Butylcholinesterase   GOT Glutamic oxaloacetic transaminase   GPT Glutamic pyruvate transaminase
LDH Lactic dehydrogenase   ALP Alkaline phosphatase.

Table 6.4 Effects of sex, age and reproductive condition on enzyme activity in the mallard

Enzyme	Sex (non-reproductive adults)	Age (young up to 58 days)	Changes during breeding season	
			Adult females	Adult males
ChE	–	+	+	+
GOT	–	–	+	+
GPT	–	+	–	+
LDH	–	–	+	–
ALP	–	+	+	+

ChE Cholinesterase GOT Glutamic oxaloacetic transaminase GPT Glutamic pyruvate transaminase LDH Lactic dehydrogenase ALP Alkaline phosphatase.

significant, and 53 out of 68 for season variation compared to only 8 out of 68 for differences between the sexes. The maximum seasonal differences were typically less than twofold, but the exceptions were LDH, GOT and GPT, which are those enzymes which have been the most widely studied. The differences were greatest in the red-winged blackbird. Another evaluation that should be read carefully by anyone interested in the clinical chemistry of wildlife is that by Fairbrother and co-workers (1990). They have examined the influence of sex, age and reproductive condition on the serum chemistry of mallards. A summary of their findings, as it relates to the enzymes considered in this section, is given in Table 6.4.

My impression from the literature available is that the study of the effect of pollutants on enzymes has lost momentum. The detailed experimental studies, such as those of Dieter (1974) and (1975), have not been followed up. However, the recent major studies of variables involved in such measurements by scientists of the US Fish and Wildlife Service (Hill and Murray, 1987) and USEPA (Fairbrother *et al.*, 1990) may be indications of resurgence of interest. The lack of correlation between simple *in vitro* and *in vivo* studies suggests that the former are not a useful method to pursue. The use of cell culture experiments is examined in section 9.4. Fortunately, most of the *in vivo* enzyme studies can be carried out by using measurements on plasma, and thus non-destructive testing is possible.

Despite the fact that the enzymes discussed in this section are highly specific – for example, GOT catalyses aspartate to glutamate – the cause of the activity change is less specific. The increase of plasma GOT occurs by leakage into the blood due to tissue damage. This does not decrease its physiological importance but does mean that we do not have a specific site in the way that we do for AChE. The possibility of using this enzyme in monitoring is demonstrated in a recent paper by Renqing (1990), who found a strong correlation between the biotic index of a diatom and the GOT activity in carp in Chinese rivers.

Nevertheless I would like to see the battery of routine biochemical tests used in human clinical medicine given a good trial in wildlife toxicology. It is imperative that these studies be carried out using rigorous quality control with as little inter-laboratory variation in methodology as possible. Only in that way will it be possible to build up a database against which wildlife toxicologists can work. The importance of keeping these variables under control is well illustrated by the work of Hill and Murray (1987).

7   *Behavioural effects: their relationship to physiological changes*

7.1 Introduction

Twenty-five years ago, at a landmark symposium at the Institute for Terrestrial Ecology, the following statements on behavioural toxicology were made (Warner *et al.*, 1966):

1. The behaviour (or activities) of an organism represent the final integrated result of a diversity of biochemical and physiological processes. Thus, a single behavioural parameter is generally more comprehensive than a physiological or biochemical parameter.
2. Behavioural patterns are known to be highly sensitive to changes in the steady state of an organism. This sensitivity is one of the key values for its use in exploring sublethal toxicity.
3. Behavioural measurements can usually be made without direct physical harm to the organism. With aquatic animals especially, implantation of detectors introduces problems of considerable complexity. Behavioural measurements can avoid this difficulty.

In this chapter, I attempt to assess the degree to which our current knowledge supports the first two statements. I believe that the third is, except for a few field studies, essentially a non-starter since, although the animals survive the experiments, it is unlikely that the experimenter will wish to keep them or be able to return them to the wild.

The word *behaviour* has rather wide meanings and some attempt needs to be made to define how it is used here. One definition given in the *Oxford Dictionary* is 'The manner in which a thing acts under specified conditions or circumstances, or in relation to other things', a definition that fits well with the first statement of Warner cited above.

7.2 Types of tests used

The dilemma facing the use of behavioural tests in wildlife toxicology is that the best studied and most easily performed and quantified are those that have the least environmental relevance.

Operant behaviour, such as conditioning to respond to a lighted or coloured key to obtain food, and visual cliff experiments to choose under artificial circumstances between the shallow or deep side of a box are too remote from real life to be capable of being related to survival or other meaningful effects on the animal. It can merely be presumed that a decrease in learning ability is an unfavourable response. Approach and avoidance behaviour is more directly related to survival, although the relationship has not been quantified. The ability to capture food is clearly important to predatory species, although the exact relationship between impaired hunting ability and survival is unknown.

Field observations are often difficult to evaluate as the organism is almost always exposed to a complex mixture of pollutants, and observations are usually difficult to quantify. A possible central nervous system response leading to behavioural changes was outlined by Ratcliffe (1970) as a possible cause in the decline of the peregrine falcon. However, observations by time-lapse at the eyries of highly contaminated peregrines revealed little in the way of abnormal behaviour (Enderson *et al.*, 1972). Decreased nest defence by prairie falcons and merlins was considered to be a factor in their reproductive success (Fyfe *et al.*, 1976), but this was not confirmed for the merlin in more detailed studies carried out subsequently (Fox and Donald, 1980). Behavioural effects were also contradictory in the case of herring gulls on the Great Lakes during the period that poor reproductive success was occurring. Fox *et al.* (1978) found that the minimum flushing distance was increased and the number of swoops by the adults on observers was decreased. In contrast, Ludwig and Tomoff (1966) noted that adults were exceptionally aggressive towards us and others. Non-pollutant-related factors, such as colony size and prior human disturbance patterns, rather than pollutants, may be the cause of these varying responses.

Studies of avoidance responses of fish to toxicants dates back to the work of Shelford and Allee (1912). There is a great deal of literature on the subject. Many pesticides and industrial chemicals have been tested, using various test systems. For a review on the earlier work in this field, the reader is referred to Cherry and Cairns (1982). The equipment used has become highly complex. The value of computerized early warning systems of this type has been discussed by Cairns and Gruber (1979). A sophisticated fish avoidance chamber with video monitor and computer-interfaced recording system has been described by Hartwell *et al.* (1987a).

In the discussion that follows, I am looking at two types of behavioural study; those in which a well-defined biomarker (using AChE, section 2.1, or biogenic amines, section 2.2, for example) was also studied and those studies which were either carried out under field conditions or at least under conditions that make extension to the real world possible.

7.3 Behavioural effects of polyhalogenated aromatic hydrocarbons

Although a great deal of work has been done on the effects of this group of chemicals on behaviour, the number of data relating behavioural changes to biochemical changes is very limited. The effects of these compounds on avian behaviour have been reviewed by Peakall (1985). The only studies that directly examined behaviour and biochemical changes appear to be those of Sharma *et al.* (1976) and McArthur *et al.* (1983), although inferences can be made from parallel experiments carried out on ring-doves.

The relationship of both the levels of biogenic amines and induction of MFOs to behavioural changes caused by dieldrin was examined in the mallard by Sharma *et al.* (1976). The results of their experiments are shown diagrammatically in Figure 7.1. The dose response of the biogenic amines – with the exception of gamma-butyric acid, which showed no significant changes – is clear. In contrast the MFOs were not significantly induced except at the highest dosage. Dieldrin produced clear changes in aggressive behaviour – pecking and avoidance – at levels lower than those at which significant changes of the biogenic amines were seen.

7.4 Behavioural effects of organophosphates

With organophosphates the reference line is the inhibition of AChE (section 2.1). There is a considerable number of avian studies in which the degree of AChE inhibition is recorded along with behaviour changes. Indeed, in a number of studies (Grue *et al.*, 1982b; King *et al.*, 1984; White *et al.*, 1983), the dosage was selected to give 50% inhibition. Avian studies for which AChE data are available are listed in Table 7.1. While these studies include several with ecological relevance, including a field study of nest defence behaviour and its effect on reproduction (King *et al.*, 1984), the behavioural changes quantified are small compared to those of AChE inhibition.

The relationship of behavioural changes in the fish, *Channa punctatus*, to inhibition of AChE and changes in biogenic amines caused by endosulfan was examined by Gopal *et al.* (1985). The results of these experiments are shown diagrammatically in Figure 7.2. Modest, but significant, increases in surface activity and distance travelled were found to increase with concentration, except at the longest time period (96 hours) when the fish were lethargic and oxygen consumption had markedly decreased. The changes in AChE are clear-cut, in regards to both time and concentration. Changes of levels of biogenic amines are less definitive, although serotonin levels are decreased by endosulfan from 24 hours onward.

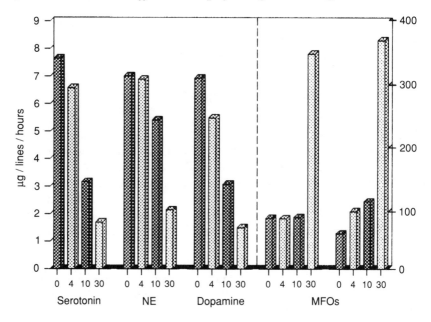

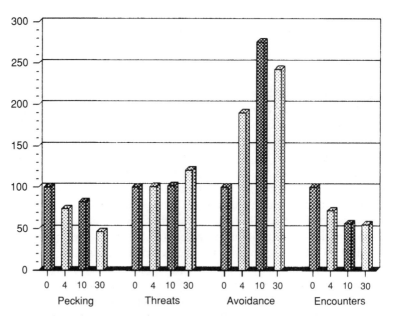

Figure 7.1 Effect of dieldrin on biogenic amines and behavioural responses in mallards. After Sharma *et al.* (1976).

Table 7.1 Relationship of behavioural effects to cholinesterase inhibition

Species Pesticide	Degree of inhibition %	Effects seen	Reference
Laughing gull Parathion	50	Incubation time decreased on days 2 and 3	White *et al.* (1983)
	46	No change in flushing distance or return time	King *et al.* (1984)
Kestrel Acephate	25	Predatory vigilance and attack behaviour not affected	Rudolph *et al.* (1984)
Kestrel Methyl parathion	50	Transient, but pronounced hypothermia	Rattner and Franson (1984)
Bobwhite quail Methyl parathion	57	Increased predation by cats	Galindo *et al.* (1985)
Starling Dicrotophos	50	Increased time perching; decreased, flying, foraging singing and displaying	Grue and Shipley (1981)
	50	Females flew fewer sorties; more time away from nest	Grue *et al.* (1982b)
White-throated Sparrow Fenitrothion	40	Slower growth rate of nestlings	Busby *et al.* (1990)

The relationship between behavioural changes and biogenic amine levels in rats exposed to dichlorvos was examined by Ali *et al.* (1980). In this experiment a daily dose of 3 mg/kg was administered for ten days; no AChE data are given but the dose can be compared to an LD_{50} value of 265 mg/kg. Marked depression of parameters measured in both the open field studies and locomotor activity was found, although there was considerable variation over the period of the experiment. Levels of dopamine were reduced in all three areas of the brain (cerebral hemispheres, cerebellum and brain stem) to the same extent, 32–37%, by day 10; norepinephrine was reduced most markedly in the cerebellum (50% compared to 21–23% in other areas), whereas serotenin was decreased more in the brain stem (50% reduction) compared to the other regions (22–36%).

The interaction of predation pressure and the effect of the insecticide abate on populations of fiddler crabs was studied by Ward *et al.* (1976). These workers followed the population densities of crabs in open marshes treated and untreated with abate and in plots that were caged over and those that were open. They found that the population of crabs in the caged areas was

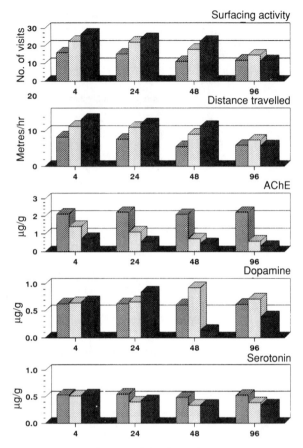

Figure 7.2 Effect of endosulfan on biogenic amines, AChE and behavioural responses of the fish, *Channa punctatus*. After Gopal *et al.* (1985).

similar in both treated and untreated areas, but that following the second application of abate (63 g/ha of active ingredient) there was a significant decrease of the population of crabs in the uncaged treated areas. Experimental studies suggest that the escape response of the crabs was impaired.

7.5 Behavioural effects of heavy metals and other pollutants

The behavioural effects of methylmercury have been reviewed by Shimai and Satoh (1985). These workers classify behavioural studies into ten groups. Most of the studies refer to mice and rats, but a few refer to primates and chickens. An overview of the findings related to mammals is given in Table 7.2. This tabulation is an oversimplification of the data as the routes and duration of dosage vary considerably. Nevertheless, it suggests that operant

Table 7.2 Effect of methylmercury on the behaviour of mammals

Test	No. of studies change/no change	Minimum dose causing effect
Development of reflexive behaviours	2/4	2 ppm in diet
Swimming ability	3/1	0.4 mg/kg daily
Spontaneous activity	3*/1	4 mg/kg
Open field behaviour	5/2	5.4 mg/kg
Maze learning	3/1	2 ppm diet
Avoidance learning	6/0	0.4 mg/kg daily
Operant learning	7/1	0.16 mg/kg
Susceptibility to convulsions	2/0	6 mg/kg
Ultrasonic vocalization	1/0	4 mg/kg
Visual function	2/0	5 mg/kg

* Two studies showed decrease; one study showed an increase
After Shimai and Satoh (1985).

learning, avoidance and swimming ability are the parameters affected at the lowest dosages. From an ecological viewpoint, none of these tests are readily related to environmental conditions. The gap between students of behavioural science from the viewpoint of assessing effects on humans and those assessing effects on wildlife is illustrated by the fact that, although some avian data are included, the detailed studies of Heinz on the effects of methylmercury on mallards are not cited.

The most detailed studies on an avian species are those of Heinz (1976a, 1976b, 1979) on the effects of methylmercury on mallards. Using an environmentally realistic dose of 0.5 ppm, he followed the effects through three generations. He found that ducklings fed methylmercury were less responsive to taped maternal calls, hyperresponsive to fright stimulus and showed no differences from controls in open field tests.

The effect of the degree of mercury contamination on the predator avoidance of the grass shrimp was studied by Kraus and Kraus (1986). These workers collected shrimp from a mercury-polluted and relatively clean estuary and exposed them in aquaria to predation by killifish. It was found that the time between captures was less for shrimp from the mercury-contaminated estuary.

A series of experiments on the effects of lead on the behaviour of the common tern and herring gull have been carried out by Burger and Gochfeld. In the first experiment, five-day-old tern chicks were injected with 0.2 mg/kg lead nitrate (Burger and Gochfeld, 1985). This dose is a quarter of the maximum tolerated dose. A series of behavioural responses was measured over a period of 24 days after dosing. Locomotion, balance, depth perception, thermoregulation and feeding behaviour were all affected by treatment with lead. In almost all cases age-related improvement was seen in control birds, but in almost all cases significant improvement was not found in lead-injected birds. In a second series of experiments (Burger and Gochfeld, 1988) the dose-response of single and duplicate injections of lead to herring gulls was examined. Fifty per cent mortality was found with a single dose of 1 mg/kg, but in the case of injection of two doses at four-day intervals, close to complete mortality was found even at the low dose (0.2 mg/kg). Behavioural tests were dose-dependent for single doses. The growth and feeding behaviour following a single dose of 0.2 mg/g was also studied (Gochfeld and Burger, 1988). Weight gain of lead-exposed chicks started to deviate from controls two weeks after injection and by three weeks experimental birds weighed only 75% of controls. Feeding efficiency also started to deviate after two weeks and was only half that of controls three weeks after injection. In a third experiment Burger (1990) examined the effects of a single dose (0.1 or 0.2 mg/g lead nitrate) on a range of behavioural tests – balance, locomotion, begging, visual cliff response and thermoregulation – in young herring gulls over a period of up to 45 days. The results are not readily summarized, as on most days most behavioural responses did not differ significantly, although over the entire period of the experiment, control birds performed better than experimentals. Balance, righting response, and individual recognition appeared to be the most sensitive parameters.

Regrettably, no tissue levels of lead are given. A rough calculation can be made, but only if assumptions are made on the degree of uptake and elimination. Burger (1990), on the basis of unpublished data, considers that the dosages used are realistic, but without detailed information it is impossible to make a comparison with environmental levels.

A critical review of the effects of heavy metals on fish behaviour has been made by Atchison *et al.* (1987). They have compiled data on acute and chronic toxicity for selected metals and have also summarized the literature on the lowest-observed-effect concentration (LOEC) for fish avoidance, attractance, fish ventilation, and cough rates. The values obtained vary considerably with water hardness and between species, which makes comparisons difficult. The only data that can be matched for both chronic toxicity and water quality are those on the brook char. These data, compiled from tables 1 and 3 of their review, are given in Table 7.3. Based on this comparison, the behaviour tests are mostly somewhat more sensitive than life cycle or early life stage tests. Atchison and co-workers consider 'that changes in certain fish behaviours, especially cough rate and avoidance reactions, are sensitive indicators of sublethal exposure to metals. Other tests involving predator avoidance, feeding behaviour, learning, social interactions and a variety of locomotor behaviours have been insufficiently studied to enable a judgement of their sensitivity or utility.' However, they consider that these tests should be seriously considered. Certainly the cough rate and avoidance reaction tests are easier to carry out than life cycle tests, although Atchison and co-workers do not see behavioural tests replacing conventional toxicity tests. They see the future direction as developing tests

Table 7.3 Comparison of LOECs for heavy metals based on life-cycle tests and behavioural tests for brook char

	LOEC (life cycle and early life stage) µg/l	LOEC (behavioural) µg/l
Cadmium	3–4	5
Copper	17	9
Lead	119	80
Mercury	0.9	3
Zinc	2000	1390

LOEC: lowest-observed-effect concentration
Compiled from Tables 1 and 3 in Atchison *et al.* (1987).

that provide ecological realism; for example, in predator-prey tests exposure of the predator to the pollutant. The correct species combinations and provision of escape cover all need to be considered. Tests must be capable of field validation. They reinforce the urging of Sprague (1981) that fish behavioural ecologists become involved in pollution-relation research.

The use of six behavioural tests to assess the sub-lethal effects in rainbow trout of six agricultural chemicals has been made by Little *et al.* (1990). The results are shown in a matrix in Table 7.4. Swimming capacity – the ability to swim against an increasing velocity of water – was found to be least sensitive, whereas swimming activity – the amount of time in motion – was a good deal more sensitive. Three parameters concerned with feeding were examined: strike frequency, amount of prey consumed and the percentage of fish feeding. There was considerable variation among chemicals, quite strikingly between two AChE inhibitors, with carbaryl not showing effects below 50% of the LC_{50}, whereas methyl parathion showed effects at 1.3% of its LC_{50}. In general, frequency of strikes was less sensitive than prey capture. Vulnerability of rainbow trout to capture by largemouth bass was a sensitive indicator, although this parameter did not always show a clear dose-response (a fact omitted from Table 7.4, which shows the lowest concentration at which a significant effect was seen). As already mentioned, there are difficulties with making predator-prey studies ecologically relevant. Examination of the matrix produced by the results of Little *et al.* (1990) clearly shows the advantages of using a battery of tests

The effect of pentachlorophenol on the actual ability of a predator (largemouth bass) to capture prey (guppy) was investigated experimentally by Brown *et al.* (1985). The aquaria used were small, but cover was provided. The bass were maintained in the experimental aquaria for a period of an hour and a half before the experiment was started. These workers found that bass had significantly lower capture success, carried out more

Table 7.4 Effect of some agricultural chemicals on behavioural parameters of the rainbow trout

Chemical	LD_{50} (96hr)	Swimming capacity	Swimming activity	Strike frequency	Daphnia consumed	% consuming daphnia	% survival from predation
Carbaryl	1.95	0.1–1	0.1–1	>1	0.1–1	0.1–1	<0.01
Chlordane	0.042	>0.02	0.002–0.02	0.002–0.02	0.002–0.02	0.002–0.02	0.002–0.02
DEF	0.66	0.05–0.1	0.005–0.05	0.005–0.05	<0.005	0.005–0.05	0.005–0.05
2,4-DMA	100	5–50	5–50	5–50	5–50	0.5–5	5–50
Methyl parathion	3.7	>0.1	<0.01	0.01–0.1	<0.1	0.01–0.1	0.01–0.1
Pentachlorophenol	0.052	>0.02	0.002–0.02	0.002–0.02	0.0002–0.002	>0.02	0.002–0.02

DEF: tributyl phosphorotrithioate
2,4-DMA: 2,4-dichlorophenoxyacetic acid
After Little *et al.* (1990).

strikes, and spent more time chasing guppies from the control and low exposure group (100 µg/l) than in the higher (500 and 700 µg/l) exposure groups. While the ecological relevance of these studies is apparent, it should be pointed out that the exposure levels causing significant effects were at 50 and 70% of the 96 hours LD_{50}. The effect of sodium pentachlorophenate on the brain levels of biogenic amines and the free amino acid tryptophan was studied by Sloley *et al.* (1986) in the rainbow trout. No major effects were seen on the levels of biogenic amines over the concentration range of pentachlorophenate studied (100–200 µg/l) nor did pre-exposure to 50 µg/l cause significant effects. The level of the tryptophan increased threefold at the highest concentration, which, however, was 90% of the lethal level.

7.6 Relationship of behavioural effects to biomarkers

I do not propose to discuss the value of behavioural studies in environmental assessment in any detail. The fish avoidance test is well established in the laboratory as a means of showing effects well below the lethal range (Hidaka and Tatsukawa, 1985), and highly automated procedures are available (Hartwell *et al.*, 1987a). Nevertheless, a note of caution should be injected. Myllyvirta and Vuorinen (1989) examined the effect of pre-exposure of fish to bleached kraft mill effluent. They found that pre-exposure to effluent reduced the avoidance behaviour, and that over the entire range of concentrations studied pre-exposed fish were observed more often in contaminated than in clean water. They concluded that the desensitization caused by pre-exposure makes it improbable that vendace can show avoidance responses in the field. Similar difficulties were found by Hartwell *et al.* (1987b) in their studies of the avoidance response of fathead minnows to a combination of metals. Control fish were highly sensitive to the presence of metals (combination of copper, chromium, arsenic and selenium), but pre-exposed fish were up to 20 times less sensitive. They considered that laboratory experiments are likely to overestimate the responsiveness of fish to metal pollution in the wild.

These difficulties do not imply that behavioural effects caused by pollution are unimportant. Studies such as those by Ward *et al.* (1976) on fiddler crabs have shown that operational levels of pesticide use can cause population effects through behavioural changes. These workers examined the effects of the pesticide temefos on caged and uncaged crabs. They found no effect on the caged crabs, indicating a lack of direct mortality, whereas the density of uncaged crabs showed a significant decrease (Figure 7.3). They considered that these effects were due to increased predation on crabs which were rendered more susceptible to capture after exposure to the pesticide.

A review by Temple (1987) confirms the widely accepted tenet that predators typically capture a disproportionally large fraction of substandard prey. But field studies of the impact of chemicals on behaviour are difficult.

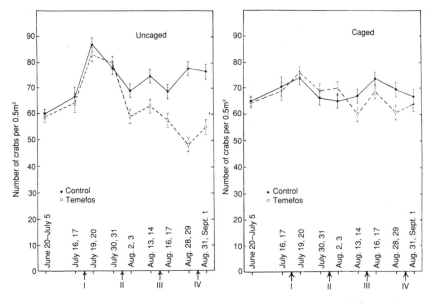

Times of temefos applications are marked with arrows and roman numerals. Bows show 90% confidence limits.

Figure 7.3 Effect of abate on the behaviour of crabs. After Ward *et al.* (1976).

Even in cases where the overall impact has been demonstrated, such as the decline of the peregrine falcon and the reproductive impairment of fish-eating birds, it has been difficult to show a behavioural component.

The overall conclusion is that behavioural parameters are not especially sensitive to exposure to pollutants and that biochemical and physiological changes are usually at least as sensitive. Furthermore, the variability of biochemical data is generally less and the dose response clearer than that obtained from behavioural. The clearest evidence is that from comparison of behavioural effects with the inhibition of AChE. In this case, enzyme inhibition is more sensitive and a much more easily determined parameter. For organochlorines and heavy metals the comparison to biomarkers is not as straightforward. Nevertheless, in general, physiological and biochemical changes are more readily measured and quantified.

It is my impression that behavioural studies have not lived up to the hopes put forward 25 years ago (Warner *et al.*, 1966). Although the behaviour of an organism does represent the integrated results of a diversity of bio-chemical and physiological processes, a single behavioural parameter is less readily quantified than the underlying biochemical and physiological processes. Nor do they seem to be more sensitive.

8   *Environmental immunotoxicology*

Steve Wong*, Michel Fournier**, Daniel Coderre**, Wanda Banska**, and Krzysztof Krzystyniak**.

*Pesticides Division, Environmental Health Directorate, Health Protection Branch, Health and Welfare Canada, Ottawa. K1A 0K9

**Département des Sciences Biologiques, Université du Québec à Montréal, C.P. 8888, Succursale "A", Montréal, Québec, H3C 3P8, Canada.

8.1 Introduction

The central role played by the immune system is recognition; it recognizes self and discriminates against non-self elements. Non-self elements represent anything that is foreign and different from an individual's own constituents. When non-self elements such as foreign cells, exogenous microorganisms (e.g. bacteria, viruses, fungi, parasites) and non-living substances (e.g. chemicals) invade the body, the immune system mounts a response in defence of the integrity of the body. The immune response can involve both humoral and cellular components.

The thymus and bone marrow are primary lymphoid organs in which T- and B-lymphocytes are differentiated respectively. The stem cell pool from which T- and B-cells arise is concentrated primarily in the bone marrow, although some stem cells circulate in the blood stream. There are several pathways of differentiation leading to mature effector cells that are capable of carrying out specialized functions such as cell-mediated immunity, or humoral immunity involving antibodies. Often the differentiation of immune cells is under the influence of thymic hormones. The major secondary lymphoid organs are the lymph nodes and the spleen. These organs process macrophages and lymphocytes that initiate and regulate immune responses. The lymph nodes and spleen provide highly dynamic support for active immune response. In general, lymph nodes respond to antigen insults in the tissues these organs drain, while the spleen responds to antigens in the blood. The precise structure of the secondary lymphoid organs governs the type of immune response. Structurally, lymph nodes can be divided into three major regions: the cortex, which is B-cell dependent; the paracortex, which is populated by both T-cells and macrophages; and the medulla, which contains various lymphocytes and macrophages. In the spleen, the lymphoid sheath surrounding arteries is largely composed of T-lymphocytes. Lying along the lymphoid sheath are lymphoid follicles with germinal centres where B-cells are populated. The immune system does not function independently of other body systems. Its close collaboration with the hormonal system is well established. Recent findings showing an association of the nervous and immune systems have been discussed in Chapter 4.

In addition to the cellular components, two major elements composed of secreted proteins in the blood play important roles in the functions of the immune system. These are the immunoglobulins (antibody) and the elements of the complement system. There are several families of immunoglobulins with specificity for antigens and having different biological functions. Immunoglobulins may bind to cell surfaces, may activate the complement system, and are able to cross the placenta. By removing free antigens from circulation, antibody molecules help to regulate the immune response. The complement system contains a group of proteins that mediate a series of interactions leading to lysis of cells.

An immune response mounted against an antigenic challenge may take several courses. Often, both cell-mediated and humoral responses are involved simultaneously. There is substantial evidence that the reserve capacity of the immune system is large, and can remain intact while other toxic manifestations are prominent. Because of the multicomponent nature and complexity of the immune system, induced changes in one part of the immune system are often accompanied by alterations in other parts, making interpretation of the significance of the changes difficult (Bellanti, 1987). On the other hand, routine parameters measured in toxicology studies, such as blood or tissue cellularity, are often considerably less sensitive indicators of toxicity than are immune function measurements (Luster *et al.*, 1988).

Immunotoxicity is defined as the adverse effects on the immune system of foreign substances (Luster *et al.*, 1988; Trizio *et al.*, 1988). A chemical substance is considered immunotoxic when it or its biotransformed product or products induce undesired events including a direct and/or indirect action on the immune system, an immunologically-based host response to the compound and/or its metabolite(s), or a modification of host structures to elements that are no longer self-recognizable. These undesirable effects of xenobiotics can be classified functionally, as for example, chemical-induced immunosuppression, allergy, immunopotentiation, autoimmunity, or altered host resistance to infections.

Because of the complexity of the immune system, there is not a single immune function assay that gives an adequate evaluation of the adverse effects of a xenobiotic on the immune system. A range of the tests available, divided into major categories, is given in Table 8.1 Tiered testing involving *in vitro* and *in vivo* assays has been proposed to assess immunotoxicity (Falchetti *et al.*, 1983; Luster *et al.*, 1988; Oytcharoy *et al.*, 1980).

Table 8.1 Types of tests available in immunotoxicology

Test area	Observations
General	Weight of lymphatic organs Cellularity of lymphoid organs Lymphocyte count Lymphocyte blastogenesis
Humoral immunity	Globulin levels Complement levels Antibody titers B-cell markers Concentration of immunoglobulin
Cell-mediated immunity	Delayed hypersensitivity T-dependent antigens T-lymphocytes
Macrophage	Phagocytosis Disposal of bacteria Peripheral monocyte count

8.2 The immune system as a target for xenobiotic interaction

It has been shown only relatively recently that various chemicals can actually affect specific components and/or functional activities of the immune system. A brief review is presented as the basis for defining the potential immunotoxic effects related to chemicals.

The phenomenon of an immune response refers to the functional characteristics of selected immune cell populations, which can be classified as humoral-mediated immunity (HMI); cell-mediated immunity (CMI); and nonspecific response (NSR). Humoral response to an antigen (HMI) is an adaptive immunity. When a foreign substance invades the body, it is recognized, taken up, and processed by antigen-presenting cells (e.g. macrophages). These cells collaborate with T-helper lymphocytes, triggering the latter to release lymphokines which stimulate a small pre-existing population of B-cells to divide, differentiate and mature into plasma cells. Plasma cells are programmed to secrete immunoglobulin molecules (IgM, IgG, IgA and IgE), which are antibodies that react specifically with the stimulating antigen, eliminating it, leaving the body primed for an even more effective response upon a subsequent encounter with the same antigen.

Adaptive immunity that involves lymphoid cells other than B-lymphocytes is usually grouped under the term *cell-mediated immunity* (CMI). Examples of CMI are delayed hypersensitivity reactions, chronic inflammation, graft rejection, immunity against cancer and infectious agents, and the up and down regulation of the immune response. Cell-mediated immunity involves the generation of cytotoxic T-lymphocytes against intracellular viruses. It also involves the activities of other T-cells and non-specific cells, such as macrophages, in the defence against foreign invasions. Cell-mediated immunity is regulated by various suppressor cells and factors.

Other important immune cell populations, including Natural Killer (NK) cells, monocytes/macrophages, and polymorphonuclear leucocytes (PMN), are involved in nonspecific immune response (NSR) with the elimination of pathogens and cancers. NK cells seem to act in primary immunosurveillance; they can eliminate transformed cells without previous contact with altered cell antigens. Macrophages and PMNs also serve as effector cells *per se*, by performing phagocytic, bactericidal and tumouricidal activities.

Invasion of a host by bacteria, viruses, or cancer stimulates the immune system to react by appropriate HMI, CMI, and/or NSR responses. A seemingly toxic effect of an agent on a particular immune response may not necessarily lead to a detectable clinical manifestation. It is possible that the immune system can adapt to the chemical stress through compensatory mechanisms and can recover from a chemical insult by a rapid regeneration of a particular cell population. On the other hand, in spite of an obvious lack of information on the mechanisms of immunotoxicity, there is a common belief that current knowledge about adverse health effects resulting from immunotoxic xenobiotics may represent only the 'tip of the iceberg'.

Chemical-induced immunotoxicity is varied and can be specific, and many immunotoxic effects have been characterized in recent years (Gleichmann *et al.*, 1989; Koller, 1987; Leibish and Moraski, 1987; Luster *et al.*, 1988; Trizio *et al.*, 1988). The various forms of chemical-induced immunotoxicity are shown in Figure 8.1.

8.2.1 Immunodeficiency and immunosuppression

Most immune deficiencies are secondary, resulting from nutritional inadequacy, intoxication due to accidental exposure as well as chemotherapy, ageing, acute and chronic infections, and cancer. Immune deficiencies involve defects in cell-mediated immunity. Direct chemically-induced injury of lymphoid organs, such as thymus atrophy, is also considered a mechanism of immunodeficiency. It is apparent that certain compounds, such as 2,3,7,8-tetrachlorodibenzo-*p*-dioxin (TCDD); 2,3,7,8-tetrachlorodibenzofuran (TCDF); polychlorinated biphenyls (PCBs); polybrominated biphenyls (PBBs); diethylstilbestrol; and some di- and tri-alkyl substituted organotins act more or less preferentially on the thymus and therefore probably affect thymus-dependent immune functions (Chastain and Pazdernik, 1985; Penninks and Seinen, 1987; Vos, 1977; Vos and Moore, 1974). In some cases, such as exposure to ethylene glycol monethyl ether (House *et al.*, 1985) or dioctyltin dichloride (Miller and Scott, 1985), a clear correlation between thymus atrophy and functional impairment of the immune system was not evident.

Any decrease of immune functions, measured through HMI, CMI, and/ or NSR, should be considered to be either immunosuppression or

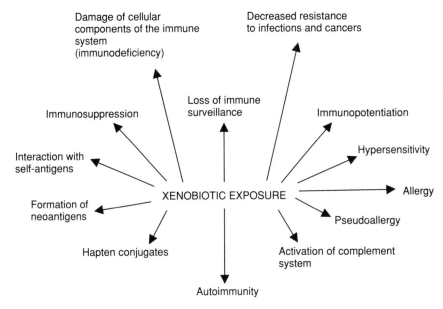

Figure 8.1 Potential immunotoxic effects of xenobiotics.

immunodepression. The decreased immune functions may have resulted from a depletion of one or more kinds of immune cell population, or an inhibition or interference of their activities. Many environmental contaminants, such as PCBs, PBBs and TCDD, have been shown to have a potential of immunosuppression in experimental animal models. Some heavy metals, alkylating agents and organochlorine insecticides are examples of immuno-suppressive compounds from different chemical groups. Chemical-induced immunosuppression, especially when long-lasting, can produce grave con-sequences, such as cancer (Dean *et al.*, 1986; Penn, 1985, 1987) or altered host resistance to infections (Bradley and Morahan, 1982; Crocker *et al.*, 1976; Krzystyniak *et al.*, 1985; Morahan *et al.*, 1984).

8.2.2 Altered host resistance to infections

Increased susceptibility to viral, bacterial and parasitic infections and to tumours is one of the most dangerous and frequently lethal effects induced by immunotoxic chemicals. Decrease of antiviral or antibacterial defence of a host exposed to immunotoxic chemicals can be usually related to the immunosuppressive potential of the chemical. Decreased antiviral resist-ance in children has been linked with massive spraying of forestry insec-ticides and/or organic chemical carriers in the insecticide formulations (Crocker *et al.*, 1976). The adverse effects of immunotoxic chemicals on the host's resistance to pathogens have been reproduced in many animal models of viral and bacterial infections, as well as in experimental cancers (Luster *et al.*, 1988; Morahan *et al.*, 1984). Organochlorine pesticides, such as DDT and dieldrin, have been shown experimentally to affect resistance to viral hepatitis in ducks (Friend and Trainer, 1974) and in mice (Krzystyniak *et al.*, 1985). Massive death of dolphins, observed recently along the Spanish coast, is suspected to be linked with increased susceptibility to infections in the animals living in contaminated waters (J.P. Revillard, personal communication).

8.2.3 Hypersensitivity, allergy and pseudoallergy

Hypersensitivity is defined as a disproportionate increase in an immune response to a previously encountered stimulus with unpleasant or sometimes damaging consequences. There are essentially four types of hypersensitivity response. Type I, or immediate hypersensitivity, commonly known as allergy (e.g. hay fever, asthma), is mediated by the anaphylactic antibody of the IgE, and sometimes IgG class with mast cells. Substances capable of eliciting type I hypersensitivity responses in pre-sensitized individuals are allergens (e.g. pollen). Allergens sensitize the anaphylactic antibodies which in turn stimulate mast cells, triggering the latter to release pharmacological mediators such as histamine. Exposure to certain chemicals and pharmacological agents has been shown to trigger type I hypersensitivity responses (Pepys, 1983). Some drugs, such as intravenous anaesthetics and radiographic contrast media, also initiate so-called pseudoallergic reactions via the formation of anaphylactic complement subcomponents (Stanworth, 1985, 1987). Type II, or cytotoxic, hypersensitivity is mediated by IgG or IgM together with complement, 'killer' (K) cells, or phagocytic cells. The antibody is directed against antigens deposited on the bodies own cells, leading to cytotoxic or phagocytic actions by K-cells and macrophages, or complement-mediated lysis. Blood transfusion reactions and many autoimmune diseases are examples of type II hypersensitivity. Type III hypersensitivity is mediated by immune complexes. When immune complexes are deposited in the tissues, complement is activated and polymorphonuclear leucocytes are attracted to the sites causing local damage. Finally, the so-called delayed-type, or type IV, hypersensitivity is mediated by a subpopulation of T-cells. When stimulated by antigens, these T-cells secrete lymphokines which mediate a range of inflammatory responses.

The term *allergy* is frequently used to define the immediate-type hypersensitivity (Pepys, 1983). In practice, an allergic mechanism can be assumed when there is a period of symptomless exposure, i.e. sensitization, followed by an overt reaction. Once sensitized, an individual may encounter allergic reactions which can be elicited by a subsequent re-exposure to a minimal and often minute amount of the allergen, far less than the amounts required for sensitization. This characteristic of allergic reactions distinguishes them from chemically-induced irritant effects (Pepys, 1983). In addition to respiratory allergic diseases and nutritional allergies, chemical-induced allergies are common. Haemolytic anaemia caused by continuous administration of chlorpromazine or phenacetin, agranulocytosis associated with amidopyrine or quinidine, contact sensitization by picryl chloride, dinitrochlorobenzene and other chemicals have been reported (Diaz and Provost, 1987; Stanworth, 1985).

There are situations where individuals show symptoms characteristic of immediate-type hypersensitivity responses, but without a proven involvement of any anaphylactic antibody of the IgE or IgG class (Stanworth, 1987). The symptoms, ranging from skin rashes to, occasionally, fatal anaphylaxis, can be induced by many chemicals. It seems likely that these chemicals, in the absence of the involvement of IgE or IgG antibodies, are able to bypass the regular two-stage mast-cell triggering process and initiate a pseudoallergic response via the formation of anaphylactic complement subcomponents.

8.2.4 Autoimmunity

Normally, the body's self-constituents are tolerated and do not stimulate a response by the immune system. Under certain circumstances, these self-structures are attacked as if they were non-self. The condition in which the body's immune system turns against itself is known as autoimmunity, sometimes called 'auto-allergy'. In autoimmunity, lymphocyte subsets reactive with self-components are produced. These lymphocytes may be B-cells which secrete autoantibodies, T-cells that attack and destroy target organs, or a combination of both. Under normal conditions, these lymphocytes are controlled by mechanisms that preclude the development of harmful autoimmunity. Principally, an aetiological agent is assumed to induce the production of altered self-structures of the host which might trigger autoimmune reactions. Certain viruses, such as Epstein-Barr, cytomegalo, and rubella, are suspected to promote the production of such non-self-structures (Gleichmann and Gleichmann, 1987). Chemicals may combine with self-structures and modify them to become non-self antigens (Kammuller *et al.*, 1989). Other chemicals may trigger a release tissue component which may not be recognized as self. All these exogenous and endogenous non-self elements may promote autoimmunity. In addition, some chemicals may accelerate autoimmunity in individuals predisposed for other reasons. It has been shown, for example, that methylcholanthrene accelerates thyroiditis in disease-predisposed rats (Bigazzi, 1985). Many drugs, such as D-penicillamine used in rheumatoid arthritis therapy (Jaffe, 1979), anticonvulsant diphenylhaydantoin (Rosenthal *et al.*, 1982; Shelby *et al.*, 1980), or antihypersensitive captopril and hydralazine (Batchelor *et al.*, 1980), have been reported to induce autoimmune symptoms in patients. It has also been postulated that many immunological and non-immunological factors, such as genetic factors and metabolic pathways of aetiological agents, contribute to the complex processes leading to autoimmune diseases (Bigazzi, 1985; Gleichmann and Gleichmann, 1987; Kammuller *et al.*, 1989).

8.2.5 Immunopotentiation

Immunorestoration and immunopotentiation are closely related. Potentiation of the immune response can be achieved in several ways, such as by raising the level of immune mediators (interferons, interleukins, cell-activating factors), or by reducing the activity of suppressor cells. Chemicals that modify immune responses are called biological response modifiers (BRMs). A variety of synthetic polymers, nucleotides, and polynucleotides are known to induce interferon production and release; many bacterial products and cytokines activate NK cells and macrophages; and some hormones, notably thymic hormones, enhance T-cell functions. Thus BRMs have the potential to enhance immune responses. Levamisole and dithiocarbamate derivatives are BMRs that have been shown to potentiate the immune response in immunopharmacological studies (Hadden, 1987; Renoux, 1985). Non-specific immunotherapy, i.e. the use of chemically defined agents to restore an altered or deficient immune system, necessitates that these agents should not be toxic. Levamisole, the levo-isomer but not the dextro-isomer of 2,3,5,6-tetrahydro-6-phenylimidazo (2, 1-b) thiazole, was the first chemically-defined agent endowed with immunoenhancing activity (Renoux and Renoux, 1980).

8.3 Relationship of immunotoxicology and other toxic effects

8.3.1 Immunotoxicity and hepatotoxicity

Systemic and organ autoimmune reactions in humans, including immuno-pathology of liver, have been linked in many cases with the autoimmune potential of drugs and chemicals. One example is liver damage in humans following halotane (1-bromo-1-chloro-2,2,2-trifluoroethane) anaesthesia (Neuberger, 1989). Clinical and serological features of patients with halotane hepatitis, such as high levels of antibody to both normal liver components and specifically to halotane-related antigens, suggested an immune-mediated cause. The immunopathological nature of halotane hepatitis is also supported by observations that halotane-related antibodies can induce lymphocytes from normal individuals to become cytotoxic to antibody-coated hepatocytes. In patients with liver failure due to acetaminophen, it has not been possible to demonstrate analogous antibodies, however (Neuberger, 1989).

8.3.2 Immunotoxicity and nephrotoxicity

Nephritis and nephropathy have been induced experimentally in laboratory animals such as rabbits, mice, rats, and guinea pigs (Druet *et al.*, 1983, 1989; Kammuller *et al.*, 1989). Mercury-induced nephritis was characterized by deposition of IgG autoantibodies on the glomerulus, causing damage to the glomerular basement membrane. Genetic factors play an important role in mercury-induced membranous glomerulonephritis; alteration of T-cell activation is suspected in the genetically affected individuals. In the brown Norway rat, the RT1-linked gene was involved in the genetic control of susceptibility to autoimmune glomerulonephritis. Renal glomerular immuno-pathology in humans has been observed in individuals using lightening creams containing mercury. Interstitial and tubular nephropathology was reported in people exposed to mercury-contaminated water (Lawrence, 1985).

8.3.3 Immunotoxicity and neurotoxicity

In healthy individuals, the central nervous system is largely isolated from the immune system by the blood-brain barrier. However, bidirectional interaction of the neuroendocrine and immune systems does occur; neuropathological features are commonly encountered in generalized immune disorders and vice versa. Neuropathological states could be attributed to an immune-mediated attack on structures of the central nervous system itself, or to immunodeficient-related infections or tumours within the central nervous system. Reye's syndrome in humans, a clinical encephalopathy with visceral fatty degeneration, has been shown to link to several factors, including viral, host genetic make-ups, and exposure to environmental chemicals such as pesticides (Crocker *et al.*, 1976). Interaction between the neuroendocrine and immune systems was first recognized in a series of stress syndrome studies over 50 years ago (Snyder, 1989). The close association of the neuroendocrine and immune systems was evidenced by the demonstration of identical cell receptors in certain cells from these systems. Several authors reported immunomodulatory effects mediated by many products of the neuroendocrine system, such as corticosteroids, cortisone, hydrocortisone (Wistar and Hildemann, 1960), opioids, and β-endorphin (Gilman *et al.*, 1982). Toxicological studies showed that the same chemicals could adversely affect both neuroendocrine and immune systems. The close relationships between these two systems provide immunotoxicologists with opportunities to apply quantitative immunotoxicity assays in the study of direct or indirect neuroendocrine toxicants (Snyder, 1989). For example, exposure to organophosphorus defoliants did not induce any clinical effects on the peripheral nervous system; the activity of an esterase common to both lymphocytes and nerve tissue was inhibited by the exposure (Lotti *et al.*, 1983). Thus, an assay for peripheral lymphocytes might be applied to monitor the extent of exposure in farm workers handling organophosphorous pesticides (Lotti *et al.*, 1983).

8.3.4 Immunotoxicity and impairment of metabolism/detoxification

Increasing evidence suggests that there is a two-directional interaction of the immune system with xenobiotic metabolism (Kaminski *et al.*, 1990; Renton, 1983; Yang *et al.*, 1986). The rate of detoxification of xenobiotics is dependent on a large number of factors, including proper functioning of the immune system. Elimination of chemicals can be impaired in immunodeficient individuals. It has been recognized that there is a major change in biotransformation of chemicals during episodes of viral or bacterial infections, or after exposure to immunopotentiating agents (Renton, 1983). Exposure of nonspecific immunostimulants often results in a decrease in hepatic microsomal activity and in the capacity of the liver to eliminate xenobiotics. Although most of the metabolizing capacity of the liver occurs in hepatocytes, it appears that nonparenchymal components, such as the lymphoid Kupffer cells, also exert an influence on xenobiotic biotransformation.

Another example of interaction of the host defence mechanisms with detoxification is an interferon-related impairment of cytochrome P-450. Involvement of the mixed function oxidase system in xenobiotic bio-oxidation and induction of cytochrome P-450 upon xenobiotic exposure are discussed in Chapter 5. As interferon *per se* can directly affect cytochrome P-450, it is speculated that the ability to depress cytochrome P-450 is a common property of all interferon inducers (Renton, 1983). On the other hand, Kaminski *et al.* (1990) demonstrated that aminoacetonitrile, a competitive inhibitor of cytochrome P-450, reversed carbon tetrachloride-mediated suppression of the T-dependent antibody response in mice. Conversely, induction of cytochrome P-450, by pretreatment of mice with ethanol prior to exposure to carbon tetrachloride, resulted in a potentiation of the immunosuppressive effects of the chemical (Kaminski *et al.*, 1990). Finally, infections of several types have been shown to alter xenobiotic biotransformation in animals. Since viruses alter metabolic activities in host cells, it seems entirely possible that metabolism of chemicals might be altered even when the infection is relatively mild. These effects can occasionally be positive, such as DDT-related induction of cytochrome P-450 in duck hepatitis; however, depression of oxidation of chemicals was noted in murine and duck hepatitis (Renton, 1983). Furthermore, increased susceptibility to parathion poisoning due to a decrease in the ability of cytomegalovirus-infected mice to detoxify the insecticide was demonstrated (Selgrade *et al.*, 1984). There is also no question that xenobiotic metabolism is impaired in humans during viral infections.

8.3.5 Immunosuppression and mutagenicity/carcinogenicity

The direct action of mutagenic/carcinogenic agents on body components, such as DNA (formation of DNA adducts with xenobiotics and DNA changes are discussed in detail in section 4.2), is a noted example among different mechanisms of carcinogenesis. It is, however, possible that potential carcinogenic effects of many chemicals of environmental concern, such as certain pesticides, can be indirect or epigenetic (Exon *et al.*, 1987). For example, benzo[a]pyrene (B[a]P) is carcinogenic and immunotoxic, whereas its noncarcinogenic analogue, benzo[e]pyren (B[e]P), does not appear to be immunotoxic (White and Holsapple, 1984). Therefore, it was proposed that epigenetic carcinogenicity of benzo[a]pyrene, or other chemical agents, could be related to its immunosuppressive potential, in addition to its potential to inflict genetic injury (Ball, 1970; Schnizlein *et al.*, 1982; Ward *et al.*, 1985; White and Holsapple, 1984). The mechanism of chemical suppression of the immune surveillance of carcinogenesis is not clear, as immunodeficiency does not necessarily result in cancer. For example, spontaneous tumour appearance in thymus-deficient nude mice is extremely rare. Overall, one possible indirect mechanism of chemical carcinogenesis can be the impairment of the immune system. Evidence supportive of immune surveillance theory is the increased incidence of cancer in immunodepressed patients (Penn, 1985, 1987). An association between long-term chemical immunosuppression and carcinogenicity has been postulated (Exon *et al.*, 1987; Penn, 1985, 1987).

8.4 Immunotoxic chemicals

As discussed before, a chemical is considered immunotoxic when it acts preferentially on the components of the immune system. In other words, elements of the immune system are specific targets of immunotoxic chemicals. Exceptionally, immunotoxic chemicals can damage elements of the immune system without any other visible signs of toxicity. Chemical classification of immunotoxic substances is extremely complicated because of their structural diversity (section 8.2).

8.4.1 Immunotoxicology of heavy metals, organometals and metalloids

Heavy metals are considered the most potent immunotoxic chemicals among the inorganics. Some heavy metals exert adverse effects on the immune system at doses that other toxicological changes are not evident (Exon, 1984). *In vitro* analysis of the immunosuppressive effects of some heavy metals on humoral immunity provided a ranking in the following order (Lawrence, 1985): mercury > copper > cadmium > cobalt > chromium > manganese > zinc > selenium. It is of interest to note that some of these heavy metals, notably zinc, cadmium, and selenium are trace elements essential for proper body functions.

Arsenic appears to be immunostimulatory at low doses, but immuno-depressive at high doses (Exon, 1984). In mice, decreased antiviral host resistance has been shown to be associated with arsenic exposure (Aranyi *et al.*, 1985a).

The effects of cadmium on the immune response varied with the route, dose, and schedule of exposure to the metal in relation to antigenic stimulation. This may explain the discordant reports on the immunotoxic properties of cadmium (Exon, 1984). Inhalation of cadmium aerosol can induce suppression of the primary humoral response in mice at a dose which is hundreds of times lower than the effective dose given by intramuscular injection. In rodents, several adverse effects of cadmium were reported, such as decreased thymus weight, increased spleen weight, decreased nonspecific host defences, phagocytosis and lymphocyte response to mitogens, and increased/decreased T-dependent antibody response (Blakley and Tomar, 1986; Cook *et al.*, 1975; Krzystyniak *et al.*, 1987; Loose *et al.*, 1978; Thomas *et al.*, 1985b).

Chromium compounds are potent skin sensitizers, ranking second to nickel among metal compounds as a cause of contact hypersensitivity (Lawrence, 1985). In rodents, chromium-induced increases of serum immunoglobulins, T-dependent antibody response, and T-lymphocyte response to mitogens were reported. Furthermore, decreased phagocytosis following chromium exposure was also observed. With regard to the mechanism of action, it was suggested that chromium itself could act as a hapten and conjugates to body proteins, inducing hypersensitivity (Exon, 1984; Lawrence, 1985). Cobalt exposure led to decreased resistance to viral infections in mice and allergy/skin sensitization in guinea pigs (Gainer, 1972; Wahlberg and Boman, 1978).

The immunotoxic effects of lead salts (acetate, chloride, carbonate, oxide) and organolead compounds (tetraethyllead, di-n-hexyllead) are well documented (Descotes, 1988). In rodents, lead-related immunosuppression of humoral and cellular responses and decreased host resistance to infections or to tumours had been demonstrated. It was also suggested that macrophages can be a target for lead, resulting in altered or impaired nonspecific immune response (Cook *et al.*, 1975; Descotes, 1988; Exon, 1984; Faith *et al.*, 1979; Hemphill *et al.*, 1971; Hillam and Ozkan, 1986; Lawrence, 1981; Neilan *et al.*, 1983). Manganese should be considered immunotoxic, as decreases in thymus weight and in host resistance to infections were reported (Maigetter *et al.*, 1976; Smialowicz *et al.*, 1985).

The immunotoxicity of inorganic (e.g. mercuric chloride) and organic (e.g. methylmercury) forms of mercury has been demonstrated (Descotes, 1988). Humoral and cellular immune responses and nonspecific host defences were reported to be markedly impaired by mercury exposure in several species. In addition, an autoimmune mechanism for mercury-related glomerulonephritis has been suggested (Bridger and Thaxon, 1983; Dieter *et al.*, 1983; Druet *et al.*, 1983, 1989, Kammuller *et al.*, 1989).

The immunotoxicity of nickel rests on its potent skin sensitization effect. Nickel-related allergy, decreased T-dependent humoral response and increased phagocytosis were reported in rats (Maurer *et al.*, 1979; Spiegelberg *et al.*, 1984). Reported immunotoxicity of selenium includes potentiation of the immune response, augmentation of phagocytosis, and activation of NK cells in humans and rats (Talcott *et al.*, 1984; Urnab and Jarstrand, 1986).

Organotins, such as di- and trialkyltin and triaryltin compounds, are used as industrial and agricultural biocides; they are neurotoxic and immunotoxic. Dialkyltins act selectively on the thymus, resulting in inhibition of thymocyte proliferation; some adverse effects of dialkyltins are reversible or are prevented by dithiol compounds. In rats, dialkyltin homologues, di-n-butyltin and di-n-octyltin dichlorides induced a dose and time-related decrease in weights of the thymus, spleen and lymph node (Descotes, 1988; Penninks and Seinen, 1987). Trialkyltins, such as tri-n-butyltin chloride and bis(tri-n-butyltin) oxide, induced thymus atrophy but had less effect on lymphoid organ weights. Diestertin compounds did not induce lymphoid atrophy, possibly because of rapid hydrolysis of their ester bond (Penninks and Seinen, 1987). Data on immunotoxicity are available for diethyltin, diheptyltin, dimethyltin, stannous, tributyltin, triethyltin, stannic, and triphenyltin chloride; di-n-butyltin, di-n-hexyltin, di-n-octyltin, di-n-pentyltin, dipropyltin, methyltin, tetra-n-octyltin, and trimethyltin dichloride; tri-n-butyltin oxide, triphenyltin acetate and triphenyltin hydroxide (Descotes, 1988; Penninks and Seinen, 1987; Seinen *et al.*, 1977; Verschuuren *et al.*, 1970).

Chronic vanadium exposure in humans can result in irritative bronchopulmonary symptoms (Descotes, 1988). Decreased phagocytosis and host resistance to infections were found in vanadium-exposed animals (Cohen *et al.*, 1986). Zinc should be considered as immunomodulating as it is moderately mitogenic for lymphocytes but impairs mitogen-induced lymphocyte proliferation (Descotes, 1988). Zinc-related decreased T-dependent antibody response and T-lymphocyte proliferation has been reported in rodents (Ehrlich, 1980; Murray *et al.*, 1983).

The conclusion reached in many immunotoxicity studies of heavy metals is that they are immunosuppressive. Exposure to heavy metals leads to increased susceptibility to infectious agents in experimental animals. Several factors, such as exposure route, dose, duration of exposure, age and genetic characteristics are of primary importance for metal-induced immunomodulation (Descotes, 1988; Exon, 1984; Lawrence, 1985). In addition, heavy metals should be considered as potential epigenetic carcinogens, acting possibly through damage of the immune system (Exon, 1984).

8.4.2 *Immunotoxicology of aromatic hydrocarbons*

Monocyclic aromatic hydrocarbons and their derivatives are potent immunotoxic substances. It appears that these aromatic hydrocarbons, such as benzene, toluene, and xylene, suppress the immune system in general. The immunotoxic potential of benzene is well documented. Reported adverse effects of benzene on the immune system of experimental animals include decreases in lymphoid organ weights, antibody production, cell-mediated immunity, host resistance to infections and to tumours, and serum complement, IgA, and IgG levels (Aoyama, 1986; Pandya *et al.*, 1986; Rosenthal and Snyder, 1987; Snyder *et al.*, 1980; Wierda *et al.*, 1981). Experimental studies in mice demonstrated the immunotoxic potential of toluene, expressed as decreased host resistance to infections (Aranyi *et al.*, 1985b; 1986).

Environmental contamination by polycyclic aromatic hydrocarbons (PAHs) is widespread, arising from energy production, motor vehicle exhaust and refuse burning. Many natural and synthetic PAHs are carcinogenic, adversely affecting the host's tumouricidal activities. Suppression of these tumoricidal mechanisms by PAHs may facilitate the metastatic outgrowth of transformed cells. Only those PAHs which are carcinogenic appear to be immunosuppressive, and the extent of immunosuppression and carcinogenicity appears to be positively correlated. Metabolism of PAHs to reactive diol-epoxides may be important in PAH-induced immunosuppression (Ward *et al.*, 1985). Like benzene and toluene, PAHs are potent immunotoxic compounds affecting both the humoral and cell-mediated immunity.

The carcinogenic and immunotoxic properties of benzo[a]pyrene (B[a]P) have been discussed in section 8.3.5. Suppression of antibody production and suppression of T-cell cytotoxicity to tumours were reported findings of benzo[a]pyrene exposure (Ball, 1970; Schnizlein *et al.*, 1982; Ward *et al.*, 1985; White and Holsapple, 1984).

Reported immunotoxic effects of 7,12-dimethylbenzanthracene were depression of thymus weight, serum antibody titers, mitogenic and allogenic pression of thymus weight, serum antibody titers, mitogenic and allogenic lymphoproliferation, T-lymphocyte cytotoxicity to tumours, and of NK cell activity, and increased susceptibility to tumours and bacteria (Ball, 1970; *et al.*, 1984).

Methylcholanthrene is carcinogenic and immunotoxic. Its adverse effects on the immune system were suppression of antibody production, mitogenic lymphoproliferation, T-lymphocyte cytotoxicity to tumours, and NK cell activity (Ball, 1970; Lill and Gangemi, 1986; Ward *et al.*, 1985; Wojdani and Alfred, 1983). Eugenol is potentially immunotoxic, as it induced increased thymus weight and decreased humoral immunity in rats (Vos, 1977).

8.4.3 *Immunotoxicology of halogenated hydrocarbons*

Several halogenated aliphatic and aromatic hydrocarbons are potent immunotoxic agents. Frequently reported immunotoxic effects of halogenated hydrocarbons in laboratory animals and humans are suppression of humoral and cellular immunity as well as host resistance to infections.

Mice exposed to trichloromethane were shown to have a lowered antibody response (Munson, 1987). Decreased T-dependent antibody production resulted from trichloroethane exposure (Sanders *et al.*, 1985). The immunotoxicity of vinyl chloride and hexachlorobenzene has been studied extensively. Vinyl chloride induced splenomegaly, depressed cell-mediated immunity and mitogenic lymphoproliferation, and possibly could be responsible for circulating immune complexes (Black *et al.*, 1983; Descotes, 1988; Sharma and Gehring, 1979). The results from studies of hexachlorobenzene on different animal species are not consistent. Hexachlorobenzene was shown to stimulate the immune system in rats but induced decreased specific immunity and nonspecific resistance in mice. No major histopathological changes of lymphoid organs were observed in either species. Impurities in technical hexachlorobenzene have been found to contribute significantly to the immunotoxic properties of the chemical (Loose *et al.*, 1977, 1979; Vos *et al.*, 1979).

Considerable information on the immunotoxic effects of halogenated aromatic hydrocarbons indicated that these compounds induce decreases in lymphoid organ weights, in resistance to infections, and in generalized immunosuppression (Descotes, 1988; Thomas and Faith, 1985). Several aromatic halogenated insecticides have been shown to be immunotoxic. Lindane-related immunosuppression of humoral and cellular immunity occurred as decreased T-dependent antibody response in mice (Andre *et al.*, 1983) and decreased human T-lymphocyte proliferation *in vitro* (Roux *et al.*, 1979). Dieldrin, recognized as an equivocal carcinogen, is potently immunotoxic. Several studies in mice showed evidence of dieldrin-associated suppression of humoral immune response, decreased phagocytosis, and reduced host resistance to infections (Friend and Trainer, 1974; Kaminski *et al.*, 1982; Krzystyniak *et al.*, 1985). Mirex-induced decreases of thymus and spleen weights, T-dependent antibody response, and serum IgG level were demonstrated in chickens and mice (Koller, 1979; Subba and Glick, 1977).

The reported effects of DDT on the immune system have been discordant. Generally, humoral immunity appears to be depressed by DDT; however, specific cell-mediated immunity does not appear to be affected. In several species (mice, rats, chickens, and rabbits) a DDT-related increased susceptibility to parasitic and viral infections, decreased delayed-type hypersensitivity, and decreased antibody response were reported. In addition, a DDT-related decreased chemotaxis, phagocytosis and allergic contact dermatitis in humans has been documented (Exon *et al.*, 1987; Faith *et al.*, 1980; Koller, 1979; Vanat and Vanat, 1971).

Polybromobiphenyls and PCBs are environmental contaminants of major concern and are considered highly immunotoxic. Studies following accidental exposure to PBBs in Michigan farms in 1973 indicated various neurological, musculoskeletal, dermatological and gastrointestinal symptoms

(Descotes, 1988). In an attempt to correlate the immunopathological changes in PBB-exposed Michigan farmers, the severity of clinical manifestations appeared to correlate well with immune abnormalities such as reduced T-lymphocyte numbers (Bekesi *et al.*, 1979). The adverse effects of PBBs on the immune system of experimental animals include decreased thymus weight, increased spleen weight and lymphocyte counts, immunosuppression expressed as decreased antibody production and decreased cellular immune response (Bekesi *et al.*, 1979; Loose *et al.*, 1977; Luster *et al.*, 1978a, b). PCBs affect the immune system by suppression of humoral response, delayed-type hypersensitivity, and lymphocyte populations. Furthermore, decreased resistance to infections was reported (Koller, 1977). Certain isomers of PCBs are known to cross the placental barrier and/or accumulate in mother's milk and thus create a potential for immunoembryotoxicity (Chang *et al.*, 1981; Descotes, 1988; Imanishi *et al.*, 1980; Thomas and Faith, 1985; Thomas and Hinsdill, 1978). The immunotoxic potential of Aroclor 1254 was studied experimentally in mice; decreased antibody production, T-dependent antibody response, and host resistance to infections and tumours were demonstrated (Koller, 1977; Thomas and Hinsdill, 1978). Immuno-toxicity data for other Aroclor chemicals such as Aroclor 1221, 1242, and 1248 show similar effects (Descotes, 1988).

The role of the Ah receptor in PHAH immunotoxicity has been reviewed by Silkworth and Vecchi (1985). These workers examined the ability of a series of PCB isomers to cause thymic atrophy and to suppress antibody formation. The isomers were selected on the basis of their hepatic microsomal enzyme-inducing capacity and on their structural configuration. Isomers 2,2',4,4'-PCB and 2,2',5,5'-PCB, which cannot take up a co-planar confirm-ation, are weakly bound to the Ah receptor and have weak immunological potential. In contrast, 3,3',4,4'-PCB can assume a planar configuration. It binds strongly to the Ah receptor and causes severe immunosuppression. The relationship of the structure of PHAHs to binding with the Ah receptor is considered in more detail in Chapter 5 (section 5.6). These compounds are immunosuppressive only in those strains that have a fully expressed Ah receptor.

Several studies of pentachlorophenol (PCP) indicate that animals exposed to the compound show an increased susceptibility to tumour growth and decreased antibody production. However, available evidence appears to suggest that immunotoxicity of technical grade PCP is due mainly to its dioxin contaminants (Exon and Koller, 1983; Forseel *et al.*, 1981; Kerkvliet *et al*, 1985a). Dichlorophenol can be considered immunotoxic, as the chemical induced increased spleen weight, increased humoral response, and decreased delayed-type hypersensitivity in rats (Exon *et al.*, 1984).

The immunotoxicity of TCDF involves suppression of humoral and cellular immunity. TCDF-induced decreased thymus weight, antibody production, and delayed-type hypersensitivity were reported in guinea pigs (Vecchi,

1987; Vos, 1977). TCDD is a highly toxic contaminant of halogenated hydrocarbons. Well-documented animal experimental studies show that TCDD exposure leads to thymic atrophy, depletion of cortical thymocytes, decreased weight of other lymphoid organs, suppression of antibody production, suppression of cell-mediated immunity in young animals, and suppression of the formation of granulocyte-macrophage colonies in the bone marrow (Chastain and Pazdernik, 1985; Clark *et al.*, 1981, 1983; Faith and Moore, 1977; Holsapple *et al.*, 1986; Kerkvliet *et al.*, 1985b; Nagarkatti *et al.*, 1984; Thomas and Hinsdill, 1979; Vecchi, 1987; Vos and Moore, 1974).

8.4.4 *Immunotoxicology of carbamates and organophosphates*

Ethylcarbamate (urethan) is potently carcinogenic and immunogenic. Several experimental studies in rodents show urethan-related suppression of NK cell activity, which might lead to decreased nonspecific host resistance to infections and to tumours (Exon *et al.*, 1987). It has also been shown that ethylcarbamate induced a decrease in bone-marrow precursor cells and suppression of antibody production in mice (Luster *et al.*, 1978b). It has to be noted, however, that the methyl analogue of urethan, methylcarbamate, appeared to be non-carcinogenic and non-immunotoxic (Descotes, 1988).

Carbaryl is one of the most widely used carbamate anticholinesterase insecticides. The immunotoxicity of carbaryl has been reviewed by Cranmer (1986). Generally, carbaryl induces significant suppression of the humoral response only at near-lethal doses. Epidemiological studies in geographical areas where carbaryl has been used extensively have not detected an increase in pesticide-related Reye's syndrome (Cranmer, 1986). However, immuno-modulatory effects of carbaryl, such as increased immunoglobulin level, have been demonstrated in animal studies (Andre *et al.*, 1983).

The organophosphate pesticides malathion, parathion, dichlorvos and methylparathion were demonstrated to suppress humoral immune response in laboratory animals (Casale *et al.*, 1983). Generally, organophosphate-associated immunosuppression might be a consequence of toxic chemical stress associated with anticholinesterase activity. Parathion-induced immuno-toxicity is possibly related to its moderate to severe cholinergic activity. Overall, parathion induced dermatitis, decreased lymphoid organ weight, humoral immune response, and cellular immune response, and increased susceptibility to viral infections (Casale *et al.*, 1983; Selgrade *et al.*, 1984). For technical grade malathion, impurities, such as OOS-trimethyl-phosphorothioate, might be responsible for its demonstrated immunotoxic potential (Rodgers *et al.*, 1986).

Cyclophosphamide (CPS), a known alkylating immunosuppressive agent, is widely used as the positive control in *in vivo* animal assays of chemical-induced modulation of immune response (Jakab and Warr, 1981; Luster *et al.*, 1981; Renoux and Renoux, 1980; Smialowicz *et al.*, 1985). In addition to inhibition of humoral and cellular immunity, CPS also suppresses host resistance to infections. Interestingly, CPS was ineffective in certain viral and bacterial infections in mice; unaltered resistance to HSV-2 virus and increased resistance to *C. neoformans* has been reported in CPS-exposed mice (Morahan *et al.*, 1984).

8.5 Assessment and prediction of immunotoxicity

The structural diversity of immunotoxic chemicals is most evident. The heterogeneity of immunotoxic compounds, representing many different chemical families, makes them extremely difficult to classify on the basis of their chemical-formula-related immunotoxic effects. Throughout current publications on immunotoxicity, immunotoxic chemicals are classified according to their origin (e.g. industrial chemicals), their presence in the environment (e.g. airborne pollutants), and the type of application (e.g. immunosuppressive or anti-inflammatory drugs), or as principally based on the type of immunotoxic effect that they elicit (e.g. immunopotentiation, allergy). Some chemical classes, however, can be characterized as potentially immunotoxic, and it can be predicted with high probability that a chemical belonging to such a class could exert similar immunotoxic properties. Such a chemical class–activity correlation is true especially for most known potent immunotoxic chemicals, such as halogenated hydrocarbons, aromatic polycyclic hydrocarbons, alkylating agents, and heavy metals. All major classes of environmental chemicals have members that are potentially immunotoxic (Table 8.2). It should be remembered, however, that these potentially immunotoxic chemical classes are likely to contain also immunoinert members.

Experimental toxicology is concerned with determining the overall effects of a chemical as judged by its ability to cause structural and functional damage. Such information generated during standard toxicity studies sometimes is relied on to provide an indication of possible immunotoxicity and an insight into the integrity and activity of the immune system of exposed animals. An important question is how much we may learn about undesirable immunological effects from the results of standard toxicity tests of chemicals, and whether supplementary investigations are necessary in recognizing the problem.

According to Kociba (1982), analysis of routine toxicity studies, such as a 90-day study, offers several advantages for an initial assessment of immunotoxicity. These advantages are: (a) the subchronic studies are conducted on animals that are in the prime of life when the thymus is at its maximal size; (b) the secondary lymphoid organs are not yet showing the age-related degenerative changes; (c) determination of serum protein fractions and other haematological parameters can indicate possible

Table 8.2 Environmental chemicals that are considered potentially immunotoxic

Chemical class	Immunotoxic effects*
Metals, organometals, metalloids	Suppression of HMI, CMI, NSR; induction of contact hypersensitivity; impairment of host resistance to infections and tumours
Halogenated hydrocarbons (aliphatic and aromatic), aromatic hydrocarbons	Suppression of HMI, CMI; impairment of host resistance to infections and tumours
Heterocyclic oxygen-containing compounds including epoxides, furans, dioxanes	Suppression of HMI, CMI; impairment of host resistance to infections and tumours
Carbamates	Modulation of CMI, HMI; modulation of host resistance to infections and tumours
Organophosphates	Suppression of HMI, CMI; impairment of host resistance to infections and tumours

* Abbreviations: HMI: humoral-mediated immunity, CMI: cell-mediated immunity, NSR: nonspecific response.

immunosuppression; (d) during routine toxicity studies, there is maximal opportunity to interpret a spectrum of effects in such a manner allowing the investigator to discriminate between direct effects of the chemical on immune organs and indirect effects, such as nutritional deficiencies, debilitation or endocrine changes. In addition, evaluation of adrenal weight and morphology, the measurement of serum corticosteroid levels or supplemental studies using adrenalectomized animals are useful to assess the role that stress-mediated adrenal glucocorticoid hormones may have in the induction of lymphopenia and/or lymphoid depletion, especially in the thymus.

Wachsmuth (1983) assessed the immunosuppressive effects of a corticosteroid in 90-day toxicity tests in dogs and rats. A good correlation between a dose-dependent reduction in lymphoid organ weight and lymphocyte counts was obtained. Histopathology revealed an extensive inflammatory reaction to parasites in the lungs and a decrease in gammaglobulin level, which was obviously biologically relevant as a sign of immunosuppression. Quantitative histopathological analysis of lymphoid tissues is difficult because only extensive changes are readily apparent. As a general conclusion, it was questioned whether such toxicological assessments reflect the function of the immune apparatus, although quantitative assessments could be made of dose responses in the lymphatic system.

Trizio *et al.* (1988) suggested that certain parameters that are routinely examined in standard toxicity testing could be useful in the assessment of possible immunotoxicity of chemicals. Among them are body weight, mortality, incidence of infections, haematological profile (erythrocyte, leucocyte and differential leucocyte counts), serum proteins (total protein, albumin, globulins), organ weights (spleen, thymus, endocrine organs), and histopathology. Incorporation of additional parameters such as assessment of cell types and numbers in the bone marrow, and microscopic examination of lymphoid organs provides valuable information.

The Organization for Economic Co-operation and Development (OECD) guidelines for short-term toxicity studies do not require routine examination of all the parameters necessary for the identification of immunotoxic potential. For example, the only lymphoid tissue examined microscopically in the 30-day toxicity study is the spleen. Strong suggestions of some alterations in the ability of the body to affect a normal immune response, however, might be concluded from the atrophy of lymphoid organs (as judged by size, weight, and histopathology), lymphopenia, or an increase or decrease in serum globulins. The reliability of such testing in identifying immunotoxic effects of chemicals needs to be further assessed for the significance to health. Exceptionally, however, a chemical-induced atrophy of lymphoid organs, such as thymus atrophy resulting from exposure to glycol ethers or dioctyltin dichloride, might not be correlated with any evidence of functional immunotoxicity (House *et al.*, 1985, Miller and Scott, 1985). Sometimes, damage is seen only at higher doses or longer exposure periods than those required to alter immune function significantly, e.g. suppression of cell-mediated immune function and severe atrophy of the thymus caused by TCDD (Vos, 1977).

Immunotoxic screening can be performed with automated procedures such as flow cytometry (Hudson *et al.*, 1985). In many instances, such as leucocyte phenotyping distribution, surface marker analysis, lymphocyte proliferative responses, macrophage phagocytosis, neutrophil function, bone marrow cellularity and others, flow cytometry can provide accurate, objective, and rapid results. There are a number of other flow cytometric applications which are useful for immunotoxicological assessment. For example, the immunotoxic/activating effect of the carbamate pesticide aminocarb on maturation of bone marrow cells was evaluated by monitoring changes in G_0, G_1, and S phases of the cell cycle and the corresponding shift in the density of surface IgM on mature B-cells (Bernier *et al.*, 1990).

Data taken from traditional toxicological studies can be useful for the identification of chemicals that are suspected to be immunotoxic. However, these data are insufficient for a full evaluation of the xenobiotic-induced immunotoxicity. A complex two-tier testing battery to assess chemical-induced immunotoxicity was proposed recently by the US National Toxicology Program (Luster *et al.*, 1988). Tier I includes assays for cell-mediated immunity (CMI), humoral-mediated immunity (HMI), and immunopathology; the last-named forms part of the protocol in standard short- and long-term toxicity and carcinogenicity studies. The aim of Tier I testing is to screen for potent immunotoxicants. Weaker immunotoxic agents are less likely to be detected. Therefore, a second tier is proposed to represent an in-depth evaluation of immunotoxicity. Tier II includes additional assays for CMI, HMI, and nonspecific immunity, as well as an examination of host resistance. Tier II is included only if functional changes are seen in Tier I at dose levels which are not overtly toxic (i.e. body weight changes). According to the scheme, Tier II will provide information on the mechanisms of immunotoxicity characterizing the nature of the immunotoxic effects (Luster *et al.*, 1988).

Since the most relevant end point for immune dysfunction is altered host resistance to viruses, bacteria, parasites, or tumour cells (Bradley and Morahan, 1982), emphasis has been placed on developing sensitive infection models and correlating immune aberrations with altered host resistance. *In vivo* experimental animal host resistance assays provide the best information for the assessment of the overall immune state of the animal. Therefore, an assay of host resistance to microbial infections or to tumours should, perhaps, be incorporated in the first tier testing.

8.6 Application of immunoassays in environmental studies

There is great interest in human medicine in relating changes in the immune system and the formation of DNA adducts to the subsequent development of neoplasia. Many of these data have been accumulated from studies on asbestos workers and smokers, while research in these areas in environmental toxicology is scarce. Significant case-studies with environmental implications include outbreaks of poisonings caused by PCB-contaminated rice and oil in Japan in 1968 and in Taiwan in 1979, the inadvertent release of poly-brominated biphenyls in Michigan in 1973, and cases of exposure to TCDDs, most notably the chemical accident at Seveso in Italy and in Love Canal in New York, and in urban streets in Missouri in the United States.

The immunological changes associated with high PCBs exposure in Japan and Taiwan have been summarized by Lee and Chang (1985). A modest decrease in immunoglobulin levels, especially IgM and IgA, was observed. While the total lymphocyte population was not altered, there was a significant reduction in both the total number and percentage of T-cells. The B-cells were not affected. A good correlation was found between plasma PCB level and the degree of delayed-type hypersensitivity responses. Effects were also seen in phagocytic cells with alterations in both polymorphonuclear cells and monocytes.

The immunological findings of farm residents and chemical workers exposed to PBBs indicated decreased T-lymphocyte population and various abnormalities in cell-mediated immunity (Bekesi *et al.*, 1987). Decreased lymphocyte response to T- and B-cell mitogens and reduced proliferation of T-lymphocytes were demonstrated, and multiple immune dysfunction in the exposed population was observed. The relationship between serum PBB level and the prevalence of immunological effects is weak, although better correlations have been found between the PBB level in white blood cells and the most severe immune dysfunction. Studies of TCDD are of particular interest because a vast amount of mechanistic information concerning the binding of TCDD to the Ah receptor is available. Hoffman *et al.* (1987), examining persons in Missouri, concluded that chronic TCDD exposure was associated with depression of cell-mediated immunity.

The effect of lead ingestion on immune responses of the mallard has been examined by Trust *et al.* (1990). Mallards experimentally exposed by oral gavage to a single lead shot showed a reduced antibody response to a challenge. The effect on lymphocyte populations was unclear because of large variation among recorded individuals values.

While there is a considerable number of data indicating that many environmental contaminants can affect the immune system, the use of this information in wildlife toxicology is minimal at the present time. The death of a large numbers of harbour seals in the North Sea has been suggested to be due to a suppression of the immune system (Dickson, 1988). So far no definitive studies to correlate the findings have been made. Dietz *et al.* (1989) concluded that 'it cannot be stated on the basis of the available data that environmental degradation did aggravate the disease, but studies on the impact of anthropogenic factors on the population dynamics of seal species are an important part of the present research on marine mammals.' The results of such studies are eagerly awaited, and it would be of particular interest to see if there is any relationship between immunosuppression and dioxin equivalents.

8.7 Conclusions

The following conclusions can be drawn from the overview of the current advancement in the assessment of chemical-induced immunotoxicity:

1. Classical toxicological findings, comparative analysis of the chemical structure, and other signals of adverse health effects of a chemical on an organism may allow the identification of a chemical as a suspected immunotoxic substance.
2. Although some insights into the immunotoxicity can be discerned from data generated through classical toxicity studies, the data do not allow a prediction of the nature of the chemical-induced immunotoxic effect.
3. Full evaluation of the immunotoxic potential of a chemical requires a battery of tests, including assays to assess the humoral-mediated immunity, cell-mediated immunity, nonspecific immunity involving macrophages,

NK cells, and other immune elements, as well as host resistance challenges. Recent advances in semiautomated immunoassays using flow cytometry or other techniques may provide efficient and more objective means in the assessment of immunotoxicity and should be incorporated in the testing protocols.

9   *The use of animals in*
wildlife toxicology

9.1 Introduction

Concerns over the use of animals in toxicological experiments are becoming an increasing force. Even non-extreme positions could, if passed into legislation, have considerable impact. It is, therefore, important to look at alternatives to animal testing and the availability of non-invasive techniques.

The use of animals in experiments has come under greater scrutiny in recent years. Much of this pressure is good; the universality of animal care committees, which include members from outside the immediate scientific community, is an important advance. It is fitting that the researcher should answer the question, 'Why is it important for this study to be carried out?' In earlier and less formal days I found the questions of my children salutary. This having been said, let me state that I deplore the actions of the extremists of the animal rights movement.

There is a strong phylogenetic aspect to the pressure on animal experimentation. Those experimenting with mammals, especially primates and species with strong association with man such as cats and dogs, come under the closest scrutiny. I have always felt that if regulations became too tough I could return to my first love, spiders. While there is an emotional component to this aspect of concern over animal experimentation, it does also parallel the ability of the organism to feel pain and, perhaps, to react to other aspects such as confinement. One option is the use of embryos. In the case of several classes – birds, fish, amphibians – these are readily obtained and normally develop externally to their parent. From a conservation viewpoint this is also an attractive option since the impact on a population of removing eggs is minimal. In the case of birds, eggs that are taken early in the breeding cycle are usually replaced. In all classes the fraction of eggs that result in adults is low. The use of embryos in wildlife toxicology has been discussed in section 3.2.

There is the paradox of growing public demand for greater safety regarding the effects of chemicals on human health and a demand for fewer or even no animal experiments. Toxicologists must, while being sensitive to the animals they use, point out the continuing need for animal experiments if safety concerns are to be met. In this chapter the possibilities of reducing the number of animals used in specific tests, such as the LD_{50}, the use of non-invasive techniques, and the use of cell culture techniques are briefly discussed.

9.2 The LD_{50} and related tests

There has been considerable interest in the development of alternative methods to animal experimentation and in decreasing the number of animals used in specific experiments. These efforts have, naturally, focussed largely on the need for, and modification of, tests used in support of regulations. Studies, such as those carried out in wildlife toxicology, have received less attention since the scale of such studies is small in comparison to those carried out for regulatory purposes. Nevertheless, the same principles should apply.

The value and limitations of the LD_{50} and the Draize tests in the regulatory process are outside the scope of this monograph. Some basic information is needed on chemicals, although every attempt to get more information from fewer animals should be supported. One important aspect of this is the availability of test data. Standardization or harmonization of data between groups of nations, such as the European Economic Community and the Organization for Economic Cooperation and Development, so that data generated in one country can be accepted in others, is important. It is vital to deal with problems of confidentiality of data. Again, this is not the place for such a discussion. But, both from the viewpoint of decreasing the amount of animal testing and the viewpoint of obtaining the information necessary for environmental assessment, it is critical that these data be freely available.

Since the LD_{50} test has been referred to frequently in this monograph, it seems appropriate to discuss its scientific relevance briefly. Modification of the procedure to reduce the number of animals has been put forward (Zbinden, 1984). This involves a pilot experiment using three groups of two or three animals. Each group is dosed with an order of magnitude higher than the last. This pilot experiment may determine that the compound is essentially non-toxic or too toxic to be of any interest. If an accurate LD_{50} is required, then additional animals must be used, but since a rough LD_{50} can be calculated from the pilot experiment, in the final experiment the dosages can be selected accurately. The approach of ranging experiments is certainly a valid concept for wildlife toxicology experiments. But, once the range has been established and it is found that a significant effect is seen at environmentally reasonable doses which justify a definitive experiment, then it is important that the sample size be adequate to demonstrate the validity of the change. A method based on small numbers of animals has been validated by Yamanaka *et al.* (1990). Their method is based on using three rats or mice of each sex at each dosage level. The initial dose used

Table 9.1 Inter-species variation of LD_{50} of TCDD and endrin

Species	LD_{50} TCDD[1] μg/kg	Dieldrin[2] mg/kg
Mammals		
Guinea pig, male	0.6 μg/kg	
female	2.1 μg/kg	
Rat, male	22 μg/kg	
female	45 μg/kg	
Rabbit	115 μg/kg	
Mule deer		6.25–12.5
Goat		25–50
Birds		
Sharp-tailed grouse		1.06
California quail		1.19
Pheasant		1.78
Rock dove		2–5
Mallard		5.64

[1] Source: International Registry of Potential Toxic Compounds
[2] Hudson, Tucker and Haegele (1984).

was 2000 mg/kg. If no animals died the tests were suspended; if deaths occurred, then toxicity tests at 200 and 20 mg/kg were carried out. These workers examined ten chemicals, using this method, and compared them to literature values of LD_{50}s. Good agreement was obtained. Dichlorvos came out as 20–200 for both sexes, using this method, while literature values ranged from 133 to 275 mg/kg. For fenthion, Yamanaka and co-workers found 20–200 for males and 200–2000 for females, compared with literature values of 150 and 190 respectively. Comparing all their data, these workers found that the maximum non-lethal dose was from two to two and half times the LD_{50} for both the oral and dermal routes.

What is the value of the LD_{50} in non-regulatory wildlife toxicology? On the negative side, there is considerable inter-species variation. The inter-species variation of the LD_{50} for TCDD and endrin is given in Table 9.1. The variation in mammals for TCDD is 200-fold, whereas the variation for endrin is much smaller. Although there is some evidence for phylogenetic relationships – for example, the grouse, quail, and pheasant are closely related and the endrin LD_{50} is close – there are numerous examples where this is not found. In an analysis of the data on the toxicity of OPs to birds Mineau (in press; discussed in section 2.1.3) found no correlation between the closeness of phylogenetic relationships and toxicity. The variation within a single genus was as great as that between different orders.

Assuming that the LD_{50} is available for the target species, what is the value of this information? While LD_{50}s are not the be-all and end-all, they are often a useful starting-point. Although the preferred approach to dosage is based on environmental levels, and some reasonable (say, ×5) multiple of environmental levels, many experimental studies are based on LD_{50}s. In the latter case initial doses of experiments can be set at fractions of the LD_{50}. In this monograph I have tended to ignore experiments where the dosages are close to lethality on the grounds that physiological changes in animals at doses close to lethality are not particularly meaningful.

Tucker and Leitzke (1979) in their review of various responses – biochemical, pathological, behavioural – compared to lethal levels concluded 'that no more than a six-fold difference in median effect level can be produced by employing any known biochemical, histopathological or behavioural effect as the bioassay end-point criteria in place of lethality. That is, the lethal dosage in any assay is less than six times greater than the dosage causing an equivalent per cent of any other selected effect.' There are three difficulties in examining specific examples to see if they, in fact, meet Tucker and Leitzke's sixfold rule. First, LD_{50}s have been determined for only a few species and thus data for most wildlife species are not available. Second, the parameters of the LD_{50} test, such as the time period, may not match those of the physiological experiment. LD_{50} values are usually determined only for short periods (usually 24, 48 or 96 hrs) or in the case of LC_{50}s for a few days. Thus LD_{50} or LC_{50} values for longer periods of the right time length to compare to physiological experiments are often not available. That LD_{50} values decrease with time approaching an asymptotic value is well known (Figure 1.7), but the actual values are usually not determined and the number of animals involved in obtaining this data is considerable. The third difficulty is whether a 50% change in a biochemical response is biologically significant. In the case of DDE-induced eggshell thinning, a decrease of 20% was sufficient to cause collapse of the peregrine falcon population over wide areas of the Holarctic. Conversely, a 50% increase in the activity of MFO activity would not be considered as a significant induction of the system; usually at least threefold and often much more is considered to be a minimum response. The sensitivity of biomarkers is considered later (section 10.3.2), but it often exceeds the sixfold of Tucker and Leitzke.

Nevertheless, for all its weaknesses, the LD_{50} is a useful benchmark. Another end-point would have similar limitations and would be likely to involve animal experimentation.

9.3 Comparison of measurements in blood to organs

The use of biomarkers is based largely on human clinical medicine. Here the justification for non-invasive techniques has to be clearly made, and in all but a few cases diagnosis is dependent on measurements on blood and urine. In this section the extent to which measurements on blood can be used instead of organs is examined.

In addition to being non-destructive, the use of blood samples has the advantage that serial studies can be made on the same individual. This can decrease the amount of variability of the data by allowing the use of an individual as its own control in experimental work. It also makes it possible to follow the time course of exposure. This can be important, particularly in cases where the time of onset of an effect is variable.

The current status of measuring biomarkers is summarized in Table 9.2. Blood has been used for a considerable time in the use of AChE inhibition, but one of the most detailed studies has just been published (Holmes and Boag, 1990). These workers exposed zebra finches to a single dose of fenitrothion and then followed the brain and plasma activities of AChE for 10 days. The dosages of fenitrothion used were 1.04, 3.80 and 11.36 mg/ kg, which resulted in maximum brain inhibition of 50, 70 and 75% and maximum plasma inhibition of 78, 82 and 89% respectively. Death within 12 hours occurred at the highest dosage. The time course of recovery from AChE inhibition at the lowest dosage is shown in Figure 9.1. The rate of recovery of brain AChE activity found by Holmes and Boag is fairly steady and indicates that exposure could be determined (20% reduction rule of Ludke *et al.*, 1975) for up to 50 hours. The slope of recovery is lower than that reported for five avian species dosed with dicrotophos by Fleming and Grue (1981). Holmes and Boag found that the coefficient of variation for activity in the plasma was more than three times greater than that in the brain, a finding that agrees with previous work (Fleming, 1981). Recovery of activity in plasma is more rapid, reaching the 80% level within 24 hours. The steep slope of the recovery curve means that accurate determination of the degree of inhibition is difficult after 12 hours. In conclusion, although plasma measurements can be used, it is only over a much shorter time period and has considerably less precision than determinations using the brain.

One consideration that sets the inhibition of AChE apart is that it is normally used following a single, acute event; the application of OP or carbamate pesticides. Most of the situations in which other biomarkers are used are the studies of chronic exposure to pollutants, that is, involving steady-state conditions. In the case of the inhibition of ALAD (section 6.3) and changes in the activity of a variety of enzymes involved in control of metabolic processes (section 6.4), blood is the tissue of choice.

DNA adduct formation can be measured using haemoglobin. Neumann (1984) has reviewed the use of haemoglobin as an exposure monitor to alkylating and arylating agents. He concludes that adducts are formed proportional to dose down to low exposure levels and that they are stable throughout the life of the erythrocytes. Furthermore, for a given agent, the ratio between tissue DNA and haemoglobin adducts is constant over a wide

Table 9.2 Availability of biomarkers in blood

Biomarker	Blood	Tissue of choice	Comment
AChE inhibition	+?	Brain	Effects in blood more transient
Neurotoxic esterases	–	Brain	Enzyme is limited to brain
Biogenic amines	–	Brain	Changes in blood too transient
DNA			
Strand breakage	?	Wide range	Nucleated avian red blood cells are possible
Adduct formation	+	Wide range	Haemoglobin is good substitute for DNA
SCE	+	Wide range	Blood lymphocytes can be used
Degree of methylation	?	Wide range	Nucleated avian red blood cells are possible
MFO	–	Liver	Western blotting technique on leucocytes is possible
Thyroid	+	Thyroid	Circulating levels of T_3 and T_4 are sensitive
Retinols	+	Liver	Advances to use plasma are being made
Porphyrins	+?	Liver	Advances to use plasma are likely
ALAD	+	Blood	Tissue of choice
Enzymes	+	Blood	Tissue of choice
Immunotoxic	–	Lymphatic cells, bone marrow	Limited number of tests available for blood

range of doses. The relationship between adduct formation in skin and haemoglobin has been studied in mice exposed to benzopyrene (Shugart and Kao, 1985) and between haemoglobin and liver in sunfish (Shugart *et al.*, 1987). In both cases, the dose response was similar (Figure 9.1) although the absolute values for adduct formation in haemoglobin were considerably lower. Sister chromatid exchange has been widely used on blood samples (Das, 1988).

The most sensitive indicator of alteration of thyroid function is changes

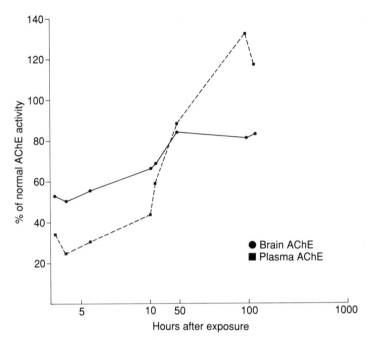

Figure 9.1 Inhibition and recovery of brain and plasma AChE activity in zebra finches following exposure to fenitrothion. After Holmes and Boag (1990).

to the plasma T_3/T_4 ratio. The limitation of purely haematological studies is that changes to the histology of the thyroid cannot be measured. Most of the retinol work has been carried out on the liver, but recent studies (Brouwer *et al.*, 1989a) show that plasma can be used. Studies on porphyrins have been carried out on the liver and kidney, but the possibility exists of using plasma.

The most serious limitation in using blood is the fact that there is virtually no activity of the mixed function oxidases in this medium, although there are possibilities of using monoclonal antibodies (section 5.7). There are severe limitations also in immunotoxicology. Some studies, such as antibody reactions, can be carried out using blood; but many studies require the removal of the lymph nodes and spleen.

One of most exciting areas of molecular biology that is beginning to become important in toxicology is the field of receptors. A list of the possible plasma receptors is given in Table 9.3. So far as wildlife toxicology is concerned, the Ah receptor is the only one that has been looked at in considerable detail (section 5.6). The muscarinic cholinergic receptor that binds acetylcholine is easily separated and has been well characterized. The use of this receptor in determination of changes of AChE activity has been discussed in section 2.1.7. Certain cyclodienes, hexachlorocyclohexane and

Table 9.3 Receptor systems associated with leucocytes

ACTH	Insulin
Acetylcholine	Parathyroid
a- and b-Adrenergic	Prolactin
Ah	Prostaglandin
Dopamine	Serotonin
Glucocorticoid	Thyroxin
Growth hormone	Vasopressin
Histamine	

Compiled from Conn (1984) (1985) (1986a,b).

some isomers of toxaphene bind with the GABA receptor (Lawrence and Cassida, 1984; Eldefrawi *et al.*, 1985) and cause changes in chloride permeability. The affinity of the dopamine receptor for dopamine has been shown to be decreased by lead and mercury (Bondy and Agrawal, 1980). An anti-oestrogenic effect on the rat pituitary cell cultures of kepone has been reported (Huang and Nelson, 1986). Oestrogenic action of a number of other pesticides (o,p'-DDT, o,p'-DDE, methoxychlor) has been previously discussed in section 3.3.1. While much detailed work on the pharmacodynamics of the interaction of environmentally important compounds with receptors and characterization of receptors is needed, it can be expected that major advances will be made in the last decade of the century.

9.4 The role of tissue culture experiments

Experiments can run the entire gamut from those on intact animals to those using established cell lines. No attempt is made to discuss the various *in vitro* test systems in detail. For additional information, the reader is referred to such works as Plaa and Hewitt (1982) and Rauckman and Padilla (1987), to name but two of the many works on the subject.

A tabulation of the advantages and disadvantages of various approaches is given in Table 9.3. Whether a particular feature is an advantage or a disadvantage depends on the question being asked. For example, in the intact animal experiments the second point in both columns can be an advantage or a disadvantage depending on the question being studied.

While the use of perfused organs has been of great value in fundamental physiological and pharmacological investigations, these techniques have not been widely used in toxicological studies. The main disadvantages are the time-consuming nature of the preparation and the short time for which they are viable. Tissue slices, while easier to carry out, also have short viability and have, largely, been superseded by primary tissue cultures. Established cell cultures suffer from a loss of organ-specific functions. For example, hepatocytes lose their ability to maintain cytochrome P450 activity needed for the pathways involved for the transformation of xenobiotics. The brief discussion of this major field here will focus on the comparison between experiments using primary cell cultures and those on intact animals.

A great deal of progress has been made in the technique of *in vitro* experiments, a term meaning literally 'in glass', over the last two decades. Early *in vitro* experiments on the inhibition of enzymes, in which a variety of metal compounds were added to solutions containing enzymes, are very limited. The lack of relationship frequently is not only magnitude but also direction. Inhibition of such enzymes as LDH and GOT is found *in vitro*, but increased levels are found *in vivo* (section 6.4). The more sophisticated approach of using *in vitro* experiments based on tissue cultures has much more promise.

The fundamental difference between *in vivo* and *in vitro* experiments is that the latter are designed to reduce the number of variables by eliminating the influence of regulation such as those imposed by the endocrine and nervous systems. This enables individual components of the system to be examined separately and in detail, but ultimately they require integration. For example, the metabolism of a xenobiotic by the P450 system can be examined in one series of experiments, but another series of experiments is needed to examine for the effect of altered steroid levels caused by P450 induction. The integration of this type of data to give an overall picture has been examined by Gillette (1986). He concludes that 'well integrated programs encompassing all of the *in vitro* approaches will be required if we are ever going to understand fully the mechanisms of action of drugs.' The importance of consistency in these various types of studies is stressed so that 'we can gain confidence that the conclusions drawn from *in vitro* experiments are relevant to events observed *in vivo*'.

Good progress has been made in culturing hepatocyte cells. This can substantially reduce the number of animals needed. Enough tissue can be obtained from two or three animals to conduct studies that would otherwise require 20 to 40 animals. Primary cultures of hepatocytes can be used to study hepatotoxicants over a period of several days. In some cases, i.e. neonatal rat hepatocytes, primary cultures have been used for biochemical studies for periods of up to 25 days and have been maintained up to 80 days with no apparent change in morphology (Acosta *et al.*, 1978). Overall, the capability of isolated hepatocytes to metabolize xenobiotics is quite similar to the metabolic patterns observed in the intact liver. Although the liver has been the target organ most widely used for *in vitro* studies, there are numerous studies on preparations of other organs. For example, Cherian (1980) has studied the effect of a variety of metallic ions on the synthesis of metallothionein caused by cadmium in cultured kidney cells, and Balazs and Ferrans (1978) the effects of a number of chemicals on myocardial cell cultures. Techniques are available to culture a wide variety of cells from different species and organs (Barnes *et al.*, 1984), so that investigators now have a wide range of systems from which to choose.

Studies in our laboratories at the National Wildlife Research Centre are, I think, a good example of the combination of intact animal and primary tissue culture work. Much of our work on the effect of specific PCB congeners on induction of porphyria (section 6.3.2) has been carried out in tissue culture experiments. When a large number of compounds, in this case congeners of PCBs, need to be studied, both separately and in combination, the number of animals that would be required for intact animal studies becomes prohibitive. An additional factor is the cost of the congeners since the amount of chemical needed per experiment is much less than that required for experiments on whole animals.

The techniques of *in vitro* testing have greatly improved in recent years. Of fundamental importance is the correspondence of *in vitro* and *in vivo* events. There have been many studies on the strength of *in vitro* tests, such as the Ames, to predict carcinogenesis. Five years ago, Stark and co-workers of the Laboratory Animal Research Center at Rockefeller University commented that 'today in vitro cytotoxicological research has a potential applicability to science similar to that held by in vitro genotoxicology 15 to 20 years ago.' For the type of biomarkers considered in this book, where the correspondence is beset with difficulties caused by pharmacokinetics outside the target organ, a useful approach is the ranking of *in vitro* and *in vivo* responses. Tyson and Green (1987) found good correlation for eight halogenated aliphatics between the concentration causing the release of 50% of the intracellular LDH and the maximally tolerated dose from whole animal studies. The *in vitro* EC_{50} value for LDH release was divided by the air–water partition coefficients to allow for the pharmacokinetic differences of comparing whole animal and closed flask systems for chemicals of different volatility. When these manipulations were carried out, an r value of 0.92 was obtained.

The close relationship of MFO induction and binding to the Ah in cultured chicken embryo hepatocytes to thymic atrophy has already been discussed (section 5.6).

While the cautions in the statement, 'it is unlikely that either cyto- or genotoxicological in vitro systems will ever provide toxicity data that will be totally equivalent to those derived from in vivo systems' (Stark *et al.*, 1986), will be accepted by most toxicologists, it is necessary to move forward. The view that 'it is not too soon to begin planning ways to integrate in vitro testing into toxicity testing as a whole' put forward by workers at the Center for Alternatives to Animal Testing at Johns Hopkins (Goldberg and Frazier, 1989) seems to express the current situation well. For both practical and moral reasons it is important that the two main streams of toxicology, *in vivo* and *in vitro*, do not remain two solitudes.

9.5 Statistical considerations

This is a completely non-technical discussion of this important subject. Those requiring technical information should consult such works as Sokal and Rohlf (1981) and Snedecor and Cochran (1980). The most important point to be made is that the biostatistical consideration should be fully examined before the experimentation starts.

The most wasteful experiments, from all points of view, are those with too few animals for a definite conclusion to be reached. When I first joined the Canadian Wildlife Service, I was asked to take part in an experiment to determine the effects of immersing polar bears in oil. My part in the experiment would have been to carry out some biochemical measurement on the blood. I declined because I considered that the sample size was too small. Since it was likely to be a sensitive issue, the number of polar bears to be exposed to oil had been reduced to two with one control. As might have been expected, a storm broke over the immersing of the polar bears in oil, and indeed the data collected had little value.

It is important to determine the variance of the parameters being studied and then to calculate the sample size that is needed to determine a specific decrease of change. This concept, for residue level studies, has been discussed in Chapter 1. Margolin (1988) examined 12 studies to determine if there is a sex difference in the rate of sister chromatid exchange (section 4.6) in humans. Sample sizes ranged down to a study that had only two and four individuals, and five studies had sample sizes in single figures. These studies led to the conclusion that no difference exists, but a large study (based on 479 individuals) established a clear statistical difference. The solution is first to determine the sample size needed to decide whether or not an effect is occurring and then to decide whether or not to do the experiment.

At the risk of being thought inconsistent, let me also make a plea for not being too hidebound by statistics. A lawyer looking at the use of biological markers in tort litigation (Johnson, 1988) quotes a court decision that discounts the traditional concept of statistical significance: 'The cold statement that a given relationship is not "statistically significant" cannot be read to mean "there is no probability of a relationship." Whether a correlation between cause and a group of effects is more likely than not – particularly in the legal sense – is a different question from that answered by tests of statistical significance, which often distinguish narrow differences in degree of probability.' There comes a time when it is necessary to make decisions on insufficient evidence. These problems are considered in more detail in the last chapter.

10   *The role of biomarkers in environmental assessment*

10.1 Biomarkers and the epidemiological approach

Epidemiology can be defined as the study of the prevalence of disease in a community at a special time and which is produced by special causes not generally present in the affected community. Classical epidemiology has been reactive rather than proactive. The earliest examples, such as nervous disorders among makers of hats ('mad as a hatter') and cancer in chimney sweeps, were cases where an abnormally high incidence of disease was noted and traced back to a specific cause.

The basic criteria of epidemiology are:

> Time sequence
> Strength of association
> Specificity of association
> Consistency of replication
> Biological plausibility

The DDE-induced eggshell thinning story is classical epidemiology. The time sequence of events that elucidated this phenomenon is given in Table 10.1. If one excludes the first, tentative speculation that there might be a problem, which appeared as a note in the magazine *British Birds*, the events occurred in a ten-year period following the Madison meeting. At this meeting peregrine falcon enthusiasts from many parts of the world discovered that the declines in their area was matched in other parts of the world. That pesticides were likely to be the common cause was immediately realized, but it was the prescient observation that the eggshells were thinner that enabled the link to reproductive failure and population changes to be made. This pointed to the direction for the studies on the physiological mechanism. The latter are discussed briefly in section 6.4.4.

While the epidemiological approach has generally been reactive rather than proactive, investigations can be launched to see if a potential hazard is causing problems. Epidemiological studies produce correlations rather than cause-and-effect relationships and as the focus moves to more subtle,

Table 10.1 Time sequence of events in the epidemiology of DDE-induced eggshell thinning in the peregrine falcon

Date	Event	Epidemiological criteria	Reference
1958	First note of abnormal breakage of peregrine eggs in UK		Ratcliffe (1958)
1965	Information on dramatic declines throughout the Holarctic is brought together at the Madison meeting	Time sequence Strength of association Consistency of replication	Hickey (1969)
1967	Demonstration that eggshell thinning first occurred in the UK in the mid-1940s	Time sequence Biological plausibility	Ratcliffe (1967)
1968	Demonstration of eggshell thinning in US in mid-1940s Association of population declines in many species with 20% eggshell thinning	Time sequence Strength of association Consistency of replication Biological plausibility	Hickey and Anderson (1968)
1970	Association of eggshell thinning with DDE residues in eggs collected in Alaska	Specificity of association Biological plausibility	Cade *et al.* (1971)
	Demonstration of DDE-induced eggshell thinning in experiments on kestrels (Falco species)	Specificity of association	Wiemeyer and Porter (1970)
1973	Mechanistic studies show enzymatic changes caused by DDE in the avian oviduct	Specificity of association Biological plausibility	Peakall *et al.* (1973)
1974	Demonstration of the presence of DDE in eggs collected in 1940s	Time sequence Specificity of association	Peakall (1974)

Table 10.1 (cont.)

Date	Event	Epidemiological criteria	Reference
1975	Demonstration of equivalent effects in both field studies and controlled experiments in the kestrel	Specificity of association Biological plausibility	Lincer (1975)
	Correlation of eggshell thinning with DDE residues in eggs from several areas	Strength of association Specificity of association Consistency of replication	Peakall *et al.* (1975a)

but widespread, problems, it becomes increasingly difficult to detect change with statistical certainty. Nevertheless, it provides an important safety net. Despite the improved testing systems that are in place today it is unlikely that pre-market testing would have caught DDT on the grounds of eggshell thinning. There are wide differences in species sensitivity to DDE-induced eggshell thinning (reviewed in Peakall (1975)), and none of the commonly used test species are sensitive. This is not to suggest that current procedures would have been inadequate to prevent the registration of DDT as an insecticide. Its high octanol/water partition coefficient, indicating its ability to bioaccumulate, and its toxicity to fish should have been sufficient to indicate the problems that could be expected.

A critical review of the role of biomarkers in clinical investigations has recently been made by Cullen (1989). Although he is looking at the problem of human occupational and environmental medicine, his conclusions are relevant to the wildlife toxicology field. His starting-point is that the easy cases are quickly found and that 'the outstanding challenge for the future is to elucidate factors which increase risk for disease modestly.' Certainly we now have the tools to investigate rapidly a situation such as that which occurred widely in the Great Lakes in the early 1970s. The question is, 'do we have the tools to evaluate the current situation in the Great Lakes?' The present problem in the Great Lakes is to assess the possible impact of chronic exposure to contaminants at levels below those causing overt harm to wildlife. It is the wildlife toxicology equivalent of Cullen's point on the difficulty of searching for weaker (but possibly more important because of the number of individuals involved) associations. While he concludes that biomarker research is fundamentally sound and that no other new approach has emerged which offers nearly the same scientific opportunity, he puts forward ideas for the future direction of the field. Basically, he considers that the most fruitful approach is the research centre-based clinical investigation. The limitation caused by the smaller sample size is more than counterbalanced by the detail that can be obtained by studying individuals known to be exposed. This approach parallels that approach used in examining the physiological and biochemical status of target species, such as the herring gull, for areas of particular concern.

The epidemiology of many North American wildlife toxicology problems was reviewed at a recent meeting hosted by the International Joint Commission (Gilbertson, 1989). The focus was on the Great Lakes, but a similar meeting could be held on the problems of the Baltic. Sophisticated monitoring programmes that should be capable of tracking environmental conditions and measuring long-term changes have been put forward. One of the most detailed is the Swedish Environmental Monitoring Programme (Bernes *et al.*, 1986). The parameters in such programmes range from chemical residue work to changes in population. The possible role of biomarkers is discussed below.

10.2 Current status of monitoring based on biomarkers

There is agreement among those interested in the field that the current status justifies the immediate implementation of pilot projects to use biomarkers in environmental monitoring. In fact, few biomarkers are considered to be sufficiently well established to use in assessing damage. The use that has been made of this approach in the US Comprehensive Environmental Response, Compensation, and Liability Act (CERCLA) is discussed in section 10.4.1.

In the concluding remarks of a conference held in 1988 by the American Chemical Society entitled 'Biomarkers of Environmental Contamination' McCarthy (1990) states that the challenges and obstacles to be addressed in an environmental monitoring programme based on biomarkers are:

1. The quantitative and qualitative relationships between chemical exposure, biomarker response and adverse effects must be established.
2. Responses due to chemical exposure must be able to be distinguished from natural sources of variability (ecological and physiological variables, species-specific differences, and individual variability) if biomarkers are to be useful in evaluating contamination.
3. The validity of extrapolating between biomarker responses measured in individual organisms and some higher-level effect at a population of community level must be established.
4. The use of exposure biomarkers in animal surrogates to evaluate the potential for human exposure should be explored.

At this point of the monograph I hope that we are aware of several biomarkers for which we can handle the types of variation listed in McCarthy's second point. It is of fundamental importance to the establishment of biomarker-based environmental monitoring programmes. A lot of work will be needed to ensure comparability between results produced by different laboratories and to put in place the necessary quality assurance/quality control protocols. The biomarker approach must parallel the work being done in monitoring programmes based on residue level measurements. The biomarker approach must be carried out as rigorously as the best of those in the field of analytical chemistry. McCarthy's fourth point can also be readily dismissed here. I have no doubt that as the biomarker field progresses it will be increasingly used to evaluate the potential for human exposure, but such uses are outside the scope of this monograph.

It is the concerns raised in McCarthy's first and third points that need to be considered in more detail. I prefer to break them down and then recombine them. It seems more convenient to consider the first of them, viz. that 'the quantitative and qualitative relationships between chemical exposure, biomarker responses and adverse effects must be established' in two parts: first, the relationship between chemical exposure and biomarker response; second, the relationship between biomarker response and adverse effects. To this second part I would add the validity of extrapolating between responses measured in individual organisms to those at a population or community level.

10.3 Relationship between chemical exposure and biomarker response

This relationship can also be considered in two parts. Are biomarkers available that respond to the main classes of pollutants that are of concern? Are these responses of the required degree of sensitivity?

10.3.1 Availability of biomarkers that respond to major classes of pollutants

A matrix summarizing the response of biomarkers considered in this monograph against the main classes of pollutants is given in Figure 10.1. Obviously, a single check mark is an oversimplification, but it does enable one to see certain patterns. Looking vertically at the matrix, one sees that alterations to DNA integrity and the immune system occur with all the major classes of pollutants. In contrast the inhibition of esterases and aminolevulinic acid dehydratase (ALAD) is much more specific. Indeed the latter is specific to a single metal, lead. The use of a suite of biomarkers with different degrees of specificity is an important aspect of environmental monitoring based on biomarkers. The relative specificity of various biomarkers is listed in Table 10.2.

Looking horizontally at the matrix in Figure 10.1, one notes that the PHAHs are the class best represented. Again the check mark system is an oversimplification. There are many different families of P450-dependent mixed function oxidase enzymes, so that some differentiation is possible, since porphyrins are strongly induced by some PHAHs and not by others, and so on.

	AChE inhibition	NTE inhibition	Biogenic amine response	DNA integrity	DNA adducts	MFO induction	Thyroid function	Retinol changes	Porphyrin profile	ALAD inhibition	Metal-binding proteins	Serum enzymes	Immune responses
Toxic metals			✓							✓	✓	✓	✓
Polyaromatic hydrocarbons				✓	✓	✓							✓
Polyhalogenated aromatic hydrocarbons			✓	✓	✓	✓	✓	✓	✓			✓	✓
Organophosphates and carbamates	✓	✓	✓									✓	✓

Figure 10.1 Biomarkers available for different classes of pollutants.

Table 10.2 Specificity of biomarkers

Highly specific	Inhibition of ALAD by lead
Moderately specific	Inhibition of AChE by OPs and carbamates
	Induction of porphyria by some PHAHs
Relatively nonspecific	Induction of MFO enzyme systems
	Sister chromatid exchange

10.3.2 Sensitivity of response of biomarkers

To be useful, biomarkers should respond to pollutants in a dose-dependent manner over a concentration of the pollutant that is environmentally meaningful. Having said that, it should be pointed out that the dose-response model is based on controlled laboratory studies. In field studies, organisms are exposed to multiple chemicals and multiple stresses (changing temperature, nutrition status, etc.), which can confound a simple dose-response relationship. In nature, it is more reasonable to think in terms of a family of dose-response relationships, based on the combination of chemical and environmental stresses to which the organism is exposed.

The inhibition of acetylcholine esterase (AChE) is, at least at the moment, a special case. In this case it is the biological response, rather than the

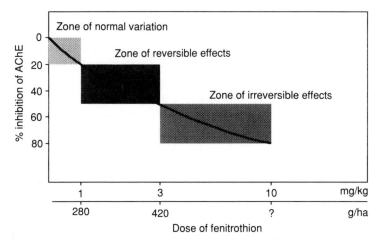

Figure 10.2 Dose response of AChE inhibition.

measurement of chemical residues, that is the 'gold' standard. In fact, in plotting the dose response of AChE, one has to plot the degree of inhibition against the dose rather than the chemical residue (Figure 10.2) since data on the latter are not available. Although analyses of residues of organophosphates have been made to confirm the exact chemical used, they are not routinely carried out. Perhaps, a few years from now, this will be the case for other chemicals and other biomarkers.

Furthermore, the inhibition of AChE is used to probe the effects of toxicants – the organophosphate and carbamate pesticides – deliberately added to the environment rather than in the broad assessment of environmental pollution. Indeed, esterases do not appear in the index of McCarthy and Shugart's recent (1990) book *Biomarkers of Environmental Contamination*. The dose response for AChE inhibition in Figure 10.2 is given in two ways. First the normal approach in terms of mg/kg given to the individual is given. This has been done, using the data of Holmes and Boag (1990). Second, the response has also been plotted against the rate of release of the pesticide into the environment. This has been based on the compilations of Mineau and Peakall (1987). It seems unlikely that similar calculations could be made for any other class of chemicals.

The other biomarker that is close to being on the 'gold' standard is the inhibition of aminolevulinic acid dehydratase (ALAD). The dose response of the ALAD activity ratio against the concentration of lead in the blood, based on the work of Scheuhammer (1989), is given in Figure 6.9. The degree of correlation is high ($r = 0.94$ for this particular data set), and the assay is easier to run than the determination of lead. The difference in the acceptance of this assay compared to AChE inhibition is only one of degree. Analysis of lead is a good deal easier than analysis for the residues of OP pesticides, and the relationship between harm caused by inhibition of AChE is more straightforward than is the case of inhibition of ALAD. Examination of Figure 6.9. shows that ALAD can be used as a reliable index of exposure to lead over a range of blood lead levels of 10 to 1000 µg/ml, which is the range of interest for environmental studies. Only at trace levels is the chemical analysis more reliable, as here one is moving into the zone of normal biological variation of the enzyme activity. However, in the case of lead, it should be pointed out that there is a background level for residue levels since lead is a naturally occurring element.

Highly sensitive methods exist for the measurement of DNA adducts. A variety of techniques are available (section 4.3). For example, using enzyme-linked immunosorbent assays, it is possible to detect one adduct in 10^8 normal nucleotides (Poirier, 1984). In a sense measurement of adducts can be looked at as a specialized means of chemical residue analysis rather than a biomarker. It represents that fraction of the pollutant that is covalently bound to a biologically important site. Thus it is a more sophisticated measurement than is usually achieved by chemical analysis. If one measures the concentration of DDE in liver, then the resultant figure of × mg/kg does not give any indication of the fraction of the DDE stored, temporarily inertly, in lipid and that which is bound to an active enzyme site. DNA adducts provide a convenient means of determination of exposure, especially for the PAHs, which lack the stability of some other classes of pollutants. Nevertheless, it is arguable whether adduct formation is a biological response or merely a possible precursor to a response.

Induction of mixed function oxidases has been widely studied, although the number of studies giving good dose-response data is quite limited. In this case there is no argument that it is a biological response, although the significance to the health of the organism can certainly be debated. One of the best of the dose-response studies comes from an early paper of Gillett (1968). In this study, rats were dosed with DDT for 14 days and the hepatic activity of aldrin epoxidase was measured. The results are shown in Figure 10.3. Using a log-log plot, the response breaks into two parts. The activity is essentially constant at low doses (although somewhat confounded by the fact that the control value itself is higher than those in the sub-ppm ranges) and then increases rapidly. The intersection of the two lines gives a 'no-effect' level of approximately 2 ppm DDT.

The dose-response curves of different MFO enzyme systems can be quite different. The responses of four such systems to a single dose of Prudhoe Bay crude oil in the liver of the herring gull are shown in Figure 10.4 (Peakall *et al.*, 1989). The activity of aminopyrine N-demethylase has reached saturation at the lowest dose used; the curve for aldrin epoxidase reaches its maximum at a dosage of 0.5 ml. In contrast, EROD increases steadily

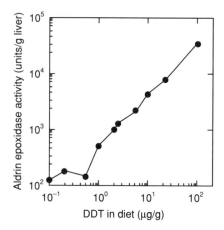

Figure 10.3 Dose response of aldrin epoxidase activity to DDT in the rat. After Gillett (1968).

throughout the range of doses used, while benzo(a)pyrene dehydrogenase shows little induction.

An interesting approach to sensitivity of biomarkers was taken many years ago by Dinman. He calculated that the minimum number of mercury atoms per cell required to cause a 20% inhibition of enzymes rich in SH groups was 4×10^4 (Dinman, 1972). This figure was based on the most sensitive *in vitro* inhibition found for any enzyme and assumed that *in vivo* effects would be as sensitive and that no binding to other sites occurred. Neither of these conditions is likely to be met, but as a first approximation it allows the calculation of the lower limit for the occurrence of biologically significant intracellular molecular interactions. The weight of 4×10^4 molecules of mercury is 4×10^{-17} g, and Dinman gives the weight of a mammalian liver cell as 7×10^{-9} g. The concentration of this minimum amount of mercury is thus 5 µg/kg. While one should not take the absolute values generated by this type of calculation too seriously, it does suggest that we are in the right ballpark for the minimum concentration, i.e. the sub- to low-ppm range.

This type of calculation can now be applied to the interaction of pollutants with receptors. Receptors have been mentioned several times although there is not a specific section devoted to them. While in the process of writing this book, I visited the British Trust for Ornithology. Standing in the crowded kitchen drinking a cup of tea, I was asked, 'What do you think will be the most significant advance in wildlife toxicology in the '90s?' Usually, for me, that is the type of question for which one thinks of a good answer on the following day, but for once it came spontaneously: 'The effect of pollutants on receptors'. Whether or not this was a correct prediction we do not yet know, but it is an area of increasing importance in toxicology. In their

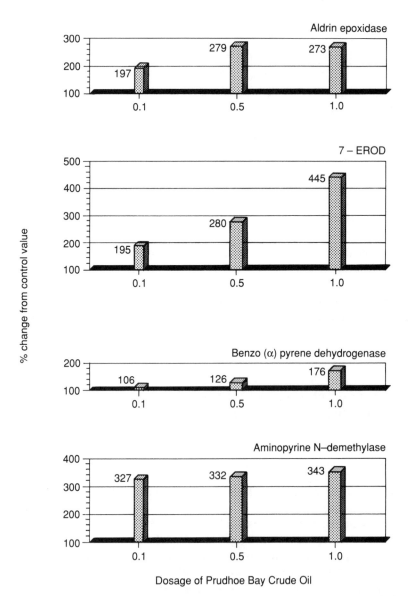

Figure 10.4 Dose response of four MFO systems to induction by crude oil in the herring gull. After Peakall *et al.* (1989).

review of the biochemistry and biology of the Ah receptor, Greenlee and Neal (1985) state that 'the identification of the Ah receptor for PHAH-inducible microsomal MFOs by Poland *et al.* (1976) was a significant event in the maturing of toxicology as a scientific discipline. This event gave the disciplines of toxicology and receptor biology a common focus.' A fruitful line of investigation might well be to examine the extent that known receptors can be blocked by environmentally important compounds and to examine the implications of the positive findings.

The concentration of the Ah receptor is given by Poland *et al.* (1976) as 81 fmol/mg protein. If one assumes that protein constitutes 10% of the tissue, then there are 8×10^{14} receptors per gram of tissue. The weight of a molecule of TCDD is 5×10^{-22} g, so that the weight of TCDD required to saturate the Ah receptors in a gram of tissue is 4×10^{-7} g.

The receptor approach is also being used in the case of the rodenticide flucoumafen. It has been found that when the hepatic binding sites become saturated, the anticoagulant effect becomes lethal. Receptor concentrations of 2.2 nmol/g liver have been reported for the rat and 1.4 nmol/g for the quail (Huckle *et al.*, 1989a,b). This approach of binding to receptors is now being used to investigate the possible impact of secondary poisoning of barn owls by rodenticides.

Without belabouring the point further, there is, I think one can safely say, sufficient evidence that individual biomarkers are sensitive enough to detect pollutants at realistic concentrations. The differential sensitivity of biomarkers is something for which we need a much better database before firm conclusions can be drawn. While there are studies involving several different biomarkers, for example, those of Andersson and coworkers (Andersson *et al.*, 1988, section 5.7) on the effects of pulp mill effluent, the evaluation of utility of the various biomarkers is something that lies in the future, but not, I would predict, in the distant future.

10.4 Relationship between responses of biomarkers to adverse effects

In an editorial in *Environmental Toxicology and Chemistry*, one of the most experienced wildlife toxicologists, Gary Heinz, asks, 'How lethal are sublethal effects?' (Heinz, 1989). He points out that in order to have rigorous proof, we need to show that a contaminant will also cause the sublethal effect in the field and proof that the sublethal effect will lead to death or reproductive failure. He points out that despite all the experimental work on sublethal effects there is very little proof at all of harm. The only example that Heinz gives that meets his criteria is DDE-induced eggshell thinning, and he asks how many others there are. The inhibition of ALAD caused by lead and the subsequent decline of the population of swans (section 6.3.7) in the United Kingdom is another example. But, unquestionably, there are not many.

I would agree with the need for a shift of research effort into this field.

At the moment we do not have the information so that we can use the projection of the important of sublethal effects in the regulation of chemicals. The detailed approach whereby we examine the entire gamut from receptors to ecosystems is the ideal to be strived for but is unlikely to be achieved except for a few of the most important pollutants. A framework for the valuation of biological markers, from the viewpoint of human medicine, was put forward by Schulte (1989). He divides the continuum from exposure to prognostic significance into seven components and examines the 21 inter-relationships between these components. This approach, which essentially parallels that of Heinz, works on a chemical-by-chemical basis.

10.4.1 Extrapolation to harm at the individual level

The degree of extrapolation that can be made from changes in biomarkers to harm to the individual in which the changes have occurred varies greatly from biomarker to biomarker. The reasons are partly in the nature of the biomarkers and partly due to limitations of knowledge. The status of the biomarkers, grouped under the headings of the main classes of pollutants that affect them, modified and expanded from a listing of Shugart *et al.* (1989), is given in Table 10.3.

The difference in the nature of biomarkers is clearly shown by looking at inhibition of esterases in comparison with induction of mixed function oxidases. Acetylcholine esterase is an essential part of the transmission of nerve impulses. Fifty per cent inhibition of AChE on the chronic basis or 80% inhibition on an acute basis causes death. In contrast, the induction of MFO enzymes is a major defensive system to rid the body of xenobiotics. Elevated levels indicate that the organism has been exposed and that physiological responses to handle the problem have occurred. There is no indication that harm has occurred. It is possible to relate MFO induction to other manifestations of harm. The calculation of dioxin equivalents (section 5.6) is the clearest example.

In the case of alterations to structure of the genetic material and alterations to the immune system there is a prima facie case for concern. Both systems are fundamental to the well-being of the animal, but nevertheless our knowledge of the actual harm caused is limited. Sister chromatid exchange (section 4.6) is a useful means of examining for the occurrence of mutational events, but the DNA has remained morphologically intact. Similarly, a properly functioning immune system is critical to the health of an individual, but that system has considerable reserve capacity so that demonstration of change cannot be taken, *per se*, as evidence of harm.

Johnson (1988), in a discussion of the use of biological markers in litigation, makes the basic point that although epidemiological studies can provide information, sometimes quantitative information, on the role of a risk factor to populations, the biostatistical data generated by such studies are not sufficient to apportion risk at an individual level. He continues, 'to the

Table 10.3 Biomarkers for different classes of pollutants

Environmental pollutant	Biomarker	Reliability index*
Toxic metals	DNA integrity	s
	Metal-binding proteins	s,d
	ALAD inhibition	s,d,p
	Immune response	s
	Levels of serum enzymes	s
PAHs	DNA/haemoglobin adducts	s,d,p
	DNA integrity	s
	MFO induction	s,d
	Immune response	s
PHAHs	Biogenic amines response	s
	DNA/haemoglobin adducts	s,d
	DNA integrity	s
	MFO induction	s
	Porphyrin profile	s
	Retinol changes	s
	Immune response	s
OPs	AChE inhibition	s,d,p
	Neuroesterases inhibition	s,d,p
	DNA integrity	s
	Enzyme profiles	s
	Immune responses	s

* s = signal of potential problem
  d = definitive indicator of type or class of pollutant
  p = predictive indicator of long-term adverse effect
  Expanded from Shugart *et al.* (1989).

extent that biologic markers become a scientifically acceptable and legally reliable means of proving that exposure to a particular risk factor caused a specific disease, judicial decisions regarding disease causation can be made with scientific certainty and without subjunctive reference to the defendant's purported negligence.'

In the environmental field, the US Comprehensive Environmental Response, Compensation, and Liability Act of 1980 (CERCLA) is an interesting use of this approach. This Act provides that, in addition to cost-recovery for response and clean-up action, natural resource trustees may recover damages for injury to natural resources, including costs of assessing such injury.

In a technical information document (USDI, 1987) various physiological responses of wildlife to evaluation of assessment of injury are considered. I did not happen to come across this piece of 'grey' literature until this monograph was almost completed. Thus the comparison with the systems selected for evaluation there and those discussed here was made after the event. Four acceptance criteria were used in this report: (1) is the response known to be the result of exposure to hazardous substances? (2) Is it known that this response is caused by hazardous substances in free-ranging organisms? (3) Is it known to cause this response in controlled experiments? (4) And is it practical to produce scientifically valid results using this method? It is interesting that there is no direct reference to harm to the organism resulting from these changes.

The first comparison that, as an author, one rushes to make is to see if one has covered those responses that are highly ranked. In this I was content, as all of the four responses that passed the criteria had been covered, although eggshell thinning has been considered as one of the best (in my personal, highly biased view, the best) examples of relating all the way from biochemical responses to population declines rather than as a specific response. When one gets to those responses that were considered to meet only two or three of the four criteria, the coverage in this monograph is, as one would expect, less consistent. Several of the more generalized responses listed in the report, e.g. adrenal and pituitary dysfunction, have been mentioned only in passing (section 3.3) and others, such as impaired renal function and thermoregulation, have not been considered here at all.

The reverse comparison, of what one has included but which has been excluded in a compilation by one's peers, is also of interest. Here, the main biochemical response that I have covered but that is not evaluated is that of the retinols. I would contend that it passed all four criteria, although I am not sure that the compilers of the USDI report would consider that it passed the first criterion. It is with this criterion that I have the greatest difficulty. In the notes in the report that support the decisions listed in the tables, there are statements, such as 'noncontaminant-related physiological factors can markedly affect the basal rate' and 'these functional changes are often evoked by a variety of other noncontaminant-related physiological factors', which are used to disqualify responses from acceptance. The difference on this criterion may well be a question of outlook. The report is looking at those responses that can be used legally to help define damage, whereas I am looking at those responses which can indicate the impact of environmental agents and which are thus worth including in a battery of biomarkers. Thus, my criterion is, 'can the biological variation be worked around?' rather than the criterion used in the report, which is, I think, 'are there serious natural confounding factors?' The definition of 'serious' is a difficult one, since all biological responses are affected to some extent by natural factors. I would argue yes for some clinical enzymes, and no for steroid hormone levels.

With these reservations in mind, I would like to look at the few disagreements that I have with the evaluation of biological responses for physiological malfunction injury for wildlife given in the report. This has been reproduced with only slight compression of wording in Table 10.4.

Table 10.4 Evaluation of biological responses to assess physiological injury caused by pollutants

	Response is often result of exposure to pollutant	Exposure to pollutant is known to cause response in free-ranging organisms	Exposure to pollutant is known to cause response in controlled experiments	Response is practical to measure	Reference in this monograph (section or chapter number)
1. Responses that meet all criteria					
Eggshell thinning	Yes	Yes	Yes	Yes	Table 10.1
Reduced avian reproduction	Yes	Yes	Yes	Yes	3
ChE inhibition	Yes	Yes	Yes	Yes	2
ALAD inhibition	Yes	Yes	Yes	Yes	6.3
2. Responses that met three criteria					
MFO induction	Yes	No	Yes	Yes	5
NTE inhibition	Yes	No	Yes	Yes	2
Thyroid dysfunction	No	Yes	Yes	Yes	6.1
Alteration in glutathione	No	Yes	Yes	Yes	5

Table 10.4 (cont.)

	Response is often result of exposure to pollutant	Exposure to pollutant is known to cause response in free-ranging organisms	Exposure to pollutant is known to cause response in controlled experiments	Response is practical to measure	Reference in this monograph (section or chapter number)
3. Responses that met two criteria					
Increased blood porphyrin	No*	No*	Yes	Yes	6.3
Alterations to neurotransmitter enzymes	*No*	No	Yes	Yes	2
Metallothionein induction	*No*	No	Yes	Yes	
Release of organ-specific enzymes	*No*	No	Yes	Yes	6.4.2
Reduced mammalian reproduction	No	Yes	Yes	No	3
ATP inhibition	No	No	Yes	Yes	6.4.4
Adrenal dysfunction	No	No	Yes	Yes	
Gonadal dysfunction	No	No	Yes	Yes	3

Pituitary dysfunction	No	No	Yes	Yes	
MFO inhibition	No	No	Yes	Yes	
Alterations to carbohydrate, lipid and protein metabolism	No	No	Yes	Yes	
Changes in DNA/RNA content or synthesis	No	No	Yes	Yes	4.2
Changes in basal metabolic rate	No	No	Yes	Yes	
Impaired thermoregulation	No	No	Yes	Yes	
Impaired intestinal transport	No	No	Yes	Yes	
Impaired renal function	No	No	Yes	Yes	
Clinical blood chemistry	No	No	Yes	Yes	7.4
Haematological alterations	No	No	Yes	Yes	

Proposed alterations to acceptance or otherwise are given in capitals
* Alterations are based on the use of liver rather than blood
After USDI (1987).

The acceptance of criteria is that given in the report. In the few cases that I disagree, the verdict is capitalized. The use of the single word verdict is something that tends to cause difficulties when the legal and scientific systems are involved with each other. Scientists are more comfortable in Scotland where the verdict 'non-proven' is allowed in addition to the stark 'guilty' and 'not guilty'.

The first response that I wish to discuss under this heading of upgrading its classification is that of induction of mixed function oxidases. Here, both evidence which should have been available to the panel (such as Kurelec *et al.*, 1977 and Spies *et al.*, 1982) and more recent work (such as that of Luxon *et al.*, 1987 and Andersson *et al.*, 1988) clearly demonstrate that there is good evidence to document the induction of mixed function oxidase activity in free-ranging wildlife. The use of MFO activity as a biomarker is discussed in section 5.5.7, and I do not think I would have any difficulty in persuading the committee on this point.

Looking at the other three responses that made a score of three out of four, I have no further disagreements with the verdicts. In the case of the inhibition of neurotoxic esterases, I do not feel that the verdict of 'No' on occurrence in free-ranging wildlife is likely to be changed. The test is useful to exclude those esterase inhibitors which cause neurological damage; on these grounds they are then excluded for widespread use as pesticides. In one sense this is the ideal test system, since it excludes substances which are likely to cause serious harm. Conversely, it is not useful as a biomarker since, by definition, compounds that cause NTE inhibition are excluded from use. The case of thyroid dysfunction is more controversial. Even within our own group at NWRC, you would not get agreement. Nevertheless, I side with the USDI report that the complexities of the responses caused by noncontaminant factors and the lack of clear relationships to pollutant loading give us a 'fail' on this response. The use of thyroid as a biomarker is discussed in more detail in section 6.1. Regarding the last of the responses that fail a single criterion, glutathione, it seems to be the case that it has not been studied in enough detail. Certainly Phase II of xenobiotic transformations has been much less studied than Phase I (induction of MFOs).

By far the largest category (18 out of 26) are those responses that meet only two of the four criteria. As the attorney for the response, I would plead for changes in only four of them. The first is for porphyrins. Here the biological response is listed as increased blood porphyrin concentration. If one accepts the inclusion of the word *blood*, then the grounds for my appeal fail. However, since the responses on the whole are not limited to those that can be carried out on blood (certainly MFO measurements are usually made in liver and NTE in brain, and neither in blood), it does not seem that the responses are limited to those that can be carried out non-invasively (chapter 9). If measurements in liver are accepted, then there is good evidence of increases caused by a range of PHAHs (especially HCB and certain congeners of PCBs; the evidence is summarized in section 6.3.2); the baseline levels of porphyrins are normally stable and increases have been demonstrated in herring gulls in the Great Lakes (Fox *et al.*, 1988). Thus, if the response is broadened to 'increased porphyrin concentration', I would argue for inclusion in category one. The case for retinols, not considered by the committee, is less strong since normal physiological mechanisms are a confounding factor. Nevertheless, recent work on the ratio of retinol to retinyl palmitate (section 6.2) indicates that there is a case to be argued.

The other three changes would not cause promotion to category one; only from category three to category two. In all three it is the first criterion, that the biological response is often the result of exposure to pollutants, with which I would argue. I have combined the two categories on alterations in neurotransmitters. Here I feel that there is evidence that biogenic amines are affected by a variety of pollutants (section 2.2) and that it is possible to measure these against the normal variation. However, I agree that this response fails the second criterion, and I have reservations that, in fact, it passes the last criterion – that of being practical to perform and producing scientifically valid results. Although the levels of biogenic amines can be readily measured, the problem of the lack of heterogeneity of brain tissue is a serious one. Studies in which the levels of amines are measured in half of the brain and the levels of environmental contaminants in the other half, although designed to allow for this heterogeneity, give an oversimplified view of the effect of the pollutant. Nevertheless, this is a system worth additional studies.

The second response in this category is that of release of organ-specific enzymes into blood. Here I would argue that changes in levels of transaminases (GOT and GPT) and lactic dehydrogenase (data are summarized in Tables 6.1 and 6.2 respectively) are quite adequate for inclusion, but the criterion of demonstration in free-ranging wildlife is not well documented. The incidence of elevated GPT activity in ducks (Dieter *et al.*, 1976) and GOT and GPT in fish (Weiser and Hinterleitner, 1980) was not definitely related to pollutant levels.

The third and last response to be considered is the induction of metallothionein. Here I would argue that the induction by heavy metals is proven and that it can be measured against the normal background variation. The exclusion of a consideration of metallothionein from this monograph was based on my view that it is easier to measure the metals themselves than the response of the induction of metallothionein.

In addition to physiological responses the report (USDI, 1988) also evaluated neoplasia injury to fish and wildlife, genetic mutation and behavioural effects. Both liver and skin neoplasia in fish were considered to pass all of the tests (section 4.7), but not in other groups of organisms. Neither of the genetic alterations – chromosome aberrations or sister chromatid exchange – was considered to meet the criteria. Both failed on the criterion

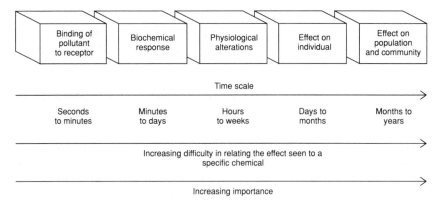

Figure 10.5 Linkages between biochemical, physiological, individual and population responses to pollutants.

of occurrence in free-ranging organisms. I would agree in both cases (sections 4.4 and 4.6). A considerable number of behaviour abnormalities are considered, and only one of the 'clinical behavioural signs of toxicity' is considered to have passed. The notes on this rather vague heading suggest that AChE-related effects are being considered. Certainly these are well documented (section 7.4). Otherwise I would agree that conclusive proof of behavioural effects is generally lacking.

Overall, my belated discovery of the existence of the CERCLA legislation and the report (USDI, 1987) on the use of physiological parameters to implement it have given me confidence. The agreement on the selection of biomarkers has been encouraging, but much more significant is the practical use to which biomarkers are being put in the tackling of environmental problems.

10.4.2 Extrapolation to harm at the population and community level

The basic objective of ecotoxicology is to ensure that the structure and function of the ecosystem are preserved. My version of the linkages between biochemical, physiological, individual and population responses is given in Figure 10.5. On the far right there are changes to the structure and function of ecosystems. The chasm that separates this from the rest of the diagram is too wide to jump or to get onto a reasonably sized piece of paper. The dilemma is that while the ultimate aim is to ensure that the ecosystem remains intact, most of the hard data lie at the opposite end.

To put things in more practical terms, let us apply the diagram to the reproductive failure of fish-eating birds in the North American Great Lakes (Figure 10.6). For me, it has the dual advantage of being one of the best-studied cases of the widespread impact of pollution and being one of which I have the best knowledge. It is an example where two initially entirely

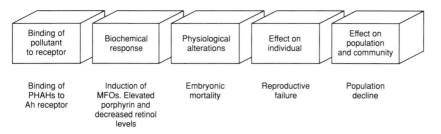

Figure 10.6 Linkages in the reproductive failure of fish-eating birds and the Ah receptor.

different lines of research – molecular biology and investigations by field biologists – eventually blended to provide a comprehensive answer to the problem.

The initial observations in the early 1970s (Gilbertson, 1975) were at the level of effects on the individual. Subsequent investigations showed serious reproductive impairment of the herring gull in Lake Ontario (Gilman *et al.*, 1977) and severe reproductive problems and widespread population declines of the double-crested cormorant (Price and Weseloh, 1986; Weseloh *et al.*, 1983).

The hypothesis that these severe effects over a wide area were caused by pollutants, notably the polyhalogenated aromatic hydrocarbons (PHAHs), was immediately put forward. It was based on a correlation between total PHAH levels and reproductive effects. The problem that was much harder to tackle was the relating of these effects to a specific chemical or chemicals. Apart from the difficulties inherent in tackling the effects of complex mixtures, there were two additional problems. First of all, the levels of most of PHAHs cross-correlated strongly to each other, and, second, the effects seen – embryotoxicity, oedema, structural abnormalities, behavioural changes – were known from laboratory studies to be caused by a wide range of PHAHs. Only in the case of eggshell thinning in cormorants was it possible to assign a specific cause with a high degree of certainty.

Detailed field studies, which included egg injection, egg exchange experiments and nest attentiveness studies, were undertaken (reviewed in Mineau *et al.*, 1984). These investigations were made more difficult by the then (late 1970s) rapidly decreasing levels of most PHAHs, following bans and restrictions on their use. All that could be established by the late 1970s was a generalized correlation with pollutant levels. With hindsight, the problem was that neither the analytical chemistry nor the environmental toxicology was sophisticated enough. The confirmation that the chlorinated dioxins (PCDDs) and chlorinated dibenzofurans (PCDFs) were present in the Great Lakes was not made until 1980 (Norstrom *et al.*, 1982). Isomer-specific analysis was soon developed and now our analytical chemical division routinely reports on the levels of 80 PHAHs compared to a dozen or so a decade ago. Similarly, although it had long been known that the number of isomers of PCBs was large, almost all the toxicological work was on the commercial mixtures. The work on specific isomers soon revealed large differences in their activity, and studies on structure-activity relationships were undertaken.

Meanwhile the molecular biologists were, without a thought of the impact their studies would eventually have on pollution problems, working on the isolation and characterization of receptors. The key finding which brought molecular biology into the realm of toxicology, the stereospecific, high-affinity binding of 2,3,7,8-TCDD to the Ah receptor (Poland *et al.*, 1976), has already been referred to in this chapter. The application of this complex biochemistry to field investigations has been made by means of expressing the complex mixtures of PHAHs as 'dioxin equivalents'. This concept is discussed in section 5.6.

Looking at the situation on the Great Lakes as we enter the 1990s, and I believe it is typical of problems elsewhere, there is good news and bad news. The good news is that the bans and restrictions put on pollutants, particularly the PHAHs, in the mid-1970s have resulted in a marked decrease of the levels of these compounds in the biota. The bad news is that the levels of these compounds are no longer decreasing, and levels have remained essentially constant during the last half of the 1980s (Figures 1.3 and 1.4). Adverse effects seen in fish-eating birds are now confined to a few specific areas, rather than occurring on a lake-wide basis as they had a decade before. Detailed studies on the reproductive failure of fish-eating birds in two highly polluted sites in the Great Lakes (Green and Saginaw bays) were undertaken in 1983 (Hoffman *et al.*, 1987; Kubiak *et al.*, 1989). Decreased reproductive failure, especially late embryonic death and increased congenital anomalies, was found. Elevated levels of AHH activity were found in double-crested cormorants and Caspian terns. The overall reproductive success of colonies of these species has been plotted against TCDD-equivalents of eggs from each colony; a high degree of correlation was found. The data for the Caspian tern are plotted in Figure 5.7. Besides these 'hot-spots', we are faced with determining the chronic effects of the complex mixture of pollutants that remain throughout the Great Lakes ecosystem.

10.5 Strategy for using biomarkers

> Between two hawks, which flies the higher pitch,
> Between two girls, which hath the merrier eye,
> I have, perhaps, some shallow spirit of judgement.
> But in these nice sharp quillets of the law,
> Good faith, I am no wiser than a daw.

> Henry IV Part 1
> Act 2, Scene 4

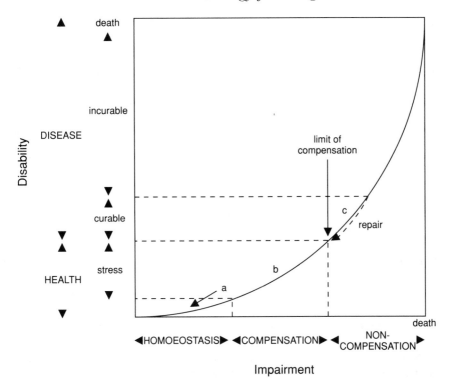

Figure 10.7 Impairment versus disability as indicators of pollutant toxicity. After Depledge (1989).

Recently Depledge (1989) has put forward a basis for using physiological indicators to detect the early effects of pollutants, with particular reference to heavy metals. He relates impairment to disability diagrammatically (Figure 10.7). Initially there is compensation by the activation of excretory and detoxification mechanisms until the limit of compensation is reached. Beyond this point physiological and behavioural processes fail.

There is a major difficulty and an unresolved problem in Depledge's approach. The difficulty is that while we can determine what this critical point is for single metals, it cannot be done for the almost infinite number of complex mixtures of pollutants that occur. Depledge's diagram is quite similar to mine of the dose response for AChE, given in Figure 10.3. I think that Depledge's three categories – homoeostasis, compensation and non-compensation – are essentially similar to my zones of normal variation, reversible effects and irreversible effects. Depledge's thesis seems to be – it is not explicitly stated – that the most critical thing is that the point marked 'the limit of compensation' not be passed. Despite the scientific soundness, there is a major difficulty and an unresolved problem in Depledge's approach from the practical viewpoint of deciding when remedial action is needed.

The difficulty is that, while we can determine what this critical point of the limit of compensation is for single compounds or metals, it cannot be done for the almost infinite number of complex mixtures of pollutants that occur in the environment. At the moment, the data are available only in a few well-studied cases for single pollutants to put an actual numerical value on this critical point. The problems involved in relating these changes to effects on populations or possible compensation due to genetic changes are even more complex.

The unresolved problem for the application of any assessment criteria is, how large an area are we prepared to pollute? In terms of metals in the marine environment, for example, are we prepared to tolerate some effects of organotin in the immediate area of the slip? This is an area already profoundly altered by man and almost certainly polluted with other compounds such as oil.

The concept of an 'exclusion zone' is tricky. It is difficult to see how an impartial scientific basis for it could be established, but it is equally difficult to see how pollution control can proceed without it. Take the example of an industrial port. One cannot expect the entire area of the port to maintain the same populations of the same animals under the same physiological conditions as a pristine area. Nor, on the other hand, is it acceptable for the port to pollute the entire estuary and nearby lake or ocean.

To come up with a solution, one has to define both the 'end-point' and the 'exclusion zone.' By using a combination of these two factors, I would like to suggest that we can use what is by environmental standards a rigorous and reasonably practical end-point. The end-point suggested here is 'that the physiological functions of organisms, outside the exclusion zone, should be within normal limits'. This approach allows for the natural variation of these functions. The actual limits would depend on these variations and would be expected to be similar to the approach put forward to assess exposure to OPs, i.e. 20% or two standard deviations of mean AChE activity (Ludke *et al.*, 1975).

The major advantages of the approach are:

1. It is, at least initially, independent of the pollutants involved and thus avoids the problem of mixtures and unknown substances. Only if the studies outside the exclusion area reveal abnormalities will detailed investigations be needed.
2. It does not require, again at least initially, proof that the effect seen is deleterious. If the area over which the effect is seen is large, it may be argued that such proof is necessary, but it is not a basic precondition.
3. Philosophically, it is a defensible position. Although pollutants are present, and analytical chemistry is too good for zero to be an objective, the functioning of the animals living in the area is normal.

There are, naturally, limitations in this approach, the most important being:

1. It implies that we have a good enough battery of tests to be confident that the physiology is indeed normal and that this battery of tests covers all the major classes of pollutants.
2. A difficult question to answer is which and how many species should be tested. However, the problem of inter-species differences exists with all other approaches.
3. It does not tackle the question, 'Is harm caused by the abnormal physiological state?'

The question of what is physiologically normal is a critical part of the approach. In many areas the cleanest material comes from the marine environment. In the case of biomarkers involving the thyroid, the fact is that this is not a true control, because the marine environment is iodide rich. But other, more subtle, differences may occur between marine and fresh-water environments. Thus, I would argue the need for the controls to come from an environment as physiologically similar as possible. As far as contamination is concerned, it is not only that it is not feasible to find control material with zero contamination but that it may not even be desirable. In practical terms broad areas of low-level contamination are now the norm, and the argument can be made that, since this is the best that we can expect in an industrialized world, this should be the control.

In using such material as the control, there is an implication of acceptance of this level of pollution. Pragmatically, there seems little option. It would be valuable to determine the biological variation of biomarker measurements under ideal (i.e. lowest contamination) conditions, not as a working control but to enable us to know how much our practical baseline has been altered from ideality.

The exclusion zone/physiological normality approach is a mixture of politics and science. This mixture, although difficult, is an inevitable part of finding the solution to environmental problems. An obvious problem occurs if the 'exclusion zone' is too large. After I had proposed the concept at a lecture at Wageningen, my old friend Professor Jan Koeman said that the trouble with the approach was that the whole of The Netherlands would be an exclusion zone. At the International Ornithology Congress in Ithaca, New York, in 1962, I took some Dutch visitors to the Adirondack Park in upstate New York. They asked about the size. When I told them it was a million hectares, they exclaimed, 'But that is bigger than Holland.' While in actual fact this is not correct, it does illustrate the difference in the concept of size between the Old and New Worlds.

The criticism that Koeman raised is crucial. Once the exclusion zone becomes large, then the pressure to question the harm caused by the physiological changes increases. At the moment we do not have the body of data on biomarkers, systematically collected over wide areas, that will answer this question. In our own laboratories we are undertaking a programme that will obtain these data for the Great Lakes and other areas of eastern Canada. Venturing to 'look into the seeds of time and say which grain will grow and which will not', I think that it is likely that we shall find some physiological abnormalities in top predator birds in considerable areas of the Great Lakes basin. I would argue that this gives a case for further clean-up rather than a case for a large exclusion zone. This would agree with the fact that there are fish advisories in place over broad areas of the lakes. If these two independent, although both somewhat arbitrary, approaches give the same answer, it will add considerable weight to the conclusion. If biomarkers, over a range of species, are outside the normal biological variation over wide areas, e.g. the whole of the Netherlands, we shall have to rethink the approach.

Certainly I would agree with the conclusion of a recent conference on biomarkers that 'Current understanding and application of biomarkers justifies their immediate implementation in an environmental program at a pilot-scale. However, the full potential of this methodology will be realized only after a larger data base of field and laboratory studies can be accumulated and analyzed' (McCarthy, 1990). Eventually the database will, hopefully, also be available to prove the harm of these changes to the organism and even to populations and communities, but until that time I would venture to suggest that the concept of physiological normality coupled with an exclusion zone is the most practical approach.

Appendix 1
Latin names of species referred to in text

The Latin names given are those used by the authors of the paper cited. These have been checked against standard texts, but no attempt has been made to bring the list up to date taxonomically. Similarly, the English names used are those used by the author even if it is clear that a name does not completely define the species, i.e. fox. Only if two different authors use the same common name for different species are more precise names used rather than sticking with the name used by the original author. However, major differences in the English names between European and North American usage, e.g. guillemot and murre, are noted. In the text the name 'quail' refers to the Japanese quail (*Coturnix coturnix*) unless otherwise stated. The bobwhite quail is referred to as the bobwhite. Similarly, the name 'kestrel' is used for the American kestrel unless otherwise stated.

Mammals

Quokka (*Setonix brachyurus*)
Short-tailed shrew (*Blarina brevicauda*)
Rhesus monkey (*Macaca mulatta*)
Mole (*Talpa europaea*)
Grey squirrel (*Sciurus carolinensis*)
White-footed mouse (*Peromyscus leucopus*)
Harvest mouse (*Reithrodontomys megalotis*)
Cotton rat (*Sigmodon hispidus*)
Chinese hamster (*Cricetulus griseus*)
Golden hamster (*Mesocricetus auratus*)
House mouse (*Mus musculus*)
Guinea pig (*Cavia aperea*)
Beluga whale (*Delphinapterus leucas*)
Ermine (*Mustela erminea*)
Mink (*M. vison*)

Polecat or ferret (*M. putorius*)
Badger (*Meles meles*)
Otter (*Lutra lutra*)
Polar bear (*Thalarctos maritimus*)
Fox (*Vulpes vulpes*)
Cat (*Felis domesticus*)
Common or harbour seal (*Phoca vitulina*)
Rabbit (*Oryctolagus cuniculus*)
Pig (*Sus scrofa*)
Roe deer (*Capreolus capreolus*)
Mule deer (*Odocileus hemionus*)
Goat (*Capra aegagrus*)

Birds

Wedge-tailed shearwater (*Puffinus pacificus*)
Leach's storm-petrel (*Oceanodroma leucorhoa*)
Brown pelican (*Pelecanus occidentalis*)
Double-crested cormorant (*Phalacrocorax auritus*)
Great egret (*Casmerodius albus*)
Snowy egret (*Leucophoyx thula*)
Black-crowned night heron (*Nycticorax nycticorax*)
Mute swan (*Cygrus olor*)
Canada goose (*Branta canadensis*)
White-fronted goose (*Anser albifrons*)
Mallard (*Anas platyrhynchos*)
Canvasback duck (*Aythya valisineria*)
Sparrowhawk (*Accipiter nisus*)
Bald eagle (*Haliaeetus leucocephalus*)
Osprey (*Pandion haliaetus*)
Peregrine falcon (*Falco peregrinus*)
Prairie falcon (*F. mexicanus*)
Merlin (*F. columbarius*)
Kestrel, or American kestrel (*Falco sparverius*)
Quail, or Japanese quail (*Coturnix coturnix*)
Burrowing owl (*Athene cunicularia*)
Chukar (*Alectoris graeca*)
Bobwhite quail (*Colinus virginianus*)
California quail (*Lophortyx california*)
Sharp-tailed grouse (*Pediocetes phasianollus*)
Grey partridge (*Perdix perdix*)
Chicken (*Gallus domesticus*)
Pheasant (*Phasianus colchicus*)
Turkey (*Meleagris gallopavo*)

Glaucous-winged gull (*Larus glaucescens*)
Lesser black-backed gull (*L. fuscus*)
Western gull (*L. occidentalis*)
Herring gull (*L. argenatatus*)
California gull (*L. californicus*)
Ring-billed gull (*L. delawarensis*)
Red-billed gull (*L. novaehollandiae*)
Laughing gull (*L. atricilla*)
Common tern (*Sterna hirundo*)
Forster's tern (*S. fosteri*)
Caspian tern (*Hydroprogne caspia*)
Common guillemot, or murre (*Uria aalge*)
Black guillemot (*Cepphus grylle*)
Puffin (*Fratercula arctica*)
Feral pigeon/Rock dove (*Columba livia*)
Ring-dove (*Streptopelia risoria*)
Barn swallow (*Hirundo rustica*)
Redstart (*Phoenicurus phoenicurus*)
Starling (*Sturnus vulgaris*)
House sparrow (*Passer domesticus*)
Red-billed quelea (*Quelea quelea*)
Red-winged blackbird (*Agelaius phoeniceus*)
Common grackle (*Quiscalus quiscula*)
White-throated sparrow (*Zonotrichia albicollis*)
Bengalese finch (*Lonchura domestica*)
Zebra finch (*Taeniopygia guttata*)

Amphibians

Southern leopard frog (*Rana sphenocephala*)
Grey tree frog (*Hyla chrysoscelis*)
Hensel toad (*Bufo arenarum*)

Fish

White sucker (*Catostomus commersoni*)
Sculpin (*Cottus gobio*)
Largemouth bass (*Micropterus salmoides*)
Bluegill, or bluegill sunfish (*Lepomis macrochirus*)
Scup (*Stenotomus chrysops*)
Striped mullet (*Mugil cephalus*)
Flagfish (*Jordanella floridae*)
Sailfin mollie (*Poecilia latipinna*)
Guppy (*P. reticulata*)

Cod (*Gadus morhua*)
Eel (*Anguilla rostrata*)
Carp (*Cyprinus carpio*)
Fathead minnow (*Pimephales promelas*)
Golden shiner (*Notemigonus crysoleucas*)
Catfish (*Ictalurus punctatus*)
Brown bullhead (*I. nebulosus*)
Herring (*Clupea harengus*)
Flounder (*Platichys flesus*)
Starry flounder (*P. stellatus*)
Winter flounder (*Pseudopleuronectes americanus*)
English sole (*Paraphrys vetulus*)
Rainbow trout (*Salmo gairdneri*)
Lake trout (*S. trutta*)
Brook trout (*Salvelinus fontinalis*)
Brook char (*S. alpinus*)
Coho salmon (*Oncoryhnchus kisutch*)
Pink salmon (*O. gorbuscha*)
Vendace (*Coregonus albula*)
Pike (*Esox lucius*)
Mudminnow (*Umbra limi*)
Spot (*Leiostromus xanthurus*)
Killifish (*Fundulus heteroclitus*)

Bivalves

Mussel (*Mytilus edulis*)

Crustaceans

Grass shrimp (*Palaemonetes pugio*)
Lobster (*Homarus americanus*)
Crayfish (*Astacus astacus*)
Fiddler crab (*Uca pugnax*)

Appendix 2
Abbreviations

Abate	O,O,O′,O′-tetramethyl O,O′-thiodi-p-phenylene bis phosphorothioate
Acephate	N-[methyl (methylthio)phosphinoxyl] acetamide
AChE	Acetylcholinesterase
ADP	Adenosine diphosphate
AEC	Adenylate energy charge
AHH	Aryl hydrocarbon hydroxylase
Aldicarb	2-methyl-2-(methylthio)propionaldehyde O-methyl-carbamyloxime
ALAD	Aminolevulinic acid dehydratase
ALAS	Aminolevulinic acid synthesase
ALP	Alkaline phosphatases
Aroclor	Mixtures of polychlorinated biphenyls of different degree of chlorination
Aminocarb	4-dimethylamino-m-tolyl methylcarbamate
AMP	Adenosine monophosphate
APC	Antigen-presenting cells
ASAL	Argininosuccinate lyase
ATPase	Adenosine triphosphatase
BaP	Benzo[a]pyrene
BaPH	Benzo[a]pyrene hydroxylase
B-cells	Bone marrow cells
BChE	Butylcholinesterase
Bidrin	O-(2, 4-dichlorophenyl) O-ethyl S-propyl
BRMs	Biological response modifiers
Carbaryl	1-napthalenyl methylcarbamate
Carbofuran	2,3-dihydro-2,2-dimethyl-7-benzofuranol methyl-carbamate
Carbophenothion	S[[(4-chlorophenyl)thio]methyl] O.O-diethyl phosphoro-dithioate
ChE	Cholinesterase

CHI	Cell mediated immunity
Chlorfenvinphos	2-chloro-1-(2,4-dichlorophenyl) ethenyl diethyl phosphate
CMI	Cell-mediated immunity
COMT	Catechol-O-methyltransferase
COPRO	Coproporphyrinogen
CPS	Cyclophosphamide
CWS	Canadian Wildlife Service
Cypermethrin	Cyano (3-phenoxyphenyl) methyl 3-(2,2-dichloroethenyl)-2,2-dimethylcyclopropanecarboxylate
DA	Dopamine
DDA	2,2-bis(p-chlorophenyl)acetic acid
DDD	2,2-bis(p-chlorophenyl)-1,1-dichloroethane
DDE	2,2-bis(p-chlorophenyl)-1,1-dichloroethylene
ddRA	3,4,didehydroretinoic acid
DDT	1,1,1-trichloro-2,2-bis (chlorophenyl) ethane
Deltamethrin	cis-3-(2,2-dibromvinyl)-2,2-dimethylcyclopropane carboxylate
Demeton	O,O-diethyl S-2-ethylthioethyl phosphorothioate
Diazinon	O,O-diethyl O-[6-methyl-2-(1-methylethyl)-4-pyrimidinyl] phosphorothioate
Dicofol	2,2,2-trichloro-1-,1-di-(4-chlorophenyl) ethanoil
Dicrotophos	3-hydroxy-N, N-dimethyl-cis-scrotonamide dimethyl phosphate
Dichlorvos	2,2-dichloroethenyl dimethyl phosphate
Dieldrin	1a,2,2a,3,6, 6a,7,7a-octahydro-2,7:3,6-dimethanonapth [2,3-b] oxirene
Dimethoate	O,O-dimethyl S-[2-(methylamino)-2-oxoethyl] phosphorothioate
DNA	Deoxyribonucleic acid
DOPA	Dihydroxyphenylalanine
DOPAC	3,4, dihydroxy phenylacetic acid
EC_{50}	Effective concentration for 50% of population
ECOD	7-ethyoxycoumarin-O-deethylase
ED_{50}	Effective dose for 50% of the population
ELISA	Enzyme-linked immunosorbent assays
Endrin	1a,2,2a,3,6,6a,7,7a-octahydro-2,7:3,6-dimethanonapth [2,3-b] oxirene
EPN	Phenyl phosphonthioic acid-O-ethyl-O-[4-nitrophenyl] ester
EROD	7-ethoxyresorufin O-deethylase
Famphur	O-[4-[(dimethylamino)sulphonyl] phenyl] O,O-dimethyl phosphorothioate
Fenitrothion	O,O-dimethyl O-(3-methyl-4-nitrophenyl) phosphorothioate

Fenthion	O,O-dimethyl O-[3-methyl-4-(methylthio)phenyl] phosphorothioate
Fenvalerate	Cyano(3-phenoxyphenyl)methyl 4-chloro-a-(1-methylethyl) benzeneacetate
FSH	Follicle-stimulating hormone
GABA	Gamma aminobutyric acid
G6P	Glucose-6-phosphatase
G6PD	Glucose-6-phosphatase dehydrogenase
GDH	Glutamate dehydrogenase
GHRH	Gonadotrophic hormone-releasing hormone
GOT	Glutamic oxaloacetic transaminase
GPT	Glutamic pyruvate transaminase
HBBs	Hexabromo biphenyls
HCB	Hexachlorobenzene
HCP	Highly carboxylated porphyrins
HMI	Humoral mediated immunity
Leptophos	O-(4-bromo-2,5-dichlorophenyl) O-methyl phenylphosphonothioate
Leptophos-oxon	Oxon derivative & leptopos
LD_{50}	Lethal dose for 50% of population
LDH	Lactic dehydrogenase
LH	Luteinizing hormone
Lindane	1',2',3',4',5',6'-1,2,3,4,5,6-hexachlorocyclohexane
LOEC	Lowest-observed-effect concentration
Kepone	1,1a,3,3a,4,5,5,5a,5b,6-decachloro-octahydro-1,3,4-metheno-2H-cyclobuta[cd]penalen-2-one
Krenite	Ethyl hydrogen (aminocarbonyl) phosphonate
MAB	Monoclonal antibodies
Malathion	Diethyl [(dimethyloxyphosphinothioyl)thio] butanedioate
Malaoxon	Oxon derivative & malathion
MAO	Monoamine oxidase
MC	3-methylcholanthrene
mCBR	Muscarinic cholinergic receptor
Methidathion	S-[(5-methoxy-2-oxo-1,3,4-thiadiazol-3(2H)-yl)methyl] O,O-dimethyl phosphorodithioate
Methyl malathion	Dimethyl [(dimethyloxyphosphinothioyl)thio] butanedioate
Methyl parathion	O,O-dimethyl O-4-(4-nitrophenyl) phosphorothioate
Methyl pirimiphos	O-[2-(diethylamino)-6-methyl-4-pyrimidinyl] O,O-dimethyl phosphorothioate
MFO	Mixed function oxidases
MHPG	3-methoxy-4-hydroxylphenylalanine
Mirex	1,1a,2,2,3,3a,4,5,5,5a,5b,6-dodecachloro-octahydro 1,3–4-methene-1 H=cyclobuta[cd]pentalene

NE	Norepinephrine (noradrenaline)
NK	Natural killer cells
NSR	Nonspecific elimination
NTE	Neurotoxic esterase or neuropathy target esterase
OCs	Organochlorines
OPIDN	Organophosphorous compound-induced delayed neurotoxicity
OPs	Organophosphates
Oxyamyl	N,N-dimethyl-*-[[9 methylcarbmoyl)oxy]imino]-*-(methylthio) acetamide
PAH	Polynuclear aromatic hydrocarbons
Paraoxon	O,O-diethyl-O-(4-nitrophenyl) phosphate
Paraquat	1,1'-dimethyl-4,4'-biphridyldiylium
Parathion	Phosphorthioic acid O,O-diethyl O-(4-nitro-phenyl)ester
PB	Phenobarbital
PBB	Polybrominated biphenyl
PBCO	Prudhoe Bay crude oil
PCBs	Polychlorinated biphenyls
PCDDs	Polychlorinated dibenzodioxins
PCDFs	Polychlorinated dibenzodioxins
PCP	Pentachlorophenol
Permethrin	(3-phenoxyphenyl)methyl 3-(2,2-dichloroethylenyl)-2,2-dimethyl cyclopropane-carboxylate
PHAH	Polyhalogenated aromatic hydrocarbons
Phosphamidon	2-chloro-3-(diethylamino)-1-methyl-3-oxo-1-propenyl dimethyl phosphate
Photomirex	8-monohydromirex
PMN	Polymorphonuclear cells
Prochloraz	N-propyl-N-[2-(2,4,6-trichlorophenoxy)-1-H-ethyl] imidazole-1-carboxamide
RBP	Retinol-binding protein
RNA	Ribonucleic acid
SCE	Sister chromatid exchange
SCOPE	Scientific Committee on Problems in the Environment
SKF-525A	2-dimethylaminoethyl 2,2-diphenyl valerate
Sulprofos	O-ethyl O-[4-(methylthio)phenyl]S-propyl phosphoro-dithioate
T_3	Thyroxine
T_4	Triiodothyronine
TBPA	Thyroxine-binding prealbumin
TBTO	Tributyl tin oxide
TCB	3,4,3',4'-tetrachlorobiphenyl
TCDD	2,3,7,8-tetrachlorodioxin
T-cells	Thymus cells

Thiofanox	3,3-dimethyl-1-(methylthio)-2-butanone O-[(methyl-amino)carbonyl]oxime
Toxaphene	Mixture of chlorinated camphenes containing 67–69% chlorine
TRH	Thyrotrophin-releasing hormone
TSH	Thyroid-stimulating hormone
UDPGT	UDPglucuronosyltransferase
URO	Uroporphyrin
UROD	Uroporphyrin decarboxylase
Warfarin	4-hydroxy-3-(3-oxo-1-phenylbutyl)-2H-1-benzopyran-2-one
2,4-D	2,4,-dimethylammonium
5-HIAA	5-hydroxyindleacetic acid
5-HT	5-hydroxyltryptamine (serotonin)
5-MeC	5-methylcytosine

References

By jove, I am not covetous for gold,
Nor care I who doth feed upon my cost:
It yearns me not if men my garments wear:
Such outward things dwell not in my desires.
But if it be a sin to covet honour
I am the most offending soul alive.
Henry IV, Act 4, Scene 3

Abiola, F., Lorgue, G., Benoit, E., Soyez, D. and Riviere, J.L. (1989) Effects of PCBs on plasma enzymes, testosterone levels, and hepatic xenobiotic metabolism in the Grey Partridge, *Perdix perdix. Bull. Environ. Contam. Toxicol.*, **43**, 473–80.

Acosta, D., Anuforo, D.C. and Smith, R.V. (1978) Primary monolayer cultures of postnatal rat liver cells with extended differentiated functions. *In Vitro*, **14**, 428–36.

Acosta, D., Sorensen, E.M.B., Anuforo, D.C., Mitchell, D.B., Ramos, K., Santone, K.S. and Smith, M.A. (1985) An in vitro approach to the study of target organ toxicity of drugs and chemicals. *In Vitro Cell. Dev. Biol.*, **21**, 495–504.

Adams, S.M., Burtis, C.A. and Beauchamp, J.J. (1985) Integrated and individual biochemical responses of rainbow trout (*Salmo gairdneri*) to varying durations of acidification stress. *Comp. Biochem. Physiol.*, **82C**, 301–10.

Addison, R.F., Zinck, M.E. and Willis, D.E. (1981) Time- and dose-dependence of hepatic mixed function oxidase activity in brook trout *Salvelinus fontinalis* on polychlorinated biphenyl residues: Implications for biological effects monitoring. *Environ. Pollut. (Ser. A)*, **25**, 211–18.

Ahokos, J.T., Karki, N.T., Oikari, A. and Soivio, A. (1976) Mixed function mono-oxygenase of fish as an indicator of pollution of aquatic environment by industrial effluent. *Bull. Environ. Contam. Toxicol.*, **16**, 270–4.

Akoso, B.T., Sleight, S.D., Nachreiner, R.F. and Aust, S.D. (1982) Effects of pu-rified polybrominated biphenyl congeners on the thyroid and pituitary glands of rats. *J. Am. Coll. Toxicol.*, **1**, 23–36.

Aldous, C.N., Farr, C.H. and Sharma, R.P. (1982) Effects of Leptophos on rat brain levels and turnover rates of biogenic amines and their metabolism. *Ecotoxicol. Environ. Saf.*, **6**, 570–6.

Aldridge, W.N. (1953) Serum esterases. 1. Two types of esterase (A and B) hy-drolysing p-nitrophenyl acetate, propionate and butyrate and a method for their determination. *Biochem. J.*, **53**, 110–17.

Ali, S.F., Chandra, O. and Hasan, M. (1980) Effects of an organophosphate (Dichlorvos) on open field behavior and locomotor activity: Correlation with regional brain monoamine levels. *Psychopharmacology*, **68**, 37–42.

Anderson, D.W. and Hickey, J.J. (1972) Eggshell changes in certain North American birds. *Proc. Int. Ornithol. Congr.*, **15**, 514–40.

Andersson, T., Forlin, L., Hardig, J. and Larsson, A. (1988) Physiological disturbances in fish living in coastal water polluted with bleached kraft pulp mill effluents. *Can. J. Fish Aquat. Sci.* **45**, 1525–36.

Andre, F., Gillon, F., Andre, C., Lafont, S. and Jourdan, G. (1983) Pesticide-containing diet augments anti-sheep red blood cell non-reaginic antibody responses in mice but may prolong murine infection with *Giardia muris*. *Environ. Res.*, **32**, 145–50.

Aoyama, K. (1986) Effects of benzene inhalation on lymphocyte subpopulations and immune response in mice. *Toxicol. Appl. Pharmacol.*, **85**, 92–101.

Aranyi, C., Bradof, J.N., O'Shea, W.J., Graham, J.A. and Miller, F.J. (1985a) Effects of arsenic trioxide inhalation exposure on pulmonary antibacterial defenses in mice. *J. Toxicol. Environ. Health*, **15**, 163–72.

Aranyi, C., O'Shea, W.J., Sherwood, R.L., Graham, J.A. and Miller, F.J. (1985b) Effects of toluene inhalation on pulmonary host defenses of mice. *Toxicol. Lett.*, **25**, 103–10.

Aranyi, C., O'Shea, W.J., Graham, J.A. and Miller, F.J. (1986) The effects of inhalation of organic chemical air contaminants on murine lung host defenses. *Fundam. Appl. Toxicol.*, **6**, 713–20.

Arcos, J.C., Conney, A.H. and Buu-Hoi, N.P. (1961) Induction of microsomal enzyme synthesis by polycyclic aromatic hyrocarbons of different molecular sizes. *J. Biol. Chem.*, **236**, 1291–6.

Astroff, B. and Safe, S. (1990) 2,3,7,8-tetrachlorodibenzo-p-dioxin as an antiestrogen: Effect on rat uterus peroxidases activity. *Biochem. Pharmacol.*, **39**, 485–8.

Asztalos, B., Nemcsok, J., Benedeczky, I., Gabriel, R., Szabo, A. and Refaie, O.J. (1990) The effects of pesticides on some biochemical parameters of carp (*Cyprinus carpio* L.). *Arch. Environ. Contam. Toxicol.*, **19**, 275–82.

Atchison, G.J., Henry, M.G. and Sandheinrich, M.B. (1987) Effects of metals on fish behavior: A review. *Environ. Biol. Fish*, **18**, 11–25.

Atkinson, D.E. and Walton, G.M. (1967) Adenosine triphosphate conservation in metabolic regulation. Rat liver citrate cleavage enzyme. *J. Biol. Chem.*, **242**, 3239–41.

Aulerich, R.J., Bursian, S.J., Breslin, W.J., Olson, B.A. and Ringer, R.K. (1985) Toxicological manifestations of 2,4,5,-2′,4′,5′-,2,3,6,2′,3′,6′-, and 3,4,5,3′,4′,5′-hexachlorobiphenyl and Aroclor 1254 in mink. *J. Toxicol. Environ. Health*, **15**, 63–79.

Aulerich, R.J., Bursian, S.J., Evans, M.G., Hochstein, J.R., Koudele, K.A., Olson, B.A. and Napolitano, A.C. (1987) Toxicity of 3,4,5,3′,4′,5′-hexachlorobiphenyl to mink. *Arch. Environ. Contam. Toxicol.*, **16**, 53–60.

Aulerich, R.J., Bursian, S.J. and Napolitano, A.C. (1988) Biological effects of epidermal growth factor and 2,3,7,8-tetrachlorodibenzo-p-dioxin on developmental parameters of neonatal mink. *Arch. Environ. Contam. Toxicol.*, **17**, 27–31.

Aulerich, R.J. and Ringer, R.K. (1977) Current status of PCB toxicity to mink, and effect on their reproduction. *Arch. Environ. Contam. Toxicol.*, **6**, 279–92.

Azais, V., Arand, M., Rauch, P., Schramm, H., Bellenand, P., Narbonne, J.-F., Oesch, F., Pascal, G. and Robertson, L.W. (1987) A time-course investigation of vitamin A levels and drug metabolizing enzyme activities in rats following a single treatment with prototypic polychlorinated biphenyls and DDT. *Toxicology*, **44**, 341–54.

Balazs, T. and Ferrans, V.J. (1978) Cardiac lesions induced by chemicals. *Environ. Health Perspect.*, **26**, 181–91.

Ball, J.K. (1970) Immunosuppression and carcinogenesis: Contrasting effects with 7,12-dimethylbenz[a]anthracene, benz[a]pyrene, and 3-methylcholanthrene. *J. Natl. Canc. Inst.*, **44**, 1–10.

Bandiera, S., Sawyer, T., Romkes, M., Zmudka, B., Safe, L., Mason, G., Keys, B. and Safe, S. (1984) Polychlorinated dibenzofurans (PCDFs): Effects of structure on binding to the 2,3,7,8-TCDD cytosolic receptor protein, AHH induction and toxicity. *Toxicology*, **32**, 131–44.

Barnes, D.W., Sirbasku, D.A. and Sato, G.H. (1984) *Cell Culture Methods for Molecular and Cell Biology: Methods for Preparation of Media, Supplements, and Substrata for Serum-free Animal Cell Culture.* Four volumes. A.R. Liss, New York.

Barron, M.G. and Adelman, I.R. (1984) Nucleic acid, protein content, and growth of larval fish sublethally exposed to various toxicants. *Can. J. Fish. Aquat. Sci.*, **41**, 141–50.

Batchelor, J.R., Welsh, K.I., Mansilla-Tinoco, R., Dollery, C.T., Hughes, G.R.V., Bernstein, R., Ryan, P., Maish, P.F., Aber, G.M., Bing, R.F., and Russel, G.I. (1980) Hydralazine-induced systemic lupus erythematosus: influence of HLA-DR and sex on susceptibility. *Lancet 1*, 1107–9.

Baumann, P.C., Smith, W.D. and Parland, W.K. (1987) Tumor frequencies and contaminant concentrations in brown bullheads from an industrialized river and a recreational lake. *Trans. Am. Fish. Soc.*, **116**, 79–86.

Bekesi, J.G., Anderson, H.A., Roboz, J.P., Roboz, J., Fischbein, A., Selikoff, I.J. and Holland, J.F. (1979) Immunologic dysfunction among PBB-exposed Michigan dairy farmers. *Annals N.Y. Acad. Sci.*, **320**, 717–28.

Bekesi, J.G., Roboz., J.P., Fishbein, A. and Selikoff, G.I., 1987 Clinical immunology studies in individuals exposed to environmental chemicals. In: *Immunotoxicology.* (Eds A. Berlin. J. Dean. M.H. Draper, E.M.B. Smith and F. Spreafico), pp. 347–61. Martinus Nijhoff, Dordrecht.

Bellanti, J.A. (1987) Summary, findings and recommendations. In: Environmental Chemical Exposures and Immune System Integrity. *Adv. Modern Environ. Technol.*, **13**, ix–xiv.

Bellrose, F.C. (1959) Lead poisoning as a mortality factor in waterfowl populations. *Ill. Nat. Hist. Surv. Bull.*, **27**, 235–88.

Bennett, E.L., Rosenzweig, M.R., Krech, D., Karlsson, H., Dye, N. and Ohlander, A. (1958) Individual, strain and age difference in cholinesterase activity of the rat brain. *J. Neurochem.*, **3**, 144–152.

Bernes, C., Giege, B., Johansson, K. and Larsson, J.E. (1986) Design of an integrated monitoring programme in Sweden. *Environ. Monit. Assess.*, **6**, 113–26.

Bernier, J., Rola-Pleszczynski, M., Flipo, D., Krzystyniak, K. and Fournier, M. (1990) Immunotoxicity of Aminocarb. II. Evaluation of the effects of sublethal exposure to aminocarb on bone marrow cells by flow cytometry. *Pest. Biochem. Physiol.*, **36**, 35–45.

Biessmann, A. (1982) Effects of PCBs on gonads, sex hormone balance and reproductive processes of Japanese Quail *Coturnix coturnix japonica* after ingestion during sexual maturation. *Environ. Pollut.*, **27A**, 15–30.

Bigazzi, P.E. (1985) Mechanisms of chemical-induced autoimmunity. In: *Immunotoxicology and Immunopharmacology* (Eds J.H. Dean, M.I. Luster, A.E. Munson, H. Amos), pp. 277–90. Raven Press, New York.

Bird, D.M., Peakall, D.B. and Miller, D.S. (1983) Enzymatic changes in the oviduct associated with DDE-induced eggshell thinning in the Kestrel, *Falco sparverius*. *Bull. Contam. Environ. Toxicol.*, **31**, 22–4.

Birge, W.J., Black, J.A. and Westerman, A.G. (1979) Evaluation of aquatic pollutants using fish and amphibian eggs as bioassay organisms. In: *Animals as Monitors of Environment Pollution* (Eds S.W. Neilsen, G. Magaki and D.G. Scarpelli), pp. 108–18. National Academy of Science.

Bitman, J. and Cecil, H.C. (1970) Estrogenic activity of DDT analogs and polychlorinated biphenyls. *Agr. Food Chem.*, **18**, 1108–12.

Bitman, J., Cecil, H.C., Harris, S.J. and Fries, G.F. (1968) Estrogenic activity of o, p'-DDT in the mammalian uterus and avian oviduct. *Science*, **162**, 371–2.

Black, J. (1983) Field and laboratory studies of environmental carcinogenesis in Niagara River fish. *J. Great Lakes Res.*, **9**, 326–34.

Black, J.J. (1988) Fish tumors as known field effects of contaminants. In: *Toxic Contamination of Large Lakes. Vol. 1.* (Ed. N.W. Schmidtke), pp. 55–81.

Black, C.M., Walker, A.E. and Catoggio, J.L. (1983) Genetic susceptibility to scleroderma-like syndrome induced by vinyl chloride. *Lancet 1*, 53–5.

Blakley, B.R. and Tomar, R.S. (1986) The effect of cadmium on antibody responses to antigens with different cellular requirements. *Int. J. Immunopharmac.*, **8**, 1009–15.

Blanck, H., Wallin, G. and Wangberg, S.-A. (1984) Species-dependent variation in algal sensitivity to chemical compounds. *Ecotoxicol. Environ. Saf.*, **8**, 339–51.

Bleavins, M.R., Bursian, S.J., Brewster, J.S. and Aulerich, R.J. (1984) Effects of dietary hexachlorobenzene exposure on regional brain biogenic amine concentrations in mink and European ferrets. *J. Toxicol. Environ. Health*, **14**, 363–77.

Bloom, S.E. (1981) Chick embryos for detecting environmental mutagens. In: *Chemical Mutagens: Principles and Methods for their Detection. Vol. 5.* (Eds A. Hollaender and F.J. de Serres), pp. 203–32.

Bloom, S.E. (1984) Sister chromatid exchange studies in the chick embryo and neonate: actions of mutagens in a developing system. In: *Sister Chromatid Exchanges* (Eds R.R. Tice and A. Hollaender), pp. 509–33.

Bloom, S.E., Nanna, U.C. and Dietert, R.R. (1987) Targeting of chemical mutagens to differentiating B-lymphocytes in vivo: Detection by direct DNA labeling and sister chromatid exchange induction. *Environ. Mutagen.*, **9**, 3–18.

Bloomquist, J.R., Adams, P.M. and Soderlund, D.M. (1986) Inhibition of γ-aminobutyric acid-stimulated chloride flux in mouse brain vesicles by polychloro-cycloalkane and pyrethroid insecticides. *Neurotoxicology*, **7**, 11–20.

Boersma, D.C., Brownlee, L.J. and Hollebone, B.R. (1984) A total hepatic induction index of metabolism for lipophilic xenobiotics. *J. Appl. Toxicol.*, **4**, 187–93.

Boersma, D.C., Ellenton, J.A. and Yagminas, A. (1986) Investigation of the hepatic mixed-function oxidase system in Herring Gull embryos in relation to environmental contaminants. *Environ. Toxicol. Chem.*, **5**, 309–18.

Bondy, S.C. and Agrawal, A.K. (1980) The inhibition of cerebral high affinity receptor sites by lead and mercury compounds. *Arch. Toxicol.*, **46**, 249–56.

Bonkovsky, H.L., Sinclair, P.R., Bement, W.J., Lambrecht, R.W. and Sinclair, J.F. (1987) Role of cytochrome P-450 in porphyria caused by halogenated aromatic compounds. *Ann. N.Y. Acad. Sci.*, **514**, 96–112.

Borchardt, T. (1983) Influence of food quantity on the kinetics of cadmium uptake and loss via food and seawater in *Mytilus edulis*. *Mar. Biol.*, **76**, 67–76.

Bracher, G.A. and Bider, J.R. (1982) Changes in terrestrial animal activity of a forest community after an application of aminocarb (Matacil). *Can. J. Zool.*, **60**, 1981–97.

Bradley, S.G. and Morahan, P.S. (1982) Approaches to assessing host resistance. *Environ. Health Persp.*, **44**, 61–9.

Brattsten, L.B. (1979) Ecological significance of mixed-function oxidations. *Drug Metab. Rev.*, **10**, 35–58.

Bridger, M.A. and Thaxon, J.P. (1983) Humoral immunity in chicken as affected by mercury. *Arch. Environ. Contam. Toxicol.*, **12**, 45–9.

Brockes, J. (1990) Reading the retinoid signals. *Nature*, **345**, 766–8.

Brouwer, A. and van den Berg, K.J. (1984) Early and differential decrease in natural retinoid levels in C57BL/Rij and DBA/2 mice by 3,4,3′,4′-tetra-chlorobiphenyl. *Toxicol. Appl. Pharmacol.*, **73**, 204–9.

Brouwer, A. and van den Berg, K.J. (1986) Binding of a metabolite of 3,4,3′,4′-tetrachlorobiphenyl to transthyretin reduces serum vitamin A transport by inhibiting the formation of the protein complex carrying both retinol and thyroxin. *Toxicol. Appl. Pharmacol.*, **85**, 301–12.

Brouwer, A., van den Berg, K.J., Blaner, W.S. and Goodman, D.S. (1986) Transthyretin (prealbumin) binding of PCBs, a model for the mechanism of interference with Vitamin A and thyroid hormone metabolism. *Chemosphere*, **15**, 1699–706.

Brouwer, A., van den Berg, K.J. and Kukler, A. (1985) Time and dose responses of the reductions in retinoid concentrations in C57BL/Rij and DBA/2 mice induced by 3,4,3′,4′-tetrachlorobiphenyl. *Toxicol. Appl. Pharmacol.*, **78**, 180–9.

Brouwer, A., Hakansson, H., Kukler, A., van den Berg, K.J. and Ahlborg, U.G. (1989b) Marked alterations in retinoid homeostasis of Sprague-Dawley rats induced by a single i.p. dose of 10 μg/kg of 2,3,7,8-tetrachlorodibenzo-p-dioxin. *Toxicology*, **58**, 267–83.

Brouwer, A., Kukler, A. and van den Berg, K.J. (1988) Alterations in retinoid concentrations in several extrahepatic organs of rats by 3,4,3′,4′-tetra-chlorobiphenyl. *Toxicology*, **50**, 317–30.

Brouwer, A., Reijnders, P.J.H. and Koeman, J.H. (1989a) Polychlorinated biphenyl (PCB)-contaminated fish induces vitamin A and thyroid hormone deficiency in the common seal (*Phoca vitulina*). *Aquat. Toxicol.*, **15**, 99–105.

Brown, J.A., Johansen, P.H., Colgan, P.W. and Mathers, R.A. (1985) Changes in the predator-avoidance behaviour of juvenile guppies (*Poecilia reticulata*) exposed to pentachlorophenol. *Can. J. Zool.*, **63**, 2001–5.

Brunström, B. (1988) Sensitivity of embryos from duck, goose, herring gull, and various chicken breeds to 3,3′,4,4′-tetrachlorobiphenyl. *Poult. Sci.*, **67**, 52–7.

Brunström, B. (1990) Mono-ortho-chlorinated chlorobiphenyls: toxicity and induction of 7-ethoxyresorufin O-deethylase (EROD) activity in chick embryos. *Arch. Toxicol.*, **64**, 188–92.

Brunström, B. and Andersson, L. (1988) Toxicity and 7-ethoxyresorufin O-deethylase-inducing potency of coplanar polychlorinated biphenyls (PCBs) in chick embryos. *Arch. Toxicol.*, **62**, 263–6.

Brunström, B. and Lund, J. (1988) Differences between chick and turkey embryos in sensitivity to 3,3′,4,4′-tetrachlorobiphenyl and in concentration/affinity of the hepatic receptor for 2,3,7,8-tetrachlorodibenzo-p-dioxin. *Comp. Biochem. Physiol. Comp. Pharmacol. Toxicol.*, **91**, 507–12.

Brunström, B. and Reutergardh, L. (1986). Differences in sensitivity of some avian species to the embryotoxicity of a PCB, 3,3′,4,4′-tetrachlorobiphenyl, injected into the eggs. *Environ. Pollut.*, **42A**, 37–45.

Buckner, C.H. and McLeod, B.B. (1975) The impact of insecticides on small forest mammals. In: *Aerial Control of Forest Insects in Canada*. (Ed. M.L. Prebble), pp. 314–18. Department of Environment, Ottawa.

Budreau, C.H. and Singh, R.P. (1973) Teratogenicity and embryotoxicity of Demeton and Fenthion in CF#1 mouse embryos. *Toxicol. Appl. Pharmacol.*, **24**, 324–32.

Buhler, D.R. and Benville, P. (1969) Effects of feeding and of DDT on the activity of hepatic glucose 6-phosphate dehydrogenase in two salmonids. *J. Fish. Res. Board Can.*, **26**, 3209–16.

Bulow, F.J. (1970) RNA–DNA ratios as indicators of recent growth rates of a fish. *J. Fish. Res. Board Can.*, **27**, 2343–9.

Bulow, F.J., Zeman, M.E., Winningham, J.R. and Hudson, W.F. (1981) Seasonal variations in RNA–DNA ratios and in indicators of feeding, reproduction, energy storage, and condition in a population of bluegill, *Lepomis macrochirus* Rafinesque. *J. Fish Biol.*, **18**, 237–44.

Bunyan, P.J., van den Heuvel, M.J., Stanley, P.I. and Wright, E.N. (1981) An intensive field trial and a multi-site surveillance exercise on the use of aldicarb to investigate methods for the assessment of possible environmental hazards presented by new pesticides. *Agro Ecosyst.*, **7**, 239–62.

Bunyan, P.J., Jennings, D.M. and Taylor, A. (1986) Organophosphorus poisoning. Diagnosis of poisoning in pheasants owing to a number of common pesticides. *J. Agric. Food Chem.*, **16**, 332–9.

Bunyan, P.J. and Page, J.M.J. (1978) Polychlorinated biphenyls. The effect on the induction of quail hepatic microsomal enzymes. *Toxicol. Appl. Pharmacol.*, **43**, 507–18.

Burdick, G.E., Harris, E.J., Dean, H.J., Walker, T.M., Skea, J. and Colby, D. (1964) The accumulation of DDT in lake trout and the effect on reproduction. *Trans. Am. Fish Soc.*, **93**, 127–36.

Burger, J. (1990) Behavioral effects of early postnatal lead exposure in herring gull (*Larus argentatus*) chicks. *Pharmacol. Biochem. Behav.*, **35**, 7–13.

Burger, J. and Gochfeld, M. (1985) Early postnatal lead exposure: Behavioral effects on common terns (*Sterna hirundo*). *J. Toxicol. Environ. Health*, **16**, 869–86.

Burger, J. and Gochfeld, M. (1988) Lead and behavioral development: effects of varying dosage and schedule on survival and performance of young common terns (*Sterna hirundo*). *J. Toxicol. Environ. Health*, **24**, 173–82.

Burns, K.A. (1976) Microsomal mixed function oxidases in an estuarine fish (*Fundulus heteroclitus*), and their induction as a result of environmental contamination. *Comp. Biochem. Physiol.*, **53B**, 443–6.

Bursain, S.J., Polin, D., Olson, B.A., Shull, L.R., Marks, H.L. and Siegel, H.S. (1983) Microsomal enzyme induction, egg production, and reproduction in three lines of Japanese quail fed polybrominated biphenyls. *J. Toxicol. Environ. Health*, **12**, 291–307.

Busby, D.G., Pearce, P.A., Garrity, N.R. and Reynolds, L.M. (1983) Effect of an organophosphorus insecticide on brain cholinesterase activity in White-throated Sparrows exposed to aerial forest spraying. *J. Appl. Ecol.*, **20**, 255–63.

Busby, D.G., White, L.M. and Pearce, P.A. (1990) Effects of aerial spraying of fenitrothion on breeding of white-throated sparrows. *J. Appl. Ecol.*, **27**, 743–55.

Butler, R.G., Harfenist, A., Leighton, F.A. and Peakall, D.B. (1988) Impact of sublethal oil and emulsion exposure on the reproductive success of Leach's Storm-petrels: Short and long term effects. *J. Appl. Ecol.*, **25**, 125–43.

Cade, T.J., Enderson, J.H., Thelander, C.G. and White, C.M. (Eds) (1988) *Peregrine falcon populations: Their management and recovery.* The Peregrine Fund, Inc. Boise, ID. 949 pp.

Cade, T.J., Lincer, J.L., White, C.M., Roseneau, D.G. and Swartz, L.G. (1971) DDE residues and eggshell changes in Alaskan falcons and hawks. *Science*, **172**, 955–7.

Cairns, J., Jr., and Gruber, D. (1979) Coupling mini- and microcomputers to biological early warning systems. *Bio. Science*, **29**, 665–9.

Carnio, J.S. and McQueen, D.J. (1973) Adverse effects of 15 ppm of p,p′-DDT on three generations of Japanese quail. *Can. J. Zool.*, **51**, 1307–12.

Carpenter, H.M., Williams, D.E. and Buhler, D.R. (1985) Hexachlorobenzene-induced porphyria in Japanese quail: Changes in microsomal enzymes. *J. Toxicol. Environ. Health*, **15**, 431–44.

Carrano, A.V., Thompson, L.H., Lindl, P.A. and Minkler, J.L. (1978) Sister chromatid exchange as an indicator of mutagenesis. *Nature*, **271**, 551–3.

Carrington, C.D. (1989) Prophylaxis and the mechanism for the initiation of organophosphorous compound-induced delayed neurotoxicity. *Arch. Toxicol.*, **63**, 165–72.

Casale, G.P., Cohen, S.D. and Dicapua, R.A. (1983) The effects of organophosphate-induced stimulation on the antibody response to sheep erythrocytes in inbred mice. *Toxicol. Appl. Pharmacol.*, **68**, 198–205.

Casterline, J., Bradlaw, J., Puma, B. and Ku, Y. (1983) Screening of fresh water fish extracts for enzyme-inducing substances by an aryl hydrocarbon hydroxylase induction bioassay technique. *J. Assoc. Off. Anal. Chem.*, **66**, 1136–9.

Cavanaugh, K.P., Goldsmith, A.R., Holmes, W.N. and Follett, B.K. (1983) Effects of ingested petroleum on the plasma prolactin levels during incubation and on the breeding success of paired mallard ducks. *Arch. Environ. Contam. Toxicol.*, **12**, 355–41.

Cavanaugh, K.P. and Holmes, W.N. (1987) Effects of ingested petroleum on the development of ovarian endocrine function in photostimulated mallard ducks (*Anas platyrhynchos*). *Arch. Environ. Contam. Toxicol.*, **16**, 247–53.

Cecil, H.C., Harris, S.J., Bitman, J. and Fries, G.F. (1973) Polychlorinated biphenyl-induced decrease in liver vitamin A in Japanese quail and rats. *Bull. Environ. Contam. Toxicol.*, **9**, 179–85.

Chambers, J.E., Heitz, J.R., McCorkle, F.M. and Yarbough, J.D. (1979) Enzyme activities following chronic exposure to crude oil in a simulated ecosystem II. Striped Mullet. *Environ. Res.*, **20**, 140–7.

Chandra, S.V., Girja, S., Srivastava, R.S. and Gupta, S.K. (1980) Combined effect of metals on biogenic amines and their distribution in the brain of mice. *Arch. Environ. Contam. Toxicol.*, **9**, 79–85.

Chang, K.J., Hsieh, K.H., Lee, T.P., Tang, S.Y. and Tung, T.C. (1981) Immunologic evaluation of patients with polychlorinated biphenyl poisoning: Determination of lymphocyte subpopulations. *Toxicol. Appl. Pharmacol.*, **61**, 58–63.

Chastain, J.E., and Pazdernik, T.L. (1985) 2,3,7,8-tetrachlorodibenzo-p-dioxin (TCDD)-induced immunotoxicity. *Int. J. Immunopharmac.*, **7**, 849–56.

Cherian, G.M. (1980) The synthesis of metallothionein and cellular adaptation to metal toxicity in primary rat kidney epithelial cell cultures. *Toxicology*, **17**, 225–31.

Cherry, D.S. and Cairns, J., Jr. (1982) Biological monitoring. Part V – preference and avoidance studies. *Water Res.*, **16**, 263–302.

Christensen, G., Hunt, E. and Fiandt, J. (1977) The effect of methylmercuric chloride, cadmium chloride and lead nitrate on six biochemical factors of the brook trout (*Salvelinus fontinalis*). *Toxicol. Appl. Pharmacol.*, **42**, 523–30.

Christensen, G.M., Olson, D. and Riedel, B. (1982) Chemical effects on the activity of eight enzymes: A review and a discussion relevant to environmental monitoring. *Environ. Res.*, **29**, 247–55.

Chu, I., Villeneuve, D.C., Valli, V.E., Secours, V.E. and Becking, G.C. (1981) Chronic toxicity of photomirex in the rat. *Toxicol. Appl. Pharmacol.*, **59**, 268–78.

Clark, D.A., Gauldie, J., Szewczuk, M.R. and Sweeney, G. (1981) Enhanced suppressor cell activity as a mechanism of immunosuppression by 2,3,7,8-tetrachlorodibenzo-p-dioxin. *Proc. Soc. Exp. Biol. Med.*, **168**, 290–9.

Clark, D.R., Jr., Spann, J.W. and Bunck, C.M. (1990) Dicofol (Kethane)-induced eggshell thinning in captive American kestrels. *Environ. Toxicol. Chem.*, **9**, 1063–9.

Clarke, J.U. (1986) Structure-activity relationships in PCBs: Use of principal components analysis to predict inducers of mixed-function oxidase activity. *Chemosphere*, **15**, 275–87.

Cohen, J.A. and Warringa, M.G.P.J. (1957) Purification and properties of dialkylfluorophosphatase. *Biochim. Biophys. Acta*, **26**, 29–39.

Cohen, M.D., Wei, C.I., Tan, H., and Kao, K.J. (1986) Effect of metavanadate on the murine immune response. *J. Toxicol. Environ. Health*, **19**, 279–98.

Collins, W.T. and Capen, C.C. (1980) Ultrastructural and functional alterations of the rat thyroid gland produced by polychlorinated biphenyls compared with iodide excess and deficiency, and thyrotrophin and thyroxine administration. *Virchows Arch. Cell Pathol.*, **33**, 213–31.

Conn, P.M. (Ed.) (1984) *The Receptors. Vol. 1.*, Academic Press, New York, 659 pp.

Conn, P.M. (Ed.) (1985) *The Receptors. Vol. 2.*, Academic Press, New York, 509 pp.

Conn, P.M. (Ed.) (1986a) *The Receptors. Vol. 3.*, Academic Press, New York, 422 pp.

Conn, P.M. (Ed.) (1986b) *The Receptors. Vol. 4.*, Academic Press, New York, 432 pp.

Conney, A.H. (1967) Pharmacological implications of microsomal enzyme induction. *Pharmacol. Rev.*, **19**, 317–66.

Conney, A.H. (1982) Induction of microsomal enzymes by foreign chemicals and carcinogenesis by polycyclic aromatic hydrocarbons: G.H.A. Clowes Memorial Lecture. *Cancer Res.*, **42**, 4875–917.

Conney, A.H., Miller, E.C. and Miller, J.A. (1956) The metabolism of methylated aminoazo dyes. V. Evidence for induction of enzyme synthesis in the rat by 3-methylcholanthrene. *Cancer Res.*, **16**, 450–9.

Conney, A.H., Welch, R.M., Kunzman, R. and Burns, J.J. (1967) Effects of pesticides on drug and steroid metabolism. *Clin. Pharmacol. Ther.*, **8**, 2–10.

Cook, J.A., Hoffman, E.O. and Di, Luzio, N.R. (1975) Influence of lead and cadmium on the susceptibility of rats to bacterial challenge (39117). *Proc. Soc. Exp. Biol. Med.*, **150**, 741–7.

Cooke, A.S. (1973) Shell thinning in avian eggs by environmental pollutants. *Environ. Pollut.*, **4**, 85–152.

Cooke, A.S. (1981) Tadpoles as indicators of harmful levels of pollution in the field. *Environ. Pollut.*, **25A**, 123–33.

Coon, N.C., Albers, P.H. and Szaro, R.C. (1979) No. 2 fuel oil decreases embryonic survival of Great Black-backed Gulls. *Bull. Environ. Contam. Toxicol.*, **21**, 152–6.

Coulston, F. (1985) Reconsideration of the dilemma of DDT for the establishment of an acceptable daily intake. *Regul. Toxicol. Pharmacol.*, **5**, 332–83.

Cranmer, M.F. (1986) Carbaryl: a review. *Neurotoxicology*, **7**, 249–328.

Crocker, J.F.S., Ozere, R.L., Safe, S.H., Digout, S.C., Rozee, K.R. and Hutzinger, O. (1976) Lethal interaction of ubiquitous insecticide carriers with virus. *Science* (Washington, DC), **192**, 1351–3.

Crosby, D.G., Tucker, R.K. and Aharonson, N. (1966) The detection of acute toxicity with *Daphnia magna*. *Food Cosmet. Toxicol.*, **4**, 503–14.

Cullen, M.R. (1989) The role of clinical investigation in biological markers research. *Environ. Res.*, **50**, 1–10.

Custer, T.W. and Ohlendorf, H.M. (1989) Brain cholinesterase activity of nestling Great Egrets, Snowy Egrets and Black-crowned Night Herons. *J. Wildl. Dis.*, **25**, 359–63.

Das, B.C. (1988) Factors that influence formation of sister chromatid exchanges in human blood lymphocytes. *CRC Crit. Rev. Toxicol.*, **19**, 43–86.

Davies, I.M. and Pirie, J.M. (1978) The mussel *Mytilus edulis* as a bio-assay organism for mercury in seawater. *Mar. Pollut. Bull.*, **9**, 128–32.

Dawe, C.J. (1987) Oncozoons and the search for carcinogen-indicator fishes. *Environ. Health Perspect.*, **71**, 129–37.

Dean, J.H., Ward, E.C. and Murray, M.J. (1986) Immunosuppression following 7, 12-dimethylbenz(a)anthracene exposure in B6C3F1 mice. II. Altered cell-mediated immunity and tumour resistance. *Int. J. Immunopharmac.*, **8**, 189–98.

De Matteis, F., Prior, B.E. and Rimington, C. (1961) Nervous and biochemical disturbances following hexachlorobenzene intoxication. *Nature*, **191**, 363–6.

Depledge, M. (1989) The rational basis for detection of the early effects of marine pollutants using physiological indicators. *Ambio*, **18**, 301–2.

Desaiah, D. (1980) Comparative effects of chlorodecone and mirex on rat cardiac ATPases and binding of 3H-catecholamines. *J. Environ. Pathol. Toxicol.*, **4**, 237–48.

Desaiah, D., Cutkomp, L.K., Koch, R.B. and Jarvinen, A. (1975) DDT: Effect of continuous exposure on ATPase activity in the fish, *Pimephales promelas*. *Arch. Environ. Contam. Toxicol.*, **3**, 132–41.

Descotes, (1988) Immunotoxicity of chemicals, In: *Immunotoxicology of Drugs and Chemicals*, pp. 297–441, Elsevier, Amsterdam, New York.

Dewailly, E., Nantel, A., Weber, J.-P. and Meyer, F. (1989) High levels of PCBs in breast milk of Inuit women from Arctic Quebec. *Bull. Environ. Contam. Toxicol.*, **43**, 641–6.

Diaz, L.A. and Provost, T.T. (1987) Dermatologic diseases. In: *Basic and Clinical Immunology* (Eds D.P. Sittes, J.D. Stobo, and J.V. Wells), pp. 516–19. Appleton & Lange, Norwalk, CT.

Dickson, D. (1988) Mystery disease strikes Europe's seals. *Science*, **214**, 893–5.

Di Giulio, R.T., Washburn, P.C., Wenning, R.J., Winston, G.W. and Jewell, C.S. (1989) Biochemical responses in aquatic animals: A review of determinations of oxidative stress. *Environ. Toxicol. Chem.*, **8**, 1103–23.

Dieter, M.P. (1974) Plasma enzyme activities in *Coturnix* quail fed graded doses of

DDE, polychlorinated biphenyl, malathion, and mercuric chloride. *Toxicol. Appl. Pharmacol.*, **27**, 86–98.

Dieter, M.P. (1975) Further studies on the use of enzyme profiles to monitor residue accumulation in wildlife: plasma enzymes in Starlings fed graded concentrations of morsodren, DDE, Arochlor 1254, and malathion. *Arch. Environ. Contam. Toxicol.*, **3**, 142–50.

Dieter, M.P. and Finley, M.T. (1979) Delta-aminolevulinic acid dehydratase enzyme activity in blood, brain, and liver of lead-dosed ducks. *Environ. Res.*, **9**, 127–35.

Dieter, M.P. and Finley, M.T. (1978) Erythrocyte delta-aminolevulinic acid dehydratase activity in mallard ducks: duration of inhibition after lead-shot exposure. *J. Wildl. Manage.*, **42**, 621–5.

Dieter, M.P., Luster, M.I., Boorman, G.A., Jamieson, C.W., Dean, J.H. and Cox, J.W. (1983) Immunological and biochemical responses in mice treated with mercuric chloride. *Toxicol. Appl. Pharmacol.*, **68**, 218–28.

Dieter, M.P., Perry, M.C. and Mulhern, B.M. (1976) Lead and PCB's in Canvasback Ducks: Relationship between enzyme levels and residues in blood. *Arch. Environ. Contam. Toxicol.*, **5**, 1–13.

Dietz, R., Heide-Jorgensen, M.-P. and Harkonen, T. (1989) Mass deaths of harbor seals (*Phoca vitulina*) in Europe. *Ambio*, **18**, 258–64.

Din, Z.B. and Brooks, J.M. (1986) Use of adenylate energy charge as a physiological indicator in toxicity experiments. *Bull. Environ. Contam. Toxicol.*, **16**, 1–8.

Dinman, B.D. (1972) 'Non-concept' of 'No-threshold': Chemicals in the environment. *Science*, **175**, 495–7.

Dixon, D.G., Hodson, P.V., Klaverkamp, J.J., Lloyd, K.M. and Roberts, J.R. (1985a) The role of biochemical indicators in the assessment of ecosystem health – their development and validation. *Natl. Res. Counc. Can. Publ.*, **24371**.

Dixon, D.R. and Clarke, K.R. (1982) Sister chromatid exchange: A sensitive method for detecting damage caused by exposure to environmental mutagens in the chromosomes of adult *Mytilus edulis*. *Mar. Biol. Lett.*, **3**, 163–72.

Dixon, D.R., Jones, I.M. and Harrison, F.L. (1985) Cytogenetic evidence of inducible processes linked with metabolism of a xenobiotic chemical in adult and larval *Mytilus edulis*. *Sci. Total Environ.*, **46**, 1–8.

Dixon, D.R. and Prosser, H. (1986) An investigation of the genotoxic effects of an organotin antifouling compound (bis(tributyltin)oxide) on the chromosomes of the edible mussel, *Mytilus edulis*. *Aquat. Toxicol.*, **8**, 185–95.

Dragomirescu, A., Raileanu, L. and Ababei, L. (1975) The effect of carbetox on glycolysis and the activity of some enzymes in carbohydrate metabolism in the fish and rat liver. *Water Res.*, **9**, 205–9.

Druet, P., Pelletier, L., Rossert, J., Druet, E., Hirsh, F. and Sapin, C. (1989) Autoimmune reactions induced by metals. In: *Autoimmunity and Toxicology*, (Eds M.E. Kammuller, N. Bloksma and W. Seinen) pp. 347–61. Elsevier, Amsterdam.

Druet, P., Sapin, C., Druet, E. and Hirsh, F. (1983) Genetic control of mercury-induced immune response in the rat. In: *Nephrotoxic Mechanisms of Drugs and Environmental Toxins* (Ed. G.A. Porter), pp. 425–35. Plenum Medical Book Company, New York, London.

Dudai, Y. (1989) *The Neurobiology of Memory. Concepts, Findings, Trends.* Oxford University Press, 340 pp.

Dunachie, J.F. and Fletcher, W.W. (1969) An investigation of the toxicity of insecticides to birds' eggs using the egg-injection technique. *Ann. Appl. Biol.*, **64**, 409–23.

Dunn, B.P., Black, J.J. and Maccubbin, A. (1987) ^{32}P post-labeling analysis of aromatic DNA adducts in fish from polluted areas. *Cancer Res.*, **47**, 6543–8.

Duvivier, J., van Cantfort, J. and Gielen, J.E. (1981) The influence of five mono-xygenase inducers on liver cytosol estradiol receptor levels in the ovariectomized adult rat. *Biochem. Biophys. Res. Commun.*, **99**, 252–8.

Eastin, W.C., Jr., Fleming, W.J. and Murray, H.C. (1982) Organophosphate inhibition of avian salt gland Na, K-ATPase activity. *Comp. Biochem. Physiol.*, **73C**, 101–7.

Edwards, A.J., Addison, R.F., Willis, D.E. and Renton, K.W. (1988) Seasonal variation of hepatic mixed function oxidases in Winter Flounder (*Pseudopleuronectes americanus*). *Mar. Environ. Res.*, **26**, 299–309.

Ehrlich, R. (1980) Interaction between environmental pollutants and respiratory infections. *Environ. Health Persp.*, **35**, 89–100.

Eldefrawi, M.E., Abalis, I.M., Filbin, M.T. and Eldefrawi, A.T. (1985) Glutamate and GABA receptors of insect muscles: Biochemical identification and interactions with insecticides. In: *Approaches to New Leads for Insecticides*. (Ed. von Keyserlingk), pp. 101–16. Springer-Verlag, Berlin.

Elder, G.H. (1978) Porphyria caused by hexachlorobenzene and other polyhalogenated aromatic hydrocarbons. In: *Heme and Hemoproteins*, (Eds F. De Matteis and W.N. Aldridge), pp. 157–200. Springer-Verlag, Berlin.

Ellenton, J.A., Brownlee, L.J. and Hollebone, B.R. (1985) Aryl hydrocarbon hydroxylase levels in Herring Gull embryos from different locations on the Great Lakes. *Environ. Toxicol. Chem.*, **4**, 615–22.

Ellenton, J.A. and McPherson, M.F. (1983) Mutagenicity studies on herring gulls from different locations on the Great Lakes. 1. Sister chromatid exchange rates in herring-gull embryos. *J. Toxicol. Environ. Health*, **12**, 317–24.

Ellenton, J.A., McPherson, M.F. and Maus, K.L. (1983) Mutagenicity studies on Herring Gulls from different locations on the Great Lakes. II. Mutagenic evaluation of extracts of Herring-gull eggs in a battery of in vitro mammalian and microbial tests. *J. Toxicol. Environ. Health*, **12**, 325–36.

Elliott, J.E., Kennedy, S.W., Jeffrey, D. and Shutt, L. (1991) Polychlorinated biphenyl (PCB) effects on hepatic mixed function oxidases and porphyria in birds: II. American Kestrel. *Comp. Biochem. Physiol.* **99C**, 141–5.

Elliott, J.E., Kennedy, S.W., Peakall, D.B. and Won, H. (1990) Polychlorinated biphenyl (PCB) effects on hepatic mixed function oxidases and porphyria in birds: I. Japanese Quail. *Comp. Biochem. Physiol.*, **96C**, 205–10.

Elliott-Feeley, E. and Armstrong, J.B. (1982) Effects of fenitrothion and carbaryl on *Xenopus laevis* development. *Toxicology*, **22**, 319–35.

Elson, P.F. (1967) Effects on wild young salmon of spraying DDT over New Brunswick forests. *J. Fish. Res. Board. Can.*, **24**, 731–67.

Enderson, J.H. and Berger, D.D. (1970) Pesticides: Eggshell thinning and lowered production of young in Prairie Falcons. *Bioscience*, **20**, 355–6.

Enderson, J.H., Temple, S.A. and Swartz, L.G. (1972) Time-lapse photographic records of nesting Peregrine Falcons. *Living Bird*, **11**, 113–28.

Environment Canada (1991) The chemical pollution of the Great Lakes and associated effects. Synopsis. *Environment Canada*, pp. 43.

Evans, R.M. (1988) The steroid and thyroid hormone receptor superfamily. *Science*, **240**, 889–95.

Exon, J.H. (1984) The immunotoxicity of selected environmental chemicals, pesticides and heavy metals. In: *Chemical Regulation in Veterinary Medicine*, pp. 355–68. Alan Liss, Inc., New York.

Exon, J.H. and Koller, L.D. (1983) Effects of chlorinated phenols on immunity in rats. *Int. J. Immunopharm.*, **5**, 131–6.

Exon, J.H., Henningsen, G.M., Osborne, C.A. and Koller, L.D. (1984) Toxicologic, pathologic and immunotoxic effects of 2, 4-dichlorophenol in rats. *J. Toxicol. Environ. Health*, **14**, 723–30.

Exon, J.H., Kerkvliet, N.I. and Talcot, P.A. (1987) Immunotoxicity of carcinogenic pesticides and related chemicals. *J. Environ. Sci. Health Part C; Environ. Carcinog. Rev.*, **5**, 73–120.

Fairbrother, A., Craig, M.A., Walker, K. and O'Loughlin, D. (1990) Changes in Mallard (Anas platyrhynchos) serum chemistry due to age, sex, and reproductive condition. *J. Wildl. Dis.*, **26**, 67–77.

Faith, R.E. and Moore, J.A. (1977) Impairment of thymus-dependent immune functions by exposure of the developing immune system to 2,3,7,8-tetrachlorodibenzo-p-dioxin (TCDD). *J. Toxicol. Environ. Health*, **3**, 415–64.

Faith, R.E., Luster, M.I. and Kimmel, C.A. (1979) Effect of chronic developmental lead exposure on cell-mediated immune functions. *Clin. Exp. Immunol.*, **35**, 413–20.

Faith, R.E., Luster, M.I. and Vos, J.G. (1980) Effects on immunocompetence by chemicals of environmental concern. *Rev. Biochem. Toxicol.*, **2**, 173–211.

Falchetti, R., Silvestri, S., Battaglia, A. and Caprino, L. (1983) Toxicological evaluation of immunomodulating drugs. In: *Current Problems in Drug Toxicology*. (Eds G. Zbinden, F. Cohadon, J.Y. Detaille and G. Mazue), pp. 248–63. Eurotext. Paris.

Farve, R.M. (1978) *Effects of PCBs on ring dove* (Streptopelia risoria) *courtship behavior*. MS Thesis. The Ohio State University, Columbus, OH. 40 pp.

Fattah, K.M.A. and Crowder, L.A. (1980) Plasma membrane ATPases from various tissues of the cockroach (*Periplaneta americana*) and mouse influenced by toxaphene. *Bull. Environ. Contam. Toxicol.*, **24**, 356–63.

Fetterolf, P.M. (1983) Effects of investigator activity on ring-billed gull behavior and reproductive performance. *Wilson Bull.*, **95**, 23–41.

Finley, M.T. and Dieter, M.P. (1979) Toxicity of experimental lead-iron shot versus commercial lead shot in Mallards. *J. Wildl. Manage.*, **42**, 32–9.

Finley, M.T., Dieter, M.P. and Locke, L.N. (1976) Delta-aminolevulinic acid dehydratase: inhibition in ducks dosed with lead shot. *Environ. Res.*, **12**, 243–9.

Fleming, W.J. (1981) Recovery of brain and plasma cholinesterase activities in ducklings exposed to organophosphorus pesticides. *Arch. Environ. Contam. Toxicol.*, **10**, 215–29.

Fleming, W.J. and Grue, C.E. (1981) Recovery of cholinesterase activity in five avian species exposed to Dictrophos, an organophosphorus pesticide. *Pestic. Biochem. Physiol.*, **16**, 129–35.

Forseel, J.H., Shull, L.R. and Kateley, J.R. (1981) Subchronic administration of technical pentachlorophenol to lactating dairy cattle: Immunotoxicologic evaluation. *J. Toxicol. Environ. Health*, **8**, 543–58.

Fossi, C., Leonzio, C., Focardi, S. and Peakall, D.B. (1990) Avian mixed function oxidase induction as a monitoring device: The influence of normal physiological functions. In: *Biomarkers of Environmental Contamination* (Eds J.F. McCarthy and L.R. Shugart), pp. 143–9. Lewis Publ., Boca Raton, FL.

Foster, R.G. and Follett, B.K. (1985) The involvement of a rhodopsin-like photopigment in the photoperiodic response of the Japanese quail. *J. Comp. Physiol. A.*, **157**, 519–28.

Foulkes, E.C. (1990) The concept of critical levels of toxic heavy metals in target tissues. *Crit. Rev. Toxicol.*, **20**, 327–39.

Foureman, G.L., White, N.B., Jr., and Bend, J.R. (1983) Biochemical evidence that Winter Flounder (*Pseudopleuronectes americanus*) have induced hepatic cytochrome P-450-dependent monooxygenase activities. *Can. J. Fish. Aquat. Sci.*, **40**, 854–65.

Fox, G.A. and Boersma, D. (1983) Characteristics of supernormal Ring-billed Gull clutches and their attending adults. *Wilson Bull.*, **95**, 552–9.

Fox, G.A. and Donald, T. (1980) Organochlorine pollutants, nest-defense behavior and reproduction success in Merlins. *Condor*, **82**, 81–4.

Fox, G.A., Gilman, A.P., Peakall, D.B. and Anderka, F.W. (1978) Behavioral abnormalities of nestling Lake Ontario Herring Gulls. *J. Wildl. Manage.*, **42**, 477–83.

Fox, G.A., Kennedy, S.W., Norstrom, R.J. and Wigfield, D.C. (1988) Porphyria in Herring Gulls: A biochemical response to chemical contamination of Great Lakes food chains. *Environ. Toxicol. Chem.*, **7**, 831–9.

Fox, G.A., Mineau, P., Collins, B. and James, P.C. (1989) The impact of the insecticide carbofuran (Furadan 480F) on the Burrowing Owl in Canada. *Techn. Rep. Ser. No. 72.*, Canadian Wildlife Service, Ottawa, 25 pp.

Francis, B.M., Metcalf, R.L. and Fisher, S.W. (1983) Response of laboratory rodents to selected avian delayed neurotoxicants. *Arch. Environ. Contamin. Toxicol.*, **12**, 731–8.

Freeman, H.C., Uthe, J.F. and Sangalang, G. (1980) The use of steroid hormone metabolism studies in assessing the sublethal effects of marine pollution. *Rapp. P.-V. Reun. Cons. Int. Explor. Mer.*, **179**, 16–22.

Friend, M., Haegele, M.A. and Wilson, R. (1973) DDE: Interference with extra-renal salt excretion in the Mallard. *Bull. Environ. Contam. Toxicol.*, **9**, 49–53.

Friend, M. and Trainer, D.D. (1974) Experimental dieldrin-duck hepatitis virus interaction studies. *J. Wildlife Manag.*, **38**, 896–902.

Fry, D.M., Swenson, J., Addiego, L.A., Grau, C.R. and Kang, A. (1986) Reduced reproduction of Wedge-tailed Shearwaters exposed to weathered Santa Barbara crude oil. *Arch. Environ. Contam. Toxicol.*, **15**, 453–63.

Fry, D.M. and Toone, C.K. (1981) DDT-induced feminization of gull embryos. *Science*, **213**, 922–4.

Fry, D.M., Toone, C.K., Speich, S.M. and Peard, R.J. (1987) Sex ratio skew and breeding patterns of gulls: Demographic and toxicological considerations. *Studies in Avian Biology*, **10**, 26–43.

Fulton, M.H. and Chambers, J.E. (1985) The toxic and teratogenic effects of selected organophosphorus compounds on the embryos of three species of amphibians. *Toxicol. Lett.*, **26**, 175–80.

Fyfe, R., Risebrough, R.W. and Walker, W., II (1976) Pollutant effects on the reproduction of the Prairie Falcons and Merlins of the Canadian prairies. *Can. Field-Nat.*, **90**, 346–55.

Gainer, J.H. (1972) Increased mortality in encephalomyocarditis virus-infected mice consuming cobalt sulfate: Tissue concentrations of cobalt. *Amer. J. Vet. Res.*, **33**, 2067–73.

Galindo, J.C., Kendall, R.J., Driver, C.J. and Lacher, T.E., Jr. (1985) The effect of methyl parathion on susceptibility of Bobwhite Quail (*Colinus virginianus*) to domestic cat predation. *Behav. Neural Biol.*, **43**, 21–36.

Gallagher, E.P. and Di Giulio, R.T. (1989) Effects of complex waste mixtures on hepatic monooxygenase activities in brown bullheads (*Ictalurus nebulosus*). *Environ. Pollut.*, **62**, 113–28.

Gasiewicz, T.A. and Rucci, G. (1984) Cytosolic receptor for 2,3,7,8-tetrachlorodibenzo-p-dioxin. Evidence for a homologous nature among various mammalian species. *Mol. Pharmacol.*, **26**, 90–8.

Gaylord, H.R. (1910) An epidemic of carcinoma of the thyroid gland among fish. *J. Am. Med. Assoc.*, **54**, 227.

Ghiasuddin, S.M. and Matsumura, F. (1982) Inhibition of gamma-aminobutyric acid (GABA)-induced chloride uptake by gamma-BHC and heptachlor epoxide. *Comp. Biochem. Physiol.*, **73C**, 141–4.

Giguere, V., Ong, E.S., Segui, P. and Evans, R.M. (1987) Identification of a receptor for the morphogen retinoic acid. *Nature*, **330**, 624–9.

Gilbertson, M. (1975) A Great Lakes Tragedy. *Nat. Can.*, **4**, 22–5.

Gilbertson, M. (1989) *Proceedings of the workshop on cause-effect linkages*. Int. Joint Comm., Windsor, Ontario. 45 pp.

Gilbertson, M., Elliott, J.E. and Peakall, D.B. (1987) Sea birds as indicators of marine pollution. In: *The Value of Birds* (Eds A.W. Diamond and F.L. Filion). *ICBP Tech. Pub.*, **6**, 231–48.

Gilbertson, M. and Reynolds, L.M. (1972) Hexachlorobenzene (HCB) in the eggs of Common Terns in Hamilton Harbour, Ontario. *Bull. Environ. Contam. Toxicol.*, **7**, 371–3.

Gillett, J.W. (1968) 'No effect' level of DDT in induction of microsomal epoxidation. *J. Agric. Food Chem.*, **16**, 295–7.

Gillette, D.M., Corey, R.D., Helferich, W.G., McFarland, J.M., Lowenstine, L.J., Moody, D.E., Hammock, B.D. and Shull, L.R. (1987a) Comparative toxicology of tetrachlorobiphenyls in mink and rats. 1. Changes in hepatic enzyme activity and smooth endoplasmic reticulum volume. *Fund. Appl. Toxicol.*, **8**, 5–14.

Gillette, D.M., Corey, R.D., Lowenstine, L.J. and Shull, L.R. (1987b) Comparative toxicology of tetrachlorobiphenyls in mink and rats. II. Pathology. *Fund. Appl. Toxicol.*, **8**, 15–22.

Gillette, J.R. (1986) On the role of pharmacokinetics in integrating results from in vivo and in vitro studies. *Food. Chem. Toxicol.*, **24**, 711–20.

Gilman, A.P., Fox, G.A., Peakall, D.B., Teeple, S.M., Carroll, T.R. and Haymes, G.T. (1977) Reproductive parameters and egg contaminant levels of Great Lakes Herring Gulls. *J. Wildl. Manage.*, **41**, 450–68.

Gilman, S., Schwartz, J., Milner, R., Bloom, F. and Feldman, J. (1982) β-endorphin enhances lymphocyte proliferative responses. *Proc. Natl. Acad. Sci.* (US), **79**, 4226–30.

Gleichmann, H., and Gleichmann, E. (1987) Mechanisms of autoimmunity. In: *Immunotoxicology* (Eds A. Berlin, J. Dean, M.H. Drapper, E.M.B. Smith and F. Spreafico), pp. 39–60. Martinus Nijhoff Publishers, Dordrecht, Boston, MA, Lancaster.

Gleichmann, E., Kimber, I. and Purchase, I.H.F. (1989) Immunotoxicology: suppressive and stimulatory effects of drugs and environmental chemicals on the immune system. *Arch. Toxicol.*, **63**, 257–73.

Gochfeld, M. and Burger, J. (1988) Effects of lead on growth and feeding behavior of young Common Terns (*Sterna hirundo*). *Arch. Environ. Contamin. Toxicol.*, **17**, 513–18.

Godowski, P.J. and Picard, D. (1989) Steroid receptors. How to be both a receptor and a transcription factor. *Biochem. Pharmacol.*, **38**, 3135–43.

Goldberg, A.M. and Frazier, J.M. (1989) Alternatives to animals in toxicity testing. *Sci. Amer.*, **261**, 24–30.

Goldman, M., Dillon, R.D. and Wilson, R.M. (1977) Thyroid function in Pekin ducklings as a consequence of erosion of ingested lead shot. *Toxicol. Appl. Pharmacol.*, **40**, 241–6.

Goldstein, J.A., Hickman, P. and Jue, D.L. (1974) Experimental hepatic porphyria induced by polychlorinated biphenyls. *Toxicol. Appl. Pharmacol.*, **27**, 437–48.

Golter, M. and Michaelson, I.A. (1975) Growth, behavior, and brain catecholamines in lead-exposed neonatal rats: A reappraisal. *Science*, **187**, 359–61.

Gopal, K., Anand, M., Mehrotra, S. and Ray, P.K. (1985) Neurobehavioural changes in freshwater fish (*Channa punctatus*) exposed to endosulfan. *J. Adv. Zool.*, **6**, 74–80.

Grant, D.L., Inverson, F., Hatina, G.V. and Villeneuve, D.C. (1974) Effects of hexachlorobenzene on liver porphyrin levels and microsomal enzymes in the rat. *Environ. Physiol. Biochem.*, **4**, 159–65.

Greenlee, W.F. and Neal, R.A. (1985) The Ah receptor: A biochemical and biologic perspective. In: *The Receptors*. (Ed. P.M. Conn), pp. 89–129. Academic Press, New York. Vol. 2.

Gregus, Z., Watkins, J.B., Thompson, T.N., Harvey, M.J., Rozman, K. and Klaassen, C.D. (1983) Hepatic phase I and phase II biotransformations in Quail and Trout: Comparison to other species commonly used in toxicity testing. *Toxicol. Appl. Pharmacol.*, **67**, 430–41.

Greig, J.B., Francis, J.E., Kay, S.J.E., Lovell, D.P. and Smith, A.G. (1984) Incomplete correlation of 2,3,7,8-tetrachlorodibenzo-p-dioxin hepatotoxicty with Ah phenotype in mice. *Toxicol. Appl. Pharmacol.*, **74**, 17–25.

Grue, C.E., Fleming, W.J., Busby, D.G. and Hill, E.F. (1983) Assessing hazards of organophosphate pesticides to wildlife. *Trans. North Am. Wildl. Nat. Resour. Conf.*, **48**, 200–20.

Grue, C.E., Hoffman, D.J., Beyer, W.N. and Franson, L.P. (1986) Lead concentrations and reproductive success in European starlings (*Sturnus vulgaris*) nesting within highway roadside verges. *Environ. Pollut.*, **42A**, 157–82.

Grue, C.E. and Hunter, C.C. (1984) Brain cholinesterase activity in fledgling starlings: Implications for monitoring exposure of songbirds to ChE inhibitors. *Bull. Environ. Contam. Toxicol.*, **32**, 282–9.

Grue, C.E., O'Shea, T.J. and Hoffman, D.J. (1984) Lead concentrations and reproduction in highway-nesting Barn Swallows. *Condor*, **86**, 383–9.

Grue, C.E., Powell, G.V.N. and Gladson, N.L. (1981) Brain cholinesterase (ChE) activity in nestling starlings: Implications for monitoring exposure of nestling songbirds to ChE inhibitors. *Bull. Environ. Contam. Toxicol.*, **26**, 544–7.

Grue, C.E., Powell, G.V.N. and Gorsuch, C.H. (1982a) Assessing effects of organophosphates on songbirds: Comparison of a captive and a free-living population. *J. Wildl. Manage.*, **46**, 766–8.

Grue, C.E., Powell, G.V.N. and McChesney, M.J. (1982b) Care of nestlings by wild female Starlings exposed to an organophosphate pesticide. *J. Appl. Ecol.*, **19**, 327–55.

Grue, C.E. and Shipley, B.K. (1981) Interpreting population estimates of birds following pesticide applications — behaviour of male Starlings exposed to an organophosphate pesticide. *Stud. Avian Biol.*, **6**, 292–6.

Grue, C.E. and Shipley, B.K. (1984) Sensitivity of nestling and adult Starlings to Dicrotophos, an organophosphate pesticide. *Environ. Res.*, **35**, 454–65.

Guengerich, F.P. (1987) Activation of organic compounds by cytochrome P-450 isozymes. In: *Mechanisms of Cell Injury: Implications for Human Health* (Ed. B.A. Fowler), pp. 7–17. John Wiley & Sons, Chichester.

Gupta, R.C. and Randerath, K. (1988) Analysis of DNA adducts by 32P-labeling and thin layer chromatography. In: *DNA Repair. Vol. 3.* (Eds E. Friedberg and P.H. Hanawalt), pp. 399–418. Marcel Decker, New York.

Hadden, J.W. (1987) Immunorestoration in secondary immunodeficiency. In: *Immunotoxicology* (Eds A. Berlin, J. Dean, M.H. Draper, E.M.B. Smith, and F. Spreafico), pp. 104–24. Martinus Nijhoff Publishers, Dordrecht, Boston, MA, Lancaster.

Haegele, M.A. and Hudson, R.H. (1973) DDE effects on reproduction of Ring Doves. *Environ. Pollut.*, **4**, 53–7.

Haegele, M.A. and Hudson, R.H. (1977) Reduction of courtship behaviour induced by DDE in male Ringed Turtle Doves. *Wilson Bull.*, **89**, 593–601.

Hamilton, J.W. and Bloom, S.E. (1986) Correlation between induction of xenobiotic metabolism and DNA damage from chemical carcinogens in the chick embryo in vivo. *Carcinogenesis*, **7**, 1101–6.

Hamilton, J.W., Denison, D.S. and Bloom, S.E. (1983) Development of basal and induced aryl hydrocarbon{benzo(a)pyrene} hydroxylase activity in the chicken embryo in ovo. *Proc. Natl. Acad. Sci.*, **80**, 3372–6.

Hardy, A.R., Stanley, P.I. and Greig-Smith, P.W. (1987) Birds as indicators of the intensity of use of agricultural pesticides in the UK. *ICBP Tech. Publ.*, **6**, 119–32.

Harfenist, A., Power, T., Clark, K.L. and Peakall, D.B. (1989) A review and evaluation of the amphibian toxicological literature. *Can. Wildl. Ser. Techn. Rep. No.*, 61. 222 pp.

Hart, A.D.M. and Westlake, D.E. (1986) Acetylcholinesterase activity and the effects of Chlorfenvinphos in regions of the starling brain. *Bull. Environ. Contam. Toxicol.*, **37**, 774–82.

Hart, L.G., Shultice, R.W. and Fouts, J.R. (1963) Stimulatory effects of chlordane on hepatic microsomal drug metabolism in the rat. *Toxicol. Appl. Pharmacol.*, **5**, 371–86.

Hartsough, G.R. (1965) Great Lakes fish now suspect as mink food. *Amer. Fur Breeder*, **38**(10), 25–7.

Hartwell, S.I., Cherry, D.S. and Cairns, J., Jr. (1987a) Avoidance responses of schooling fathead minnows (*Pimephales promelas*) to a blend of metals during a 9-month exposure. *Environ. Toxicol. Chem.*, **6**, 177–87.

Hartwell, S.I., Cherry, D.S. and Cairns, J., Jr. (1987b) Field validation of avoidance of elevated metals by fathead minnows (*Pimephales promelas*) following in situ acclimation. *Environ. Toxicol. Chem.*, **6**, 189–200.

Heinrich-Hirsch, B., Hofmann, D., Webb, J. and Neubert, D. (1990) Activity of aldrinepoxidase, 7-ethoxycoumarin-O-deethylase and 7-ethoxyresorefufin-O-deethylase during the development of chick embryos in ovo. *Arch. Toxicol.*, **64**, 128–34.

Heinz, G.H. (1976a) Methylmercury: Second-year feeding effects on mallard reproduction and duckling behavior. *J. Wildl. Manage.*, **40**, 82–90.

Heinz, G.H. (1976b) Methylmercury: Second-generation reproductive and behavioral effects on Mallard Ducks. *J. Wildl. Manage.*, **40**, 710–15.

Heinz, G.H. (1979) Methylmercury: Reproductive and behavioral effects on three generations of mallard ducks. *J. Wildl. Manage.*, **43**, 394–401.

Heinz, G.H. (1989) How lethal are sublethal effects? *Environ. Toxicol. Chem.*, **8**, 463–4.

Heinz, G.H., Hill, E.F. and Contrera, J.F. (1980) Dopamine and norepinephrine depletion in Ring Doves fed DDE, dieldrin, and Aroclor 1254. *Toxicol. Appl. Pharmacol.*, **53**, 75–82.

Heinz, G.H. and Johnson, R.W. (1981) Diagnostic brain residues of dieldrin: Some new insights. In: *Avian and Mammalian Wildlife Toxicology: Second Conference, ASTM STP 757* (Eds D.W. Lamb and E.E. Kenaga), pp. 72–92. Am. Soc. Test. Mater.

Helle, E., Olsson, M. and Jensen, S. (1976) PCB levels correlated with pathological changes in seal uteri. *Ambio*, **5**, 261–3.

Helmy, M.M., Lemke, A.E., Jacob, P.G. and Al-Sultan, Y.Y. (1979) Haematological changes in Kuwait Mullet, *Liza macrolepis* (Smith), induced by heavy metals. *Indian J. Mar. Sci.*, **8**, 278–81.

Hemphill, F.E., Kaeberle, M.L. and Buck, W.B. (1971) Lead suppression of mouse resistance to Salmonella typhimurium. *Science*, **172**, 1031–2.

Hendrickson, C.M. and Bowden, J.A. (1976) In vitro inhibition of lactic acid dehydrogenase by insecticidal polychlorinated hydrocarbons. 2. Inhibition by dieldrin and related compounds. *J. Agric. Food Chem.*, **24**, 756–9.

Hernberg, S., Nikkkanen, J., Mellin, G. and Lilius, H. (1970) γ-aminolevulinic acid dehydrase as a measure of lead exposure. *Arch. Environ. Health*, **21**, 140–5.

Hetzel, B.S. (1989) *The Story of Iodine Deficiency: An International Challenge in Nutrition.* Oxford University Press, pp. 236.

Hickey, J.J. (1969) *The Peregrine Falcon Populations: Their Biology and Decline.* University of Wisconsin Press, Madison, WI. 596 pp.

Hickey, J.J. and Anderson, D.W. (1968) Chlorinated hydrocarbons and eggshell changes in raptorial and fish-eating birds. *Science*, **162**, 271–3.

Hidaka, H. and Tatsukawa, R. (1985) Avoidance test of a fish (*Oryzias latipes*) to aquatic contaminants, with special reference to monochloramine. *Arch. Environ. Contam. Toxicol.*, **14**, 565–71.

Hill, E.F. (1988) Brain cholinesterase activity of apparently normal wild birds. *J. Wildl. Dis.*, **24**, 51–61.

Hill, E.F. (1989) Sex and storage affect cholinesterase activity in blood plasma of Japanese Quail. *J. Wildl. Dis.*, **25**, 580–5.

Hill, E.F. and Fleming, W.J. (1982) Anticholinesterase poisoning of birds: Field monitoring and diagnosis of acute poisoning. *Environ. Toxicol. Chem.*, **1**, 27–38.

Hill, E.F. and Murray, H.C. (1987) Seasonal variation in diagnostic enzymes and biochemical constituents of captive northern bobwhite quail and passerines. *Comp. Biochem. Physiol.*, **87B**, 933–40.

Hillam, R.P. and Ozkan, A.N. (1986) Comparison of local and systemic immunity after intratracheal, intraperitoneal, and intravenous immunization of mice exposed to either aerosolized or ingested lead. *Environ. Res.*, **39**, 265–77.

Hilmy, A.M., Shabana, M.B. and Said, M.M. (1981) The role of serum transminases (SGO-T and SGP-T) and alkaline phosphatases in relation to inorganic phosphorus with respect to mercury poisoning in *Aphanius dispar rupp* (Teleostei) of the Red Sea. *Comp. Biochem. Physiol.*, **68C**, 69–74.

Hinderer, R.K. and Menzer, R.E. (1976) Enzyme activities and cytochrome P-450 levels of some Japanese Quail tissues with respect to their metabolism of several pesticides. *Pest. Biochem. Physiol.*, **6**, 161–9.

Hochstein, J.R., Aulerich, R.J. and Bursain, S.J. (1988) Acute toxicity of 2,3,7,8-tetrachlorodibenzo-p-dioxin to mink. *Arch. Environ. Contam. Toxicol.*, **17**, 33–7.

Hodson, P.V., Blunt, B.R., Spry, D.J. and Austen, K. (1977) Evaluation of erythrocyte γ-amino levulinic dehydratase activity as a short-term indicator in fish of a harmful exposure to lead. *J. Fish. Res. Board Can.*, **34**, 501–8.

Hoffman, D.J. (1988) Effects of Krenite brush control agent (Fosamine ammonium) on embryonic development in mallards and bobwhite. *Environ. Toxicol. Chem.*, **7**, 69–75.

Hoffman, D.J. (1990) Embryotoxicity and teratogenicity of environmental contaminants to bird eggs. Rev. *Environ. Contam. Toxicol.*, **115**, 39–90.

Hoffman, D.J. and Albers, P.H. (1984) Evaluation of potential embryotoxicity and teratogenicity of 42 herbicides, insecticides, and petroleum contaminants to Mallard eggs. *Arch. Environ. Contam. Toxicol.*, **13**, 15–27.

Hoffman, D.J. and Eastin, W.C., Jr. (1981) Effects of Malathion, Diazinon, and Parathion on Mallard embryo development and cholinesterase activity. *Environ. Res.*, **26**, 472–85.

Hoffman, D.J., Eastin, W.C., Jr. and Gay, M.L. (1982) Embryotoxic and biochemical effects of waste crankcase oil on birds' eggs. *Toxicol. Appl. Pharmacol.*, **63**, 230–41.

Hoffman, D.J., Franson, J.C., Pattee, O.H., Bunck, C.M. and Murray, H.C. (1985) Biochemical and hematological effects of lead ingestion in nestling American Kestrels (*Falco sparverius*). *Comp. Biochem. Physiol.*, **80C**, 431–9.

Hoffman, D.J., Pattee, O.H., Wiemeyer, S.N. and Mulhern, B. (1981) Effects of lead shot ingestion on delta-aminolevulinic acid dehydratase activity, hemoglobin concentration and serum chemistry in Bald Eagles, *Haliaeetus leucocephalus*. *J. Wildl. Dis.*, **17**, 423–31.

Hoffman, D.J., Rattner, B.A., Sileo, L., Docherty, D. and Kubiak, T.J. (1987) Embryotoxicity, teratogenicity and aryl hydrocarbon hydroxylase activity in Forster's terns on Green Bay, Lake Michigan. *Environ. Res.*, **42**, 176–84.

Hoffman, D.J. and Sileo, L. (1984) Neurotoxic and teratogenic effects of an organophosphorous insecticide (phenyl phosphonothioic acid-O-ethyl-O-[4-nitrophenyl] ester) on Mallard development. *Toxicol. Appl. Pharmacol.*, **73**, 284–94.

Hoffman, D.J., Sileo, L. and Murray, H.C. (1984) Subchronic organophosphorus ester-induced delayed neurotoxicity in mallards. *Toxicol. Appl. Pharmacol.*, **75**, 128–36.

Hoffman, R.E., Stehr-Green, P.A., Webb, K.B., Evans, G., Knutsen, A.P., Schramm, W.F., Staake, J.L., Gibson, B.B. and Steinberg, K.K. (1987) Health effects of long-term exposure to 2,3,7,8-tetrachlorodibenzo-p-dioxin. *Adv. Modern Environ. Toxicol.*, **13**, 217–35.

Hohman, W.L., Pritchert, R.D., Pace III, R.M., Woolington, D.W. and Helm, R. (1990) Influence of ingested lead on body mass of wintering canvasbacks. *J. Wildl. Manage.*, **54**, 211–15.

Holdway, D.A., Sloley, B.D., Downer, R.G.H. and Dixon, D.G. (1986) Effect of pulse exposure to methoxychlor on brain serotonin levels in American Flagfish (*Jordanella floridae*, Goode and Bean) as modified by time after exposure, concentration and gender. *Environ. Toxicol. Chem.*, **5**, 289–94.

Holmes, S.B. and Boag, P.T. (1990) Inhibition of brain and plasma cholinesterase activity in zebra finches orally dosed with fenitrothion. *Environ. Toxicol. Chem.*, **9**, 323–34.

Holmes, W.N. and Cavanaugh, K.P. (1990) Some evidence for an effect of ingested petroleum on the fertility of the mallard drake (*Anas platyrhynchos*). *Arch. Environ. Contam. Toxicol.*, **19**, 898–901.

Holmes, W.N., Cavanaugh, K.P. and Cronshaw, J. (1978) The effects of ingested petroleum on oviposition and some aspects of reproduction in experimental colonies of mallard ducks (*Anas platyrhynchos*). *J. Reprod. Fert.*, **54**, 335–47.

Holsapple, M.P., McCay, J.A. and Barnes, D.W. (1986) Immunosuppression without liver induction by subchronic exposure to 2,7-dichlorodibenzo-p-dioxin in adult female B6C3F1 mice. *Toxicol. Appl. Pharmacol.*, **83**, 445–55.

Hose, J.E., Cross, J.N., Smith, S.G. and Diehl, D. (1987) Elevated circulating erythrocyte micronuclei in fishes from contaminated sites off southern California. *Mar. Environ. Res.*, **22**, 167–76.

House, R.V., Lauer, L.D., Murray, M.J., Ward, E.C. and Dean, J.H. (1985) Immunological studies in B6C3F1 mice following exposure to ethylene glycol monomethyl ether and its principal metabolite methoxyacetic acid. *Toxicol. Appl. Pharmacol.*, **77**, 358–62.

Hrdina, P.D., Peters, D.A.V. and Singhal, R.L. (1976) Effects of chronic exposure to cadmium, lead and mercury on brain biogenic amines in the rat. *Res. Commun. Chem. Pathol. Pharmacol.*, **15**, 483–93.

Huang, E., Nelson, S.-R. and F.R. (1986) Anti-estrogenic action of chlordecone in rat pituitary gonadotrophs in vitro. *Toxicol. Appl. Pharmacol.*, **82**, 62–9.

Huckle, K.R., Hutson, D.H., Logan, C.J., Morrison, B.J. and Warburton, P.A. (1989a) The fate of the rodenticide flocoumafen in the rat: Retention and elimination of a single oral dose. *Pestic. Sci.*, **25**, 297–312.

Huckle, K.R., Warburton, P.A., Forbes, S. and Logan, C.J. (1989b) Studies on the fate of flocoumafen in the Japanese Quail (*Coturnix coturnix japonica*). *Xenobiotica*, **18**, 51–62.

Hudson, P.M., Chen, P.H., Tilson, H.A. and Hong, J.S. (1985) Effects of p,p'-DDT on the rat brain concentrations of biogenic amine and amino acid neurotransmitters and their association with p,p'-DDT-induced tremor and hyperthermia. *J. Neurochem.*, **45**, 1349–55.

Hudson, J.L., Duque, R.E. and Lovett, E.J. (1985) Application of flow cytometry in immunotoxicology. In: *Immunotoxicology and Pharmacology* (Eds J.H. Dean, M.I. Luster, A.E. Munson, and H. Amos), pp. 159–77. Raven Press, New York.

Hudson, P.M., Tilson, H.A., Chen, P.H. and Hong, J.S. (1986) Neurobehavioral effects of permethrin are associated with alterations in regional levels of biogenic amine metabolites and amino acid neurotransmitters. *Neurotoxicology*, **7**, 143–54.

Hudson, R.H., Tucker, R.K. and Haegele, M.A. (1984) Handbook of toxicity of pesticides to wildlife. *U.S. Fish and Wildl. Ser. Resource Publ.*, **153**, pp. 90.

Hunt, G.L. and Hunt, M.W., Jr. (1977) Female–female pairing in western gulls (*Larus occidentalis*) in southern California. *Science*, **196**, 1466–7.

Hurst, J.G., Newcomer, W.S. and Morrison, J.A. (1974) Some effects of DDT, toxaphene and polychlorinated biphenyl on thyroid function in bobwhite quail. *Poult. Sci.*, **53**, 125–33.

Hutton, M. (1980) Metal contamination of feral Pigeons (*Columba livia*) from the London area: Part 2 – Biological effects of lead exposure. *Environ. Pollut.*, **22A**, 281–93.

Hutzinger, O., Nash, D.M., Safe, S., DeFreitas, A.S.W., Norstrom, R.J., Wildish, D.J. and Zitko, V. (1972) Polychlorinated biphenyls: Metabolic behavior of pure isomers in pigeons, rats, and brook trout. *Science*, **178**, 312–4.

IUPAC–IUB (1982) Nomenclature of retinoids. *Eur. J. Biochem.*, **129**, 1–5.

Ikeda, M., Conney, A.H. and Burns, J.J. (1968) Stimulatory effect of phenobarbital and insecticides on warfarin metabolism in the rat. *J. Pharmacol. Exp. Ther.*, **162**, 338–43.

Imanishi, J., Nomura, H., Matsubara, M., Kita, M., Won, S.J., Mizutani, T. and Kishida, T. (1980) Effect of polychlorinated biphenyl on viral infections in mice. *Infect. Immunity.*, **29**, 275–7.

Ito, Y. (1973) Studies on the influence of PCB on aquatic organisms. II. Changes in blood characteristics and plasma enzyme activities of Carp administered orally with PCB. *Bull. Jpn. Soc. Sci. Fish.*, **39**, 1135–8.

Jackim, E., Hamlin, J.M. and Sonis, S. (1970) Effects of metal poisoning on five liver enzymes in the Killifish (*Fundulus heteroclitus*). *J. Fish. Res. Board Can.*, **27**, 383–90.

Jaffe, I.A. (1979) Penicillamine in rheumatoid arthritis: Clinical pharmacology and biochemical properties. *Scand. J. Rheumatol.* (Suppl), **28**, 58–64.

Jakab, G.J. and Warr, G.A. (1981) Lung defenses against viral and bacterial challenges during immunosuppression with cyclophosphamide in mice. *Am. Rev. Respir. Dis.*, **123**, 524–8.

James, M.O. and Bend, J.R. (1980) Polycyclic aromatic hydrocarbon induction of cytochrome P-450-dependent mixed-function oxidases in marine fish. *Toxicol. Appl. Pharmacol.*, **54**, 117–33.

Janicki, R.H. and Kinter, W.B. (1971) DDT: Disrupted osmoregulatory events in the intestine of the eel *Anguilla rostrata* adapted to seawater. *Science*, **173**, 1146–8.

Jefferies, D.J. (1975) The role of the thyroid in the production of sublethal effects by organochlorine insecticides and polychlorinated biphenyls. In: *Organochlorine Insecticides: Persistent Organic Pollutants* (Ed. F. Moriarty), pp. 132–230. Academic Press, London.

Jefferies, D.J. (1967) The delay in ovulation produced by pp'-DDT and its possible significance in the field. *Ibis*, **109**, 266–72.

Jefferies, D.J. and French, M.C. (1971) Hyper- and hypothyroidism in pigeons fed DDT: An explanation for the 'thin eggshell phenomenon'. *Environ. Pollut.*, **1**, 235–42.

Jefferies, D.J. and French, M.C. (1972) Changes induced in the pigeon thyroid by p,p'-DDE and dieldrin. *J. Wildl. Manage.*, **36**, 24–30.

Jefferies, D.J. and Parslow, J.L.F. (1972) Effect of one polychlorinated biphenyl on size and activity of the gull thyroid. *Bull. Environ. Contam. Toxicol.*, **8**, 306–10.

Jefferies, D.J. and Parslow, J.L.F. (1976) Thyroid changes in PCB-dosed guillemots and their indication of one of the mechanisms of action of these materials. *Environ. Pollut.*, **10**, 293–311.

Jennings, D.M., Bunyan, P.J., Brown, P.M., Stanley, P.I. and Jones, F.J.S. (1975) Organophosphorus poisoning: A comparative study of the toxicity of carbophenothion to the Canada goose, the pigeon, and the Japanese quail. *Pestic. Sci.*, **6**, 245–57.

Jensen, S. (1966) Report of a new chemical hazard. *New Sci.*, **32**, 612.

Jensen, S., Jansson, B. and Olsson, M. (1979) Number and identity of anthropogenic substances known to be present in Baltic seals and their possible effects on reproduction. *Ann. N.Y. Acad. Sci.*, **320**, 436–48.

Jimenez, B.D., Burtis, L.S., Ezell, G.H., Egan, B.Z., Lee, N.E., Beauchamp, J.J. and McCarthy, J.F. (1988) The mixed function oxidase system of Bluegill Sunfish,

Lepomis macrochirus: Correlation of activities in experimental and wild fish. *Environ. Toxicol. Chem.*, **7**, 623–34.

John, T.M. and George, J.C. (1977) Blood levels of cyclic AMP, thyroxine, uric acid, certain metabolites and electrolytes under heat-stress and dehydration in the pigeon. *Arch. Int. Physiol. Biochim.*, **85** (Fascicule 3): 571–82.

Johnson, M.K. (1975) The delayed neuropathy caused by some organophosphorus esters: Mechanism and challenge. *CRC Crit. Rev. Toxicol.*, **3**, 289–316.

Johnson, M.K. (1982) The target for initiation of delayed neurotoxicity by organophosphorus esters: Biochemical studies and toxicological applications. *Rev. Biochem. Toxicol.*, **4**, 141–212.

Johnson, M.K. (1990) Organophosphates and delayed neuropathy – is NTE alive and well? *Toxicol. Appl. Pharmacol.*, **102**, 385–99.

Johnson, R.H. (1988) Biological markers in tort litigation. *Stat. Sci.*, **3**, 367–70.

Johnston, C.A., Demarest, K.T., McCormack, K.M., Hook, J.B. and Moore, K.E. (1980) Endocrinological, neurochemical, and anabolic effects of polybrominated biphenyls in male and female rats. *Toxicol. Appl. Pharmacol.*, **56**, 240–7.

Johnston, G., Collett, G., Walker, C., Dawson, A., Boyd, I. and Osborn, D. (1989) Enhancement of malathion toxicity to the hybrid Red-legged Partridge following exposure to Prochloraz. *Pestic. Biochem. Physiol.*, **35**, 107–18.

Johnstone, G.J., Ecobichon, D.J. and Huntzinger, O. (1974) The influence of pure polychlorinated biphenyl compounds on hepatic function in the rat. *Toxicol. Appl. Pharmacol.*, **28**, 66–81.

Jones, K.G. and Sweeney, G.D. (1980) Dependence of the porphyrogenic effect of 2,3,7,8-tetrachlorodibenzo(p)dioxin upon inheritance of aryl hydrocarbon hydroxylase responsiveness. *Toxicol. Appl. Pharmacol.*, **53**, 42–9.

Kacew, S. and Singhal, R.L. (1973) The influence of p, p'-DDT, -chlordane, heptachlor and endrin on hepatic and renal carbohydrate metabolism and cyclic AMP-adenyl cyclase system. *Life Sci.*, **13**, 1363–71.

Kaminski, N.E., Barnes, D.W., Jordan, S.D. and Holsapple, M.P. (1990) The role of metabolism in carbon tetrachloride-mediated immunosuppression: In vivo studies. *Toxicol. Appl. Pharmacol.*, **102**, 9–20.

Kaminski, N.E., Roberts, J.F. and Guthrie, E.F. (1982) The effect of DDT and dieldrin on rat peritoneal macrophages. *Pest. Biochem. Physiol.*, **17**, 191–5l.

Kammuller, M.E., Bloksma, N. and Seinen, W. (1989) Autoimmunity and toxicology. Immune disregulation induced by drugs and chemicals. In: *Autoimmunity and Toxicology* (Eds M.E. Kammuller, N. Bloksma and W. Seinen), pp. 132–230. Elsevier, Amsterdam.

Kanter, P.M. and Schwartz, H.S. (1979) A hydroxylapatite batch assay for quantitation of cellular DNA damage. *Anal. Biochem.*, **97**, 77–84.

Karlsson, B., Persson, B., Sodergren, A. and Ulfstrand, S. (1974) Locomotory and dehydrogenase activities of Redstarts, *Phoenicurus phoenicurus* L. (Aves), given PCB and DDT. *Environ. Pollut.*, **7**, 53–63.

Kato, R. (1961) Modifications of the toxicity of strychnine and octomethyl-pyrophosphoramide (OMPA) induced by pretreatment with phenaglycodol and thiopental. *Arzhelm. Forsch.*, **11**, 797–8.

Kawanishi, S., Mizutani, T. and Sano, S. (1978) Induction of porphyrin synthesis in chick embryo liver cell culture by synthetic polychlorobiphenyl isomers. *Biochim. Biophys. Acta.*, **540**, 83–92.

Keil, J.E., Priester, L.E. and Sandifer, S.H. (1971) Polychlorinated biphenyl (Aroclor

1242): Effects of uptake on growth, nucleic acids and chlorophyll of a marine diatom. *Bull. Environ. Contam. Toxicol.*, **6**, 156–9.

Keith, J.O. (1978) *Synergistic effects of DDE and food stress on reproduction in Brown Pelicans and Ring Doves.* Ph.D. Thesis, Ohio State University, Columbus, OH.

Kemp, J.R. and Wallace, K.B. (1990) Molecular determinants of the species-selective inhibition of brain acetylcholinesterase. *Toxicol. Appl. Pharmacol.*, **104**, 246–58.

Kenaga, E.E. (1981) Comparative toxicity of 131 596 chemicals to plant seeds. *Ecotoxicol. Environ. Saf.*, **5**, 469–75.

Kennedy, S.W. (1991) A novel approach to the detection of organophosphate pesticide – induced inhibition of cholinesterase in wild birds.

Kennedy, S.W., Wigfield, D.C. and Fox, G.A. (1986) The delay in polyhalogenated aromatic hydrocarbon-induced porphyria: Mechanistic reality or methodological artefact? *Toxicol. Lett.*, **31**, 235–41.

Kennedy, S.W. and Wigfield, D.C. (1986) The delay in polyhalogenated aromatic hydrocarbon-induced porphyria: The effect of diet preparation. *Toxicol. Lett.*, **32**, 195–202.

Kerkvliet, I.N., Brauner, J.A. and Beacher-Steppan, L. (1985a) Effects of dietary technical pentachlorophenol exposure on T-cell, macrophage and natural cell activity in C57B1/6 mice. *Int. J. Immunopharmac.*, **7**, 239–47.

Kerkvliet, N.I., Brauner, J.A. and Matlock, J.P. (1985b) Humoral immunotoxicity of polychlorinated diphenyl ethers, phenoxyphenols, dioxins and furans present as contaminants of technical grade pentachlorophenol. *Toxicol.*, **36**, 307–24.

Kezić, N., Britvić, S., Protić, M., Simmons, J.E., Rijavec, M., Zahn, R.K. and Kurelec, B. (1983) Activity of benz[a]pyrene monooxygenase in fish from the Save River, Yugoslovia: Correlation with pollution. *Sci. Total Environ.*, **27**, 59–69.

Khera, K.S. and Lyon, D.A. (1968) Chick and duck embryos in the evaluation of pesticide toxicity. *Toxicol. Appl. Pharmacol.*, **13**, 1–15.

Kimber, I., Jones, K. and Vignali, D.A. (1986) The influence of 7,12-dimethylbenzantracene on natural killer (NK) cell function in rats. *J. Clin. Lab. Immunol.*, **20**, 193–205.

Kimbrough, R.D. (1972) Toxicity of chlorinated hydrocarbons and related compounds. A review including chlorinated dibenzodioxins and chlorinated dibenzofurans. *Arch. Environ. Health*, **25**, 125–31.

Kimbrough, R.D. and Gaines, T.B. (1968) Effect of organic phosphorus compounds and alkylating agents on the rat fetus. *Arch. Environ. Health*, **16**, 805–8.

King, K.A., White, D.H. and Mitchell, C.A. (1984) Nest defence behavior and reproductive success of laughing gulls sublethally dosed with parathion. *Bull. Environ. Contam. Toxicol.*, **33**, 499–504.

Koch, R.B., Cutkomp, L.K. and Do, F.M. (1969) Chlorinated hydrocarbon insecticide inhibition of cockroach and honeybee ATPases. *Life Sci.*, **8**, 289–97.

Kociba, R.J. (1982) Morphologic considerations in the detection of immune suppression in routine toxicity studies. In: *Immunologic Considerations in Toxicology* (Ed. R.P. Sharma), pp. 124–31. CRC Press, Boca Raton, FL.

Koeman, J.H., Van Velzen-Blad, H.C.W., De Vries, R. and Vos, J.G. (1973) Effects of PCB and DDE in cormorants and evaluation of PCB residues from an experimental study. *J. Reprod. Fert.*, **19**(suppl.), 353–64.

Kohn, K.W. (1983) The significance of DNA-damage assays in toxicity and carcinogenicity assessment. *Ann. N.Y. Acad. Sci.*, **407**, 106–18.

Koller, L.D. (1977) Enhanced polychlorinated biphenyl lesions in Moloney leukaemia virus-infected mice. *Clin. Toxicol.*, **11**, 107–16.

Koller, L.D. (1979) Effects of environmental contaminants on the immune system. *Adv. Vet. Sci. Comp. Med.*, **23**, 267–95.

Koller, L.D. (1987) Immunotoxicology today. *Toxicol. Pathol.*, **15**, 346–51.

Koss, G., Schuler, E., Arndt, B., Seidel, J., Seubert, S. and Seubert, A. (1986) A comparative toxicological study on pike (*Esox lucius* L.) from the River Rhine and River Lahn. *Aquat. Toxicol.*, **8**, 1–9.

Koval, P.J., Peterle, T.J. and Harder, J.D. (1987) Effects of polychlorinated biphenyls on mourning dove reproduction and circulating progesterone levels. *Bull. Environ. Contam. Toxicol.*, **39**, 663–70.

Kraus, M.L. and Kraus, D.B. (1986) Differences in the effects of mercury on predator avoidance in two populations of the grass shrimp *Palaemonetes pugio*. *Mar. Environ. Res.*, **18**, 277–89.

Krzystyniak, K., Hugo, P., Flipo, D. and Fournier, M. (1985) Increased susceptibility to mouse hepatitis virus 3 of peritoneal macrophages exposed to dieldrin. *Toxicol. Appl. Pharmacol.*, **80**, 397–408.

Krzystyniak, K., Fournier, M., Trottier, B., Nadeau, D. and Chevalier, G. (1987) Immunosuppression in mice after inhalation of cadmium aerosol. *Toxicol. Lett.*, **38**, 1–12.

Kubiak, T.J., Harris, H.J., Smith, L.M., Schwartz, T.R., Stalling, D.L., Trick, J.A., Sileo, L., Docherty, D.E. and Erdman, T.C. (1989) Microcontaminants and Reproductive Impairment of the Forster's Tern on Green Bay, Lake Michigan – 1983. *Arch. Environ. Contam. Toxicol.*, **18**, 706–27.

Kumar, M.V.S. and Desiraju, T. (1990) Regional alterations of brain biogenic amines and GABA/glutamate levels in rats following chronic lead exposure during neonatal development. *Arch. Toxicol.*, **64**, 305–14.

Kupfer, D. (1975) Effects of pesticides and related compounds on steroid metabolism and function. *CRC Crit. Rev. Toxicol.*, **4**, 83–124.

Kurelec, B., Britvić, S., Rijavec, M., Muller, W.E.G. and Zahn, R.K. (1977) Benz(a)pyrene monooxygenase induction in marine fish – molecular response to oil pollution. *Mar. Biol.*, **44**, 211–16.

Lack, D. (1954) *The Natural Regulation of Animal Numbers*. Oxford University Press. 343 pp.

Lambert, B., Lindblad, A., Hormberg, K. and Francesconi, D. (1982) Use of sister chromatid exchange to monitor human populations for exposure to toxicologically harmful agents. In: *Sister Chromatid Exchange*. (Ed. S. Wolff), pp. 149–82. John Wiley & Sons, New York.

Lambrecht, R., Sinclair, P.R., Bement, W.J. and Sinclair, J.F. (1988) Uroporphyrin accumulation in cultured chick embryo hepatocytes. Comparison of 2,3,7,8-tetrachlorodibenzo-p-dioxin and 3,4,3′,4′-tetrachlorobiphenyl. *Toxicol. Appl. Pharmacol.*, **96**, 507–16.

Lamont, T.G., Bagley, G.E. and Reichel, W.L. (1970) Residues of o,p′-DDD and o, p′-DDT in brown pelican eggs and mallard ducks. *Bull. Environ. Contam. Toxicol.*, **5**, 231–6.

Lane, C.E. and Scura, E.D. (1970) Effects of dieldrin on glutamic oxaloacetic transaminase in *Poecilia latipinna*. *J. Fish. Res. Board Can.*, **27**, 1869–71.

Lawrence, D.A. (1981) In vivo and in vitro effects of lead on humoral and cell-mediated immunity. *Infect. Immunity*, **31**, 136–42.

Lawrence, D.A. (1985) Immunotoxicity of heavy metals. In: *Immunotoxicology and*

Immunopharmacology (Eds J.H. Dean, M.L. Luster, A.E. Munson, and H. Amos), pp. 341–53. Raven Press, New York.

Lawrence, L.J. and Casida, J.E. (1984) Interactions of lindane, toxaphane and cyclodienes with brain-specific δ-butylbicyclophosphorothionate receptor. *Life Sci.*, **35**, 171–8.

Leatherland, J.F. (1982) Environmental physiology of the teleostean thyroid gland: A review. *Environ. Biol. Fish*, **7**, 83–110.

Leatherland, J.F. and Sonstegard, R.A. (1978) Lowering of serum thyroxine and triidothyronine levels in yearling coho salmon, *Oncorhynchus kisutch*, by dietary Mirex and PCBs. *J. Fish. Res. Board Can.*, **35**, 1285–9.

Leatherland, J.F. and Sonstegard, R.A. (1979) Effect of dietary mirex and PCB (Aroclor 1254) on thyroid activity and lipid reserves in rainbow trout *Salmo gairdneri* Richardson. *J. Fish Dis.*, **2**, 43–8.

Leatherland, J.F. and Sonstegard, R.A. (1980) Effect of dietary Mirex and PCB's in combination with food deprivation and testosterone administration on thyroid activity and bioaccumulation of organochlorines in rainbow trout *Salmo gairdneri* Richardson. *J. Fish Dis.*, **3**, 115–24.

Leatherland, J.F. and Sonstegard, R.A. (1982a) Bioaccumulation of organochlorines by yearling coho salmon (*Oncorhynchus kisutch* Walbaum) fed diets containing Great Lakes coho salmon, and the pathophysiological responses of the recipients. *Comp. Biochem. Physiol.*, **72C**, 91–9.

Leatherland, J.F. and Sonstegard, R.A. (1982b) Thyroid responses in rats fed diets formulated with Great Lakes coho salmon. *Bull. Environ. Contam. Toxicol.*, **29**, 341–6.

Leatherland, J.F. and Sonstegard, R.A. (1984) Pathobiological responses of feral teleosts to environmental stressors: Interlake studies of the physiology of Great Lakes salmon. *Adv. Environ. Sci. Technol.*, **16**, 115–49.

Leatherland, J.F., Sonstegard, R.A. and Moccia, R.D. (1981) Interlake differences in body and gonad weights and serum constitutents of Great Lake coho salmon (*Oncorhynchus kisutch*). *Comp. Biochem. Physiol.*, **69A**, 701–4.

Lee, Y.-Z., O'Brien, P.J., Payne, J.F. and Rahimtula, A.D. (1986) Toxicity of petroleum crude oils and their effects on xenobiotic metabolizing enzymes activities in the chicken embryo in ovo. *Environ. Res.*, **39**, 153–63.

Lee, T.P. and Chang, K.-J. (1985) Health effects of polychlorinated biphenyls. In: *Immunotoxocology and Immunopharmacology*. (Ed J.H. Dean, M.L. Luster, A.E. Munson and H. Amos), pp. 415–22. Raven Press, New York.

Lefkowitz, R.J. Kobilka, B.K. and Caron, M.G. (1989) The new biology of drug receptors. *Biochem. Pharmacol.* **38**, 2941–8.

Leibish, I.J., and Moraski, R.M. (1987) Mechanisms of immunomodulation by drugs. *Toxicol. Pathol.*, **15**, 338–45.

Leighton, F.A. (1990) The systemic toxicity of Prudhoe Bay crude and other petroleum oils to CD-1 mice. *Arch. Environ. Contam. Toxicol.*, **19**, 257–62.

Lehtinen, K.-J. (1990) Mixed-function oxygenase enzyme responses and physiological disorders in fish exposed to Kraft pulp-mill effluents: A hypothetical model. *Ambio*, **19**, 259–65.

Lill, P.H. and Gangemi, D. (1986) Suppressive effects of 3-methylcholanthrene on the in vitro antitumour activity of naturally cytotoxic cells. *J. Toxicol. Environ. Health*, **17**, 347–56.

Lim, C.K., Li, F. and Peters, T.J. (1988) High-performance liquid chromatography of porphyrins. *J. Chromatogr.*, **429**, 123–53.

Lima, S.L. and Dill, L.M. (1990) Behavioral decisions made under the risk of predation; a review and prospectus. *Can. J. Zool.*, **68**, 619–40.

Lincer, J.L. (1975) DDE-induced eggshell thinning in the American Kestrel: A comparison of the field situation and laboratory results. *J. Appl. Ecol.*, **12**, 781–93.

Linder, R.E., Gaines, T.B. and Kimbrough, R.D. (1974) The effect of polychlorinated biphenyls on rat reproduction. *Food Cosmet. Toxicol.*, **12**, 63–77.

Litterst, C.L. (1978) Variations in hepatic microsomal drug metabolism in C57B1/6 mice from three different suppliers. *Pharmacology*, **16**, 131–4.

Little, E.E., Archeski, R.D., Flerov, B.A. and Kozlovakaya, V.I. (1990) Behavioral indicators of sublethal toxicity in Rainbow Trout. *Arch. Environ. Contam. Toxicol.*, **19**, 380–5.

de Llamas, M.C., de Castro, A.C. and de D'Angelo, A.M.P. (1985) Cholinesterase activities in developing amphibian embryos following exposure to the insecticides dieldrin and malathion. *Arch. Environ. Contam. Toxicol.*, **14**, 161–6.

Lock, R.A.C., Cruijsen, P.M.J.M. and van Overbeeke, A.P. (1981) Effects of mercuric chloride and methylmercuric chloride on the osmoregulatory function of the gills in Rainbow Trout, *Salmo gairdneri* Richardson. *Comp. Biochem. Physiol.*, **68C**, 151–9.

Logner, K.R., Neathery, M.W., Miller, W.J. Gentry, R.P., Blackmon, D.M. and White, F.D. (1984) Lead toxicity and metabolism from lead sulfate fed to Holstein calves. *J. Dairy Sci.*, **67**, 1007–13.

Loose, L.D., Pittman, K.A., Benitz, K.F. and Silkworth, J.B. (1977) Polychlorinated biphenyl and hexachlorobenzene induced humoral immunosuppression. *J. Reticuloendothelial Soc.*, **22**, 253–71.

Loose, L.D., Silkworth, J.B. and Simpson, D.W. (1978) Influence of cadmium on the phagocytic and microbicidal activity of murine peritoneal macrophages, pulmonary alveolar macrophages and polymorphonuclear neutrophils. *Infect. Immunity*, **22**, 378–81.

Lotti, M., Becker, C., Aminoff, M., Woodrof, J., Seiber, J., Talcott, R. and Richardson, R. (1983) Occupational exposure to the cotton defoliants DEF and Merphos. A rational approach to monitoring organophosphorus-induced delayed neurotoxicity. *J. Occup. Med.*, **25**, 517–22.

Lotti, M. and Johnson, M.K. (1978) Neurotoxicity of organophosphorus pesticides: Predictions can be based on in vitro studies with hen and human enzymes. *Arch. Toxicol.*, **41**, 215–21.

Lubet, R.A., Nims, R.W., Ward, J.M., Rice, J.M. and Diwan, B.A. (1989) Induction of cytochrome P450b and its relationship to liver tumor promotion. *J. Amer. Coll. Toxicol.*, **8**, 259–68.

Ludke, J.L., Hill, E.F. and Dieter, M.P. (1975) Cholinesterase (ChE) response and related mortality among birds fed ChE inhibitors. *Arch. Environ. Contam. Toxicol.*, **3**, 1–21.

Ludwig, J.P. and Tomoff, C.S. (1966) Reproductive success and insecticide residues in Lake Michigan Herring Gulls. *Jack-Pine Warbler*, **44**(2), 77–85.

Lundholm, E. (1987) Thinning of eggshells in birds by DDE: Mode of action on the eggshell gland. *Comp. Biochem. Physiol.*, **88C**, 1–22.

Lundholm, C.E. (1988) The effects of DDE, PCB and chlordane on the binding of progesterone to its cytoplasmic receptor in the eggshell gland mucosa of birds and the endometrium of the mammalian uterus. *Comp. Biochem. Physiol.*, **89C**, 361–8.

Luster, M.I., Dean, J.H., Boorman, G.A., Archer, D.L., Lauer, L., Lawson, L.D., Moore, J.A. and Wilson, R.E. (1981) The effects of orthophenylphenol, tris(2, 3-dichloropropyl)phosphate and cyclophosphamide on the immune system and host susceptibility of mice following subchronic exposure. *Toxicol. Appl. Pharmacol.*, **58**, 252–61.

Luster, M.I., Faith, R.E. and Moore, J.A. (1978a) Effects of polybrominated biphenyls (PBB) on immune response in Rodents. *Environ. Health Persp.*, **23**, 227–32.

Luster, M.I., Faith, R.E., and Kimmel, C.A. (1978b) Depression of humoral immunity in rats following chronic developmental exposure. *J. Environ. Pathol. Toxicol.*, **1**, 397–402.

Luster, M.I., Munson, A.E. Thomas, P.T., Holsapple, M.P., Fenters, J.D., White, K.L., Lauer, L.D., Germolec, D.R., Rosenthal, G.J. and Dean, J.H. (1988) Development of a testing battery to assess chemical-induced immunotoxicity: National Toxicology Program's guidelines for immunotoxicity evaluation in mice. *Fundam. Appl. Toxicol.*, **10**, 2–19.

Luxon, P.L., Hodson, P.V. and Borgmann, U. (1987) Hepatic aryl hydrocarbon hydroxylase activity of Lake Trout (*Salvelinus namaycush*) as an indicator of organic pollution. *Environ. Toxicol. Chem.*, **6**, 649–57.

Mac, M.J. (1988) Toxic substances and survival of Lake Michigan salmonids: Field and laboratory approaches. *Adv. Environ. Sci. Technol.*, **21**, 389–401.

Maccubbin, A.E., Black, P., Trzeciak, L. and Black, J.J. (1985) Evidence for polynuclear aromatic hydrocarbons in the diet of bottom-feeding fish. *Bull. Environ. Contam. Toxicol.*, **34**, 876–82.

Mackness, M.I., Thompson, H.M., Hardy, A.R. and Walker, C.H. (1987) Distinction between 'A'-esterases and arylesterases. *Biochem. J.*, **245**, 293–6.

Mackness, M.J. and Walker, C.H. (1988) Multiple forms of sheep serum 'A'-esterase activity associated with high density lipoproteins. *Biochem. J.*, **250**, 539–45.

Maigetter, R.Z., Ehrlich, R., Fenters, J.D. and Gardner, D.E. (1976) Potentiating effects of manganese dioxide on experimental respiratory infections. *Environ. Res.*, **11**, 386–91.

Malins, D.C., Krahn, M.M., Myers, M.S., Rhodes, L.D., Brown, D.W., Krone, C.A., McCain, B.B. and Chan, S.-L. (1985) Toxic chemicals in sediments and biota from a creosote-polluted harbor: Relationships with hepatic neoplasms and other hepatic lesions in English sole (*Parophrys vetulus*). *Carcinogenesis*, **6**, 1463–9.

Malins, D.C., McCain, B.B., Landahl, J.T., Myers, M.S., Krahn, M.M., Brown, D.W., Chan, S.-L. and Roubal, W.T. (1988) Neoplastic and other diseases in fish in relation to toxic chemicals: An overview. *Aquat. Toxic.*, **11**, 43–67.

Margolin, B.H. (1988) Statistical aspects of using biologic markers. *Stat. Sci.*, **3**, 351–7.

Margolin, B.H. and Shelby, M.D. (1985) Sister chromatid exchanges: A re-examination of the evidence for sex and race differences in humans. *Environ. Mutagen.*, **7** (Suppl. 4), 63–72.

Marks, G.S. (1985) Exposure to toxic agents: The heme biosynthetic pathway and hemoproteins as indicator. *CRC Crit. Rev. Toxicol.*, **15**, 151–79.

Mason, G., Farrell, K., Keys, B., Piskorska-Pliszczynska, J., Safe, L. and Safe, S. (1986) Polychlorinated dibenzo-p-dioxins: Quantitative in vitro and in vivo structure-activity relationships. *Toxicology*, **41**, 21–31.

Mason, G., Zacharewski T., Denomme M.A., Safe, L. and Safe, S. (1987) Poly-brominated dibenzo-p-dioxins and related compounds: Quantitative in vivo and in vitro structure-activity relationships. *Toxicology*, **44**, 245–55.

Maugh, T.H., II (1984) Tracking exposure to toxic substances. *Science*, **226**, 1183–4.

Maurer, T., Thomann, N., Weinrich, E.G. and Hess, R. (1979) Predictive evaluation in animals of the contact allergenic potential of medically important substances. II. Comparison of different methods of cutaneous sensitization with weak allergens. *Contact Derm.*, **5**, 1–10.

McArthur, M.L.B., Fox, G.A., Peakall, D.B. and Philogène, B.J.R. (1983) Ecological significance of behavioral and hormonal abnormalities in breeding ring doves fed an organochlorine chemical mixture. *Arch. Environ. Contam. Toxicol.*, **12**, 343–53.

McBee, K. and Bickham, J.W. (1988) Petrochemical-related DNA damage in wild rodents detected by flow cytometry. *Bull. Environ. Contam. Toxicol.*, **40**, 343–9.

McBee, K., Bickham, J.W., Brown, K.W. and Donnelly, K.C. (1987) Chromosomal aberrations in native small mammals (*Peromyscus leucopus* and *Sigmodon hispidus*) at a petrochemical waste disposal site: I. Standard karyology. *Arch. Environ. Contam. Toxicol.*, **16**, 681–8.

McCarthy, J.F. (1990) Concluding remarks: Implementation of a biomarker-based environmental monitoring program. In: *Biomarkers of Environmental Contamination* (Ed. J.F. McCarthy and L.R. Shugart), pp. 429–39. Lewis Publishers, Boca Raton, FL.

McCarthy, J.F. and Shugart, L.R. (1990) Biological markers of environmental contamination. In: *Biomarkers of Environmental Contamination* (Eds J.F. McCarthy and L.R. Shugart), pp. 3–14. Lewis Publishers, Boca Raton, FL.

McKee, M.J. and Knowles, C.O. (1986) Protein, nucleic acid and adenylate levels in *Daphnia magna* during chronic exposure to chlordecone. *Environ. Pollut.*, **42A**, 335–51.

McKinney, J.D., Fawkes, J., Jordan, S., Chae, K., Oatley, S., Coleman, R.E. and Briner, W. (1985) 2,3,7,8-tetrachlorodibenzo-p-dioxin (TCDD) as a potent and persistent thyroxine agonist: A mechanistic model for toxicity based on molecular reactivity. *Environ. Health Perspect.*, **61**, 41–53.

McLaughlin, J., Jr., Marliac, J.-P., Verrett, M.J., Mutchler, M.K. and Fitzhugh, O.G. (1963) The injection of chemicals into the yolk sac of fertile eggs prior to incubation as a toxicity test. *Toxicol. Appl. Pharmacol.*, **5**, 760–71.

McMurtry, M.J., Wales, D.L., Scheider, W.A., Beggs, G.L. and Dimond, P.E. (1989) Relationship of mercury concentrations in lake trout (*Salvelinus namaycush*) and smallmouth bass (*Micropterus dolomieui*) to the physical and chemical characteristics of Ontario lakes. *Can. J. Fish. Aquat. Sci.*, **46**, 426–34.

Meany, J.E. and Pocker, Y. (1979) The in vitro inactivation of lactate dehydrogenase by organochlorine insecticides. *Pest. Biochem. Physiol.*, **11**, 232–42.

Meiniel, R. (1977) Activités cholinestérasiques et expression de la tératogènese axiale chez l'embryon de Caille exposé aux organophosphores. *C.R. Acad. Sci. Paris*, **285**, 401–4.

Meneely, G.A. and Wyttenbach, C.R. (1989) Effects of the organophosphate insecticides diazinon and parathion on Bobwhite Quail embryos: Skeletal defects and acetylcholinesterase activity. *J. Exp. Zool.*, **252**, 60–70.

Mentlein, R., Ronai, A., Robbi, M., Heyman, E. and Deimling, O.V. (1987) Genetic identification of rat liver carboxylesterases isolated in different laboratories. *Biochim. Biophys. Acta*, **913**, 27–38.

Metcalfe, C.D., Cairns, V.W. and Fitzsimons, J.D. (1988) Experimental induction of liver tumours in rainbow trout (*Salmo gairdneri*) by contaminated sediment from Hamilton Harbour, Ontario. *Can. J. Fish. Aquat. Sci.*, **45**, 2161–7.

Meyers, S.M., Cummings, J.L. and Bennett, R.S. (1990) Effects of methyl parathion on Red-winged Blackbird (*Agelaius phoeniceus*) incubation behavior and nesting success. *Environ. Toxicol. Chem.*, **9**, 807–14.

Miller, D.S., Kinter, W.B. and Peakall, D.B. (1976a) Enzymatic basis for DDE-induced eggshell thinning in a sensitive bird. *Nature*, **259**, 122–4.

Miller, D.S., Kinter, W.B., Peakall, D.B. and Risebrough, R.W. (1976b) DDE feeding and plasma osmoregulation in ducks, guillemots, and puffins. *Am. J. Physiol.*, **231**, 370–6.

Miller, K. and Scott, M.P. (1985) Immunological consequences of dioctyltin dichloride (DOTC)-induced thymic injury. *Toxicol. Appl. Pharmacol.*, **78**, 395–403.

Mineau, P. (1991) The regulatory assessment of cholinesterase-inhibiting insecticides. In: (Mineau, P. Ed.) *Cholinesterase-Inhibiting Insecticides – their impact on wildlife and the environment*. Elevier, Austlerdan, pp. 277–300.

Mineau, P., Fox, G.A., Norstrom, R.J., Weseloh, D.V., Hallett, D.J. and Ellenton, J.A. (1984) Using the herring gull to monitor levels and effects of organochlorine contamination in the Canadian Great Lakes. *Adv. Environ. Sci. Technol.*, **14**, 425–52.

Mineau, P. and Peakall, D.B. (1987) An evaluation of avian impact assessment techniques following broad-scale forest insecticide sprays. *Environ. Toxicol. Chem.*, **6**, 781–91.

Mineau, P. and Pedrosa, M. (1986) A portable device for non-destructive determination of avian embryonic viability. *J. Field Ornithol.*, **57**, 53–6.

Mineau, P. and Weseloh, D.V.C. (1981) Low-disturbance monitoring of Herring Gull reproductive success on the Great Lakes. *Colonial Waterbirds*, **4**, 138–42.

Miranda, C.L., Henderson, M.C., Wang, J.-L., Nakaue, H.S. and Buhler, D.R. (1986) Induction of acute renal porphyria in Japanese quail by Aroclor 1254. *Biochem. Pharmacol.*, **35**, 3637–9.

Miranda, C.L., Henderson, M.C., Wang, J.-L., Nakaue, H.S. and Buhler, D.R. (1987) Effects of polychlorinated biphenyls on porphyrin synthesis and cytochrome P-450-dependent monooxygenases in small intestine and liver of Japanese Quail. *J. Toxicol. Environ. Health*, **20**, 27–35.

Mitchell, C.A. and White, D.H. (1982) Seasonal brain acetylcholinesterase activity in three species of shorebirds overwintering in Texas. *Bull. Environ. Contam. Toxicol.*, **29**, 360–5.

Moccia, R.D., Fox, G.A. and Britton, A. (1986) A quantitative assessment of thyroid histopathology of herring gulls (*Larus argentatus*) from the Great Lakes and a hypothesis on the causal role of environmental contaminants. *J. Wildl. Dis.*, **22**, 60–70.

Moccia, R.D., Leatherland, J.F. and Sonstegard, R.A. (1977) Increasing frequency of thyroid goiters in Coho salmon (*Oncorhynchus kisutch*) in the Great Lakes. *Science*, **198**, 425–6.

Moccia, R.D., Leatherland, J.F., Sonstegard, R.A. and Holdrinet, M.V.H. (1978) Are goiter frequencies in Great Lakes salmon correlated with organochlorine residues? *Chemosphere*, **7**, 649–52.

Modak, A.T., Weintraub, S.T. and Stavinoha, W.B. (1975) Effect of chronic ingestion of lead on the central cholinergic system in rat brain regions. *Toxicol. Appl. Pharmacol.*, **34**, 340–7.

Mohamed, Z.A. and El-Sheamy, M.K. (1989) Aminotransferases, alkaline phosphatase and cholinesterases of hen tissues administered a single oral dose of alpha cyano pyrethyroids. *Egypt J. Food Sci.*, **16**, 105–10.

Moore, N.W. (1966) A pesticide monitoring system with special reference to the selection of indicator species. *J. Appl. Ecol.*, **3**, 261–9.

Morahan, P.S., Bradley, A.E., Munson, A.E., Duke, S., Fromtling, R.A., Marciano-Cabral, F. and Jesse, E. (1984) Immunotoxic effects of diethylsilbertrol (DES) and calcium chloride (CAD) on host resistance: Comparison with cyclophosphamide (CPS). *Prog. Clin. Biol. Res.*, **161**, 403–6.

Morgan, M.J., Fancey, L.L. and Kiceniuk, J.W. (1990) Response and recovery of brain acetylcholinesterase activity in Atlantic salmon (*Salmo salar*) exposed to fenitrothion. *Can. J. Fish. Aquat. Sci.*, **47**, 1652–4.

Mouw, D., Kalitis, K., Anver, M., Schwartz, J., Constan, A., Hartung, R., Cohen, B. and Ringer, D. (1975) Lead. Possible toxicity in urban vs rural rats. *Arch. Environ. Health*, **30**, 276–80.

Munson, A.E. (1987) Immunopathologic alterations observed in laboratory animals: Consideration in the choice of experimental animals for evaluation of the immune system. In: *Environmental Chemical Exposures and Immune System Integrity* (Eds E.J. Burger, R.G. Tardiff, and J.A. Bellanti), pp. 35–48. Princeton Scientific Publishing Co., Inc., Princeton, NJ.

Murphy, S.D. and Cheever, K.L. (1968) Effect of feeding insecticides. Inhibition of carboxyesterase and cholinesterase activities in rats. *Arch. Environ. Health*, **17**, 749–58.

Murray, M.J., Wilson, F.D., Fisher, G.L. and Erikson, K.L. (1983) Modulation of murine lymphocyte and macrophage proliferation by parenteral zinc. *Clin. Exp. Immunol.*, **53**, 744–9.

Myers, M.S., Landahl, J.T., Krahn, M.M., Johnson, L.L. and McCain, B.B. (1990) Overview of studies on liver carcinogenesis in English sole from Puget Sound; evidence for a xenobiotic chemical etiology. I: Pathology and epizootiology. *Sci. Total Environ.*, **94**, 33–50.

Myllyvirta, T.P. and Vuorinen, P.J. (1989) Avoidance of bleached kraft mill effluent by pre-exposed *Coregonus albula* L. *Water Res.*, **23**, 1219–27.

Nagarkatti, P.S., Sweeney, G.D., Gauldie, J. and Clark, D.A. (1984) Sensitivity to suppression of cytotoxic T-cell generation by 2,3,7,8-tetrachlorodi-benzo-p-dioxin (TCDD) is dependent on the Ah genotype of the murine host. *Toxicol. Appl. Pharmacol.*, **72**, 169–76.

Nayak, B.N., Ray, M., Persaud, T.V.N. and Nigli, M. (1989a) Relationship of embryotoxicity to genotoxicity of lead nitrate in mice. *Exp. Pathol.*, **36**, 65–73.

Nayak, B.N., Ray, M., Persaud, T.V.N. and Nigli, M. (1989b) Embryotoxicity and in vivo cytogenetic changes following maternal exposure to cadmium chloride in mice. *Exp. Pathol.*, **36**, 75–80.

Nayak, B.N. and Petras, M.L. (1985) Environmental monitoring for genotoxicity: In vivo sister chromatid exchange in the house mouse (*Mus musculus*). *Can. J. Genet. Cytol.*, **27**, 351–6.

Neal, R.A. (1967) Studies on the metabolism of diethyl 4-nitrophenyl phosphorothionate (parathion) in vitro. *Biochem. J.*, **103**, 183–91.

Nebert, D.W. and Gonzalez, F.J. (1987) P450 genes: Structure, evolution, and regulation. *Ann. Rev. Biochem.*, **56**, 945–93.

Nebert, D.W., Nelson, D.R. and Feyereisen, R. (1989) Evolution of the cytochrome P450 genes. *Xenobiotica*, **19**, 1149–60.

Neilan, B.A., O'Neil, K. and Handwerger, B.S. (1983) Effect of low-level lead exposure on antibody-dependent and natural killer cell-mediated cytotoxicity. *Toxicol. Appl. Pharmacol.*, **69**, 272–5.

Neuberger, J. (1989) Halotane hepatitis – an example of possibly immune-mediated hepatotoxicity. In: *Autoimmunity and Toxicology* (Eds M.E. Kammuller, N. Bloksma, and W. Seinen), pp. 215–36. Elsevier Science Publishers, Amsterdam.

Neumann, H.-G. (1984) Analysis of hemoglobin as a dose monitor for alkylating and arylating agents. *Arch. Toxicol.*, **56**, 1–6.

Newton, I. (1989) *Lifetime Reproduction in Birds*. Academic Press, London. 479 pp.

Noji, S., Nohno, T., Koyama, E., Muto, K., Ohyama, K., Aoki, Y., Tamura, K., Ohsugi, K., Ide, H., Taniguchi, S. and Saito, T. (1991) Retinoic acid induces polarizing activity but is unlikely to be a morphogen in the chick limb bud. *Nature*, **350**, 83–6.

Noltie, D.B. (1988) Differences in breeding Lake Superior pink salmon (*Oncorhynchus gorbuscha* (Walbaum)) associated with variation in thyroid hyperplasia. *Can. J. Zool.*, **66**, 2688–94.

Noltie, D.B., Leatherland, J.F. and Keenleyside, M.H.A. (1988) Epizootics of thyroid hyperplasia in Lake Superior and Lake Erie pink salmon (*Oncorhynchus gorbuscha* (Walbaum)). *Can. J. Zool.*, **66**, 2676–87.

Norstrom, R.J., Hallett, D.J., Simon, M. and Mulvihill, M.J. (1982) Analysis of Great Lakes herring gull eggs for tetrachlorodibenzo-p-dioxins. In: *Chlorinated Dioxins and Related Compounds – Impact on the Environment*, pp. 173–81. Pergamon Press, New York.

Norstrom, R.J., Simon, M., Muir, D.C.G. and Schweinsburg, R.E. (1988) Organo-chlorine contaminants in Arctic marine food chains: Identification, geographical distribution and temporal trends in polar bears. *Environ. Sci. Technol.*, **22**, 1063–71.

NRC (National Research Council) (1987) Committee on biological markers. *Environ. Health Persp.*, **74**, 3–9.

Oesch-Bartlomowicz, B. and Oesch, F. (1990) Phosphorylation of cytochrome P450 isoenzymes in intact hepatocytes and its importance for their function in metabolic processes. *Arch. Toxicol.*, **64**, 257–61.

Oishi, T. and Konishi, T. (1978) Effects of photoperiod and temperature on testicular and thyroid activity of the Japanese quail. *Gen. Comp. Endocrinol.*, **36**, 250–4.

Oliver, J. and Baylé, J.D. (1980) Brain photoceptors for the photo-induced testicular response in birds. *Experientia*, **38**, 1021–9.

Osborn, D. and Harris, M.P. (1979) A procedure for implanting a slow release formulation of an environmental pollutant into a free-living animal. *Environ. Pollut.*, **19**, 139–44.

Oytcharoy, R., Guentcheva, G. and Michalova, S. (1980) Some approaches to experimental testing of drugs of immunotoxicity. *Arch. Toxicol. Suppl.*, **4**, 120–31.

Pandya, K.P., Shanker, R., Gupta, A., Khan, W.A. and Ray, P.K. (1986) Modulation of benzene toxicity by an interferon inducer (6MFA). *Toxicology*, **39**, 291–305.

Park, S.S., Cha, S.-J., Miller, H., Persson, A.V., Coon, M.J. and Gelboin, H.V. (1982) Monoclonal antibodies to rabbit liver cytochrome P-450LM2 and cytochrome P-450LM4. *Mol. Pharmacol.*, **21**, 248–58.

Park, S.S., Miller, H., Klotz, A.V., Kloepper-Sams, P.J., Stegeman, J.J. and Gelboin, H.V. (1986) Monoclonal antibodies to liver microsomal cytochrome P-450E of the marine fish *Stenotomus chrysops* (scup): Cross reactivity with 3-methylcholanthrene induced rat cytochrome P-450. *Arch. Biochem. Biophys.*, **249**, 339–50.

Parke, D.V., Ioannides, C., Iwasaki, K. and Lewis, D.F.V. (1985) Microsomal enzymes and toxicity – Conclusions. In: *Microsomes and Drug Oxidations* (Eds A.R. Boobis, J. Caldwell, F. De Matteis and C.R. Elcombe), pp. 402–13. Taylor & Francis, London.

Passer, E.L., Leinaeng, R.H., Birmingham, L.W., Cruz, H., Dupervil, C., Persaud, E.J. and Dolensek, E.P. (1989) Effect of lead on blood protoporphyrin levels of a group of ring teal ducks (*Callonetta leucophrys*). *Zool. Biol.*, **8**, 357–65.

Passino, D.R.M. and Smith, S.B. (1987) Acute bioassays and hazard evaluation of representative contaminants detected in Great Lakes fish. *Environ. Toxicol. Chem.*, **6**, 901–7.

Pattee, O.H. and Hennes, S.K. (1983) Bald Eagles and waterfowl: The lead shot connection. *Trans. 48th North Am. Wildl. Conf.*, pp. 230–7.

Payne, J.F. (1976) Field evaluation of benzopyrene hydroxylase induction as a monitor for marine petroleum pollution. *Science*, **191**, 945–6.

Payne, J.F. (1984) Mixed-function oxygenases in biological monitoring programs: Review of potential usage in different phyla of aquatic animals. In: *Ecotoxicological Testing for the Marine Environment. Vol. 1*, pp. 625–55.

Payne, J.F. and Fancey, L.L. (1982) Effect of long-term exposure to petroleum on mixed function oxygenases in fish: Further support for use of the enzyme system in biological monitoring. *Chemosphere*, **11**, 207–13.

Payne, J.F., Fancey, L.L., Rahimtula, A.D. and Porter, E.L. (1987) Review and perspective on the use of mixed-function oxygenase enzymes in biological monitoring. *Comp. Pharmacol. Physiol.*, **86C**, 233–45.

Peakall, D.B. (1970) p,p′-DDT: Effect on calcium metabolism and concentration of estradiol in the blood. *Science*, **168**, 592–4.

Peakall, D.B. (1974) DDE: Its presence in Peregrine eggs in 1948. *Science*, **183**, 673–4.

Peakall, D.B. (1975) Physiological effects of chlorinated hydrocarbons on avian species. In: *Environmental Dynamics of Pesticides* (Eds R. Haque, and V.H. Freed), pp. 343–60. Plenum Press, New York.

Peakall, D.B. (1976) Effects of toxaphene on hepatic enzyme induction and circulating steroid levels in the rat. *Environ. Health Perspect.*, **13**, 117–20.

Peakall, D.B. (1983) Methods for assessment of the effects of pollutants on avian reproduction. In: *Methods for Assessing the Effects of Chemicals on Reproductive Functions* (Eds V.B. Vouk, and P.J. Sheehan), SCOPE, **20**, 345–63.

Peakall, D.B. (1985) Behavioral responses of birds to pesticides and other contaminants. *Residue Rev.*, **96**, 45–77.

Peakall, D.B. (1987) Accumulation and effects on birds. In: *PCBs and the Environment. Vol. 2.* (Ed. J.S. Waid), pp. 31–47. CRC Press, Boca Raton, FL.

Peakall, D.B. and Bart, J.R. (1983) Impacts of aerial applications of insecticides on forest birds. *CRC Crit. Rev. Environ. Control*, **13**, 117–65.

Peakall, D.B., Cade, T.J., White, C.M. and Haugh, J.R. (1975a) Organochlorine residues in Alaskan Peregrines. *Pestic. Monit. J.*, **8**, 255–60.

Peakall, D.B., Fox, G.A., Gilman, A.P., Hallett, D.J. and Norstrom, R.J. (1980b) Reproductive success of herring gulls as an indicator of Great Lakes water quality. In: *Hydrocarbons and Halogenated Hydrocarbons* (Eds B.K. Afghan, and D. Mackay), pp. 337–44. Plenum Press, New York.

Peakall, D.B., Hallett, D., Miller, D.S., Butler, R.G. and Kinter, W.B. (1980a)

Effects of ingested crude oil on black guillemots: A combined field and laboratory study. *Ambio*, **9**, 28–30.

Peakall, D.B., Lincer, J.L. and Bloom, S.E. (1972) Embryonic mortality and chromosomal alterations caused by Aroclor 1254 in ring doves. *Environ. Health Perspect.*, **2**, 103–4.

Peakall, D.B., Lincer, J.L., Risebrough, R.W., Pritchard, J.B. and Kinter, W.B. (1973) DDE-induced eggshell thinning: Structural and physiological effects in three species. *Comp. Gen. Pharmacol.*, **4**, 305–13.

Peakall, D.B., Miller, D.S. and Kinter, W.B. (1975b) Blood calcium levels and the mechanism of DDE-induced eggshell thinning. *Environ. Pollut.*, **9**, 289–94.

Peakall, D.B., Norstrom, R.J., Jeffrey, D.A. and Leighton, F.A. (1989) Induction of hepatic mixed function oxidases in the herring gull (Larus argentatus) by Prudhoe Bay crude oil and its fractions. *Comp. Biochem. Physiol.*, **94C**, 461–3.

Peakall, D.B., Norstrom, R.J., Rahimtula, A.D. and Butler, R.D. (1986) Characterization of mixed-function oxidase systems of the nestling herring gull and its implications for bioeffects monitoring. *Environ. Toxicol. Chem.*, **5**, 379–85.

Peakall, D.B. and Peakall, M.L. (1973) Effect of a polychlorinated biphenyl on the reproduction of artificially and naturally incubated dove eggs. *J. Appl. Ecol.*, **10**, 863–8.

Peakall, D.B. and Tucker, R.K. (1985) Extrapolation from single species studies to populations, communities and ecosystems. In: *Methods for Estimating Risk of Chemical Injury: Human and Non-human Biota and Ecosystems* (Eds V.B. Vouk, G.C. Butler, D.G. Hoel, and D.B. Peakall), SCOPE, **26**, 611–36.

Pearce, P.A. and Price, I.M. (1975) Effects on amphibians. In: *Aerial Control of Forest Insects in Canada* (Ed. M.L. Prebble), pp. 301–5. Department of Environment, Ottawa.

Pearce, P.A., Peakall, D.B. and Erskine, A.J. (1976) Impact on forest birds of the 1975 spruce budworm spray operations in New Brunswick. *Can. Wildl. Ser. Prog. Notes*, **62**, 7.

Pedersen, M.G., Hershberger, W.K., Zachariah, P.K. and Juchaw, M.R. (1976) Hepatic biotransformation of environmental xenobiotics in six strains of rainbow trout (*Salmo gairdneri*). *J. Fish. Res. Board Can.*, **33**, 666–75.

Penn, I. (1985) Neoplastic consequences of immunosuppression. In: *Immunotoxicology and Immunopharmacology* (Eds J.H. Dean, M.I. Luster, A.E. Munson, H. Amos), pp. 79–90. Raven Press, New York.

Penn, I. (1987) The neoplastic consequences of immunodepression. In: *Immunotoxicology* (Eds A. Berlin, J. Dean, M.H. Draper, E.M.B. Smith, F. Spreafico), pp. 69–83. Martinus Nijhoff Publishers, Dordrecht.

Penninks, A. and Seinen, W. (1987) Immunotoxicity of organotin compounds. A cell biological approach to dialkyltin induced thymus atrophy. In: *Immunotoxicology* (Eds A. Berlin, J. Dean, M.H. Draper, E.M.B. Smith, F. Spreafico), pp. 258–78. Martinus Nijhoff Publishers, Dordrecht.

Pepys, J. (1983) Allergic reactions of the respiratory tract to low molecular weight chemicals. In: *Immunotoxicology* (Eds G.G. Gibson, R. Hubbard, and D.V. Parke), pp. 107–24. Academic Press, London.

Perry, P. and Evans, H.J. (1975) Cytological detection of mutagen-carcinogen exposure by sister chromatic exchange. *Nature*, **258**, 121–4.

Petkovich, M., Brand, N.J., Krust, A. and Chambon, P. (1987) A human retinoic acid receptor which belongs to the family of nuclear receptors. *Nature*, **330**, 444–50.

Phillips, D.H., Hewer, A., Martin, C.N., Garner, R.C. and King, M.M. (1988) Correlation of DNA adduct levels in human lung with cigarette smoking. *Nature*, **336**, 790–2.

Phillips, W.E.J. (1963) DDT and the metabolism of Vitamin A and carotene in the rat. *Can. J. Biochem. Physiol.*, **41**, 1793–802.

Pisarev, D.L.K.D., Molina, M.D.C.R.D. and Viale, L.C.S.M.D. (1990) Thyroid function and thyroxine metabolism in hexachlorobenzene-induced porphyria. *Biochem. Pharmacol.*, **39**, 817–25.

Plaa, G.L. and Hewitt, W.R. (1982) *Toxicology of the Liver*. Raven Press, New York. 338 pp.

Poirier, M.C. (1984) The use of carcinogen-DNA adduct antisera for quantitation and localization of genomic damage in animal models and human population. *Environ. Mutagen.*, **6**, 879–87.

Poland, A., Glover, E. and Kende, A.S. (1976) Stereospecific, high affinity binding of 2,3,7,8-tetrachlorodibenzo-p-dioxin by hepatic cytosol. Evidence that the binding species is receptor for induction of aryl hydroxylase. *J. Biol. Chem.*, **251**, 4936–46.

Poland, A. and Knutson, J.C. (1982) 2,3,7,8-tetrachlorodibenzo-p-dioxin and related halogenated aromatic hydrocarbons: Examination of the mechanism of toxicity. *Annu. Rev. Pharmacol. Toxicol.*, **22**, 517–54.

Pough, F.H. (1976) Acid precipitation and embryonic mortality of spotted salamanders, *Ambystoma maculatum. Science*, **192**, 68–70.

Powell, G.V.N. (1984) Production by an altricial songbird, the red-winged blackbird, in fields treated with the organophosphate insecticide fenthion. *J. Appl. Ecol.*, **21**, 83–95.

Powell, G.V.N. and Gray, D.C. (1980) Dosing free-living nestling starlings with an organophosphate pesticide, Famphur. *J. Wildl. Manage.*, **44**, 918–21.

Prein, A.E., Thie, G.M., Alink, G.M., Koeman, J.H. and Poels, C.L.M. (1978) Cytogenetic changes in fish exposed to water of the River Rhine. *Sci. Total Environ.*, **9**, 287–91.

Price, I.M. and Weseloh, D.V. (1986) Increased numbers and productivity of Double-crested Cormorants, *Phalacrocorax auritus*, on Lake Ontario. *Canad. Field-Nat.*, **100**, 474–82.

Prijono, W.B. and Leighton, F.A. (1991) Parallel measurement of brain acetylcholinesterase and the muscarinic cholinergic receptor in the diagnosis of acute, lethal poisoning by anti-cholinesterase pesticides. *J. Wildl. Dis.* 27: 10–115.

Proctor, N.H. and Casida, J.E. (1975) Organophosphorus and methyl carbamate insecticide teratogenesis: Diminished NAD in chicken embryos. *Science*, **190**, 580–2.

Raha, A., Reddy, V., Houser, W. and Bresnick, E. (1990) Binding characteristics of 4S PAH-binding protein and Ah receptor from rats and mice. *J. Toxicol. Environ. Health*, **29**, 339–55.

Ralph, C.J. and Scott, J.M. (Eds) (1981) *Estimating Numbers of Terrestrial Birds*. Cooper Ornithology Soc., Lawrence, KS. 630 pp.

Ratcliffe, D.A. (1958) Broken eggs in Peregrine eyries. *Brit. Birds*, **51**, 23–6.

Ratcliffe, D.A. (1967) Decrease in eggshell weight in certain birds of prey. *Nature*, **215**, 208–10.

Ratcliffe, D.A. (1970) Changes attributable to pesticides in egg breakage frequency and eggshell thickness in some British birds. *J. Appl. Ecol.*, **7**, 67–116.

Rathke, D.E. and McRae, G. (1989) *1987 Report on Great Lakes Water Quality, Appendix B, Great Lakes Surveillance. Vol. 3.* Int. Joint Comm. Rep.

Rattner, B.A. (1982) Diagnosis of anticholinesterase poisoning in birds: Effects of environmental temperature and underfeeding on cholinesterase activity. *Environ. Toxicol. Chem.*, **1**, 329–35.

Rattner, B.A., Clarke, R.N. and Ottinger, M.A. (1986) Depression of plasma luteinizing hormone concentration in quail by the anticholinesterase insecticide parathion. *Comp. Biochem. Physiol.*, **83C**, 451–3.

Rattner, B.A. and Eastin, W.C., Jr. (1981) Plasma corticosterone and thyroxine concentrations during chronic ingestion of crude oil in Mallard ducks (*Anas platyrhynchos*). *Comp. Biochem. Physiol.*, **68C**, 103–7.

Rattner, B.A. and Franson, J.C. (1984) Methyl parathion and fenvalerate toxicity in American Kestrels: Acute physiological responses and effects of cold. *Can. J. Physiol. Pharmacol.*, **62**, 787–92.

Rattner, B.A., Sileo, L. and Scanes, C.G. (1982a) Oviposition and the plasma concentrations of LH, progesterone and corticosterone in bobwhite quail (*Colinus virginianus*) fed parathion. *J. Reprod. Fert.*, **66**, 147–55.

Rattner, B.A., Sileo, L. and Scanes, C.G. (1982b) Hormonal responses and tolerance to cold of female quail following parathion ingestion. *Pest. Biochem. Physiol.*, **18**, 132–8.

Rauckman, E.J. and Padilla, G.M. (1987) *The Isolated Hepatocyte: Use in Toxicology and Xenobiotic Transformations*. Academic Press New York.

Razin, A. and Riggs, A.D. (1980) DNA methylation and gene function. *Science*, **210**, 604–10.

Razin, A. and Szyf, M. (1984) DNA methylation patterns: Formation and function. *Biochim. Biophys. Acta*, **782**, 331–42.

Reijnders, P.J.H. (1980) Organochlorine and heavy metal residues in harbour seals from the Wadden Sea and their possible effects on reproduction. *Netherlands J. Sea Res.*, **14**, 30–65.

Reijnders, P.J.H. (1986) Reproductive failure in common seals feeding on fish from polluted coastal waters. *Nature*, **324**, 456–7.

Reinert, R.E. and Hohreiter, D.W. (1984) Adenylate energy charge as a measure of stress in fish. *Adv. Environ. Sci. Technol.*, **16**, 151–61.

Renoux, G. and Renoux, M. (1980) The effects of sodium diethyldithiocarbamate, azathioprine, cyclophosphamide, or hydrocortisone acetate administered alone or in association for 4 weeks on the immune responses of BALB/c mice. *Clin. Immunol. Immunopathol.*, **15**, 23–32.

Renoux, G. (1985) Immunomodulatory agents. In: *Immunopharmacology and Immunotoxicology* (Eds J. Dean, M.I. Luster, A.E. Munson, and H. Amos), pp. 193–205. Raven Press, New York.

Renqing, Z. (1990) Relationship between serum GOT of *Cyprinus carpio* and biotic index of diatom in the diagnosis of water quality. *Bull. Environ. Contam. Toxicol.*, **44**, 844–50.

Renton, K.W. (1983) Relationship between the enzymes of detoxication and host defence mechanisms. In: *Biological Basis of Detoxication* (Eds J. Caldwell and W.B. Jakoby), pp. 307–24. Academic Press, New York.

Richardson, H.L., Stier, A.R. and Borsos-Nachtnebel, E. (1952) Liver tumor inhibition and adrenal histologic responses in rats to which 3'-methyl-4-dimethyl-aminoazobenzene and 20-methylcholanthrene were simultaneously administered. *Cancer Res.*, **12**, 356–61.

Richert, E.P. and Prahlad, K.V. (1972) Effects of DDT and its metabolites on thyroid of the Japanese quail, *Corturnix corturnix japonica. Poult. Sci.*, **51**, 196–200.

Richie, P.J. and Peterle, T.J. (1979) Effects of DDE on circulating luteinizing hormone levels in Ring Doves during courtship and nesting. *Bull. Environ. Contam. Toxicol.*, **23**, 220–6.

Ridgway, L.P. and Karnofsky, D.A. (1952) The effects of metals on the chick embryo: Toxicity and production of abnormalities in development. *Ann. N.Y. Acad. Sci.*, **55**, 203–15.

Rifkind, A.B., Sassa, S., Reyes, J. and Muschick, H. (1985) Polychlorinated aromatic hydrocarbon lethality, mixed-function oxidase induction, and uroporphyrinogen decarboxylase inhibition in the chick embryo: Dissociation of dose-response relationships. *Toxicol. Appl. Pharmacol.*, **78**, 268–79.

Risebrough, R.W. (1986) Pesticides and bird populations. *Curr. Ornithol.*, **3**, 397–427.

Risebrough, R.W., Reiche, P., Herman, S.G., Peakall, D.B. and Kirven, M.N. (1968) Polychlorinated biphenyls in the global ecosystem. *Nature*, **220**, 1098–1102.

Robinson, S.H., Cantoni, O. and Costa, M. (1984) Analysis of metal-induced DNA lesions and DNA-repair replication in mammalian cells. *Mutat. Res.*, **131**, 173–81.

Rodgers, K.E., Immamura, T. and Devens, B.H. (1986) Organophosphorus pesticide immunotoxicity: Effects of O,O,S-trimethyl phosphorothioate on cellular and humoral immune response systems. *Immunopharmacol.*, **12**, 193–202.

Ronis, M.J.J., Andersson, T., Hansson, T. and Walker, C.H. (1989b). Differential expression of multiple forms of cytochrome P-450 in vertebrates: Antibodies to purified rat cytochrome P-450s as molecular probes for the evolution of P-450 gene families I and II. *Mar. Environ. Res.*, **28**, 131–5.

Ronis, M.J.J., Borlakoglu, J., Walker, C.H., Hanson, T. and Stegeman, J.J. (1989a) Expression of orthologues to rat P-450IA1 and IIB1 in sea-birds from the Irish Sea 1978–88. Evidence for environmental induction. *Mar. Environ. Res.*, **28**, 123–30.

Ronis, M.J.J. and Walker, C.H. (1985) Species variations in the metabolism of liposoluble organochlorine compounds by hepatic microsomal monooxygenase: Comparative kinetics in four vertebrate species. *Comp. Biochem. Physiol.*, **82C**, 445–9.

Ronis, M.J.J. and Walker, C.H. (1989) The microsomal monooxygenases of birds. In: *Reviews in Biochemical Toxicology* (Eds E. Hodgson, J.R. Bend and R.M. Philpot), pp. 301–84. Elsevier, New York.

Rosenthal, C.J., Noguera, C.A., Coppola, A. and Kapelner, S.N. (1982) Pseudolymphoma with mycosis fungoides manifestations, hyperresponsiveness to diphenylhydantoin, and lymphocyte disregulation. *Cancer*, **49**, 2305–14.

Rosenthal, C.J. and Snyder, C.A. (1987) Inhaled benzene reduces aspects of cell-mediated tumour surveillance in mice. *Toxicol. Appl. Pharmacol.*, **88**, 35–43.

Rossi, E. and Curnow, D.H. (1986) Porphyrins. In: *HPLC of Small Molecules* (Ed. C.K. Kim), pp. 261–303. IRL Press, Washington, DC.

Roux, F., Treich, I., Brun, C., Desoize, B. and Fournier, E. (1979) Effect of lindane on human lymphocyte responses to phytohemagglutinin. *Biochem. Pharmacol.*, **28**, 2419–26.

Rudolph, S.G., Zinkl, J.G., Anderson, D.W. and Shea, P.J. (1984) Prey-capturing ability of American kestrels fed DDE and acephate or acephate alone. *Arch. Environ. Contam. Toxicol.*, **13**, 367–72.

Safe, S. (1986) Comparative toxicology and mechanism of action of polychlorinated dibenzo-p-dioxins and dibenzofurans. *Annu. Rev. Pharmacol.*, **26**, 371–99.

Safe, S. (1987) Determination of 2,3,7,8-TCDD equivalent factors (TEFs): Support for the use of the in vitro AHH induction assay. *Chemosphere*, **16**, 791–802.

Sanders, O.T. and Kirkpatrick, R.L. (1977) Reproductive characteristics and corticoid levels in female white-footed mice fed ab libitum and restricted diets containing a polychlorinated biphenyl. *Environ. Res.*, **13**, 358–63.

Sanders, V.M., White, K.L. and Munson, A.E. (1985) Humoral and cell-mediated immune status of mice exposed to 1,1,2-trichloroethane. *Drug Chem. Toxicol.*, **8**, 357–72.

Sassa, S., Sugita, O., Ohnuma, N., Imajo, S., Noguchi, T. and Kappas, A. (1986) Studies of the influence of chloro-substituent sites and conformational energy in polychlorinated biphenyls on uroporphyrin formation in chick-embryo liver cell cultures. *Biochem. J.*, **235**, 291–6.

Sastry, K.V. and Sharma, K. (1980) Mercury induced haematological and biochemical anomalies in *Ophiocephalus* (Channa) *punctatus*. *Toxicol. Lett.*, **5**, 245–9.

Sawyer, T. and Safe, S. (1982) PCB isomers and congeners: Induction of aryl hydrocarbon hydroxylase and ethoxyresofin O-deethylase activities in rat hepatoma cells. *Toxicol. Lett.*, **13**, 87–94.

Sawyer, T.W. and Safe, S. (1985) In vitro AHH induction by chlorinated biphenyl and dibenzofuran mixtures: Additive effects. *Chemosphere*, **14**, 79–84.

Sawyer, T.W., Vatcher, A.D. and Safe, S. (1984) Comparative aryl hydrocarbon hydroxylase induction activities of commercial PCBs in Wistar rats and rat hepatoma H-4-IIE cells in culture. *Chemosphere*, **13**, 695–701.

Schafer, E.W. (1972) The acute oral toxicity of 369 pesticidal, pharmaceutical and other chemicals to wild birds. *Toxicol. Appl. Pharmacol.*, **21**, 315–30.

Schafer, E.W., Jr., and Brunton, R.B. (1979) Indicator bird species for toxicity determinations: Is the technique usable in test method development? In: *Vertebrate Pest Control* (Ed. J.R. Beck), pp. 157–68. ASTM STP 680.

Scheider, W.A., Jeffries, D.S. and Dillon, P.J. (1979) Effects of acidic precipitation on precambrian freshwaters in southern Ontario. *J. Great Lakes Res.*, **5**, 45–51.

Scheuhammer, A.M. (1987a) Erythrocyte γ-aminolevulinic acid dehydratase in birds. I. The effects of lead and other metals in vitro. *Toxicology*, **45**, 155–63.

Scheuhammer, A.M. (1987b) Erythrocyte γ-aminolevulinic acid dehydratase in birds. II. The effects of lead exposure in vivo. *Toxicology.*, **45**, 165–75.

Scheuhammer A.M. (1989) Monitoring wild bird populations for lead exposure. *J. Wildl. Manage.*, **53**, 759–65.

Schmidt-Nielsen, K. (1960) The salt-secreting gland of marine birds. *Circulation*, **21**, 955–67.

Schnizlein, C.T., Bice, D.E., Mitchell, C.E. and Hahn, F.F. (1982) Effects on rat lung immunity by acute lung exposure to benzo (a)pyrene. *Arch. Environ. Health*, **37**, 201–6.

Schulte, P.A. (1989) A conceptual framework for the validation and use of biologic markers. *Environ. Res.*, **48**, 129–44.

Scott, M.L., Vadehra, D.V., Mullenhoff, P.A., Rumsey, G.L. and Rice, R.W. (1971) Results of experiments on the effects of PCB's on laying hen performance. *Proc. Cornell Nutritional Conference for Feed Manufacturers*, pp. 56–64.

Sears, J. (1989) A review of lead poisoning among the River Thames Mute Swan *Cygnus olor* population. *Wildfowl*, **40**, 151–2.

Seinen, W., Vos, J.G., Van Spange, I., Snoek, M., Brands, R. and Hooykaas, H. (1977) Toxicity of organotin compounds. II. Comparative in vivo and in vitro studies with various organotin and organolead compounds in different animal species with special emphasis on lymphocyte cytotoxicity. *Toxicol. Appl. Pharmacol.*, **42**, 197–212.

Selgrade, M.K., Daniels, M.J., Illing, J.W., Ralston, A.L., Grandy, M.A., Charlet, E. and Graham, J. (1984) Increased susceptibility to parathion poisoning following murine cytomegalovirus infection. *Toxicol. Appl. Pharmacol.*, **76**, 356–64.

Sharma, R.P. (1973) Brain biogenic amines: Depletion by chronic dieldrin exposure. *Life Sci.*, **13**, 1245–51.

Sharma, R.P. and Gehring, P.J. (1979) Immunologic effects of vinyl chloride in mice. *Ann. N.Y. Acad. Sci.*, **320**, 551–63.

Sharma, R.P., Winn, D.S. and Low, J.B. (1976) Toxic, neurochemical and behaviorial effects of dieldrin exposure in Mallard ducks. *Arch. Environ. Contam. Toxicol.*, **5**, 43–53.

Sharma, S.K., Chaturvedi, L.D. and Sastry, K.V. (1979) Acute endrin toxicity on oxidases *Ophiocephalus punctatus* (Bloch). *Bull. Environ. Contam. Toxicol.*, **23**, 153–7.

Shelby, H.J., Rothman, S.J. and Buckley, R.H. (1980) Phenytoin hypersensitivity. *J. Allergy Clin. Immunol.*, **66**, 166–72.

Shelford, V.E. and Allee, W.C. (1912) The reaction of fishes to gradients of dissolved atmospheric gases. *J. Exp. Zool.*, **14**, 207–66.

Shellhammer, H.S. (1961) Variation in brain acetylcholinesterase in the wild mice *Mus musculus* and *Reithrodontomys megalotis*. *Physiol. Zool.*, **34**, 312–8.

Shimai, S. and Satoh, H. (1985) Behavioral teratology of methylmercury. *J. Toxicol. Sci.*, **10**, 199–216.

Shugart, L. (1986) Quantifying adductive modification of hemoglobin from mice exposed to benzo[a]pyrene. *Anal. Biochem.*, **152**, 365–9.

Shugart, L.R. (1988) Quantitation of chemically induced damage to DNA of aquatic organisms by alkaline unwinding assay. *Aquat. Toxicol.*, **13**, 43–52.

Shugart, L.R. (1990a) Biological monitoring: Testing for genotoxicity. In: *Biomarkers of Environmental Contamination* (Eds J.F. McCarthy and L.R. Shugart), pp. 205–16. Lewis Publishers, Boca Raton, FL.

Shugart, L.R. (1990b) 5-methyl deoxycytidine content of DNA from Bluegill Sunfish (*Lepomis macrochirus*) exposed to benzo[a]pyrene. *Environ. Toxicol. Chem.*, **9**, 205–8.

Shugart, L.R., Adams, S.M., Jiminez, B.D., Talmage, S.S. and McCarthy, J.F. (1989) Biological markers to study exposure in animals and bioavailability of environmental contaminants. *ACS Sympos. Ser.*, **382**, 86–97.

Shugart, L., Holland, J.M. and Rahn, R.O. (1983) Dosimetry of PAH skin carcinogenesis: Covalent binding of benzo[a]pyrene to mouse epidermal DNA. *Carcinogenesis*, **4**, 195–8.

Shugart, L. and Kao, J. (1985) Examination of adduct formation in vivo in the mouse between benzo(a)pyrene and DNA of skin and hemoglobin of red blood cells. *Environ. Health Perspect.*, **62**, 223–6.

Shugart, L.R., McCarthy, J.F., Jimenez, B.D. and Daniels, J. (1987) Analysis of adduct formation in the bluegill sunfish (*Lepomis macrochirus*) between benzo[a]pyrene and DNA of the liver and hemoglobin of the erythrocyte. *Aquat. Toxicol.*, **9**, 319–5.

Silbergeld, E.K. and Goldberg, A.M. (1975) Pharmacological and neurochemical investigations of lead-induced hyperactivity. *Neuropharmacology*, **14**, 431–44.

Silkworth. J.B. and Vecchi, A. (1985) Role of the Ah receptor in halogenated aromatic hydrocarbon immunotoxicity. In: *Immunotoxicology and Immunopharmacology* (Eds J.H. Dean, M.I. Luster. A.E. Munson and H. Amos), pp. 263–75. Raven Press, New York.

Silver, R. and Ball, G.F. (1989) Brain, hormone and behavior interactions in avian reproduction: Status and prospectus. *Condor*, **91**, 966–78.

Silverman, D.H.S. and Karnovsky, M.L. (1989) Serotonin and peptide immunoneuromodulators: Recent discoveries and new ideas. *Adv. Enzymol.*, **62**, 203–26.

Sinclair, P.R., Bement, W.J., Bonkovsky, H.L., Lambrecht, R.W., Frezza, J.E., Sinclair, J.F., Urquhart, A.J. and Elder, G.H. (1986) Uroporphyrin accumulation produced by halogenated biphenyls in chick-embryo hepatocytes. *Biochem. J.*, **237**, 63–71.

Sinclair, P.R., Lambrecht, R. and Sinclair, J. (1987) Evidence for cytochrome P450-mediated oxidation of uroporphyrinogen by cell-free liver extracts from chick embryos treated with 3-methylcholanthrene. *Biochem. Biophys. Res. Commun.*, **146**, 1324–9.

Singh, A.K. and Drewes, L.R. (1987) Neurotoxic effects of low-level chronic acephate exposure in rats. *Environ. Res.*, **43**, 342–9.

Singh, S.P., Sharma, L.D. and Bahga, H.S. (1989) Hemato-biochemical parameters in cockerels given Fenthion at different dietary doses. *Indian J. Vet. Med.*, **9**, 64–6.

Skåre, J.U., Stenersen, J., Kveseth, N. and Polder, A. (1985) Time trends of organochlorine chemical residues in seven sedentary marine fish species from a Norwegian fjord during the period 1972–82. *Arch. Environ. Contam. Toxicol.*, **14**, 33–41.

Sladen, W.J.L., Menzie, C.M. and Reichel, W.L. (1966) DDT residues in Adelie penguins and a crabeater seal from Antarctica. *Nature*, **210**, 670–3.

Sloley, B.D., Hickie, B.E., Dixon, D.G., Downer, R.G.H. and Martin, R.J. (1986) The effects of sodium pentachlorophenate, diet and sampling procedure on amine and tryptophan concentrations in the brain of rainbow trout, *Salmo gairdneri* Richardson. *J. Fish Biol.*, **28**, 267–77.

Slooff, W. and De Zwart, D. (1983) Bio-indicators and chemical pollution of surface waters. *Environ. Monit. Assess.*, **3**, 237–45.

Smialowicz, R.J., Luebke, R.W., Riddle, M.M., Rogers, R.R. and Lowe, D.G. (1985) Evaluation of the immunotoxic potential of chlordecone with comparison to cyclophosphamide. *J. Toxicol. Environ. Health*, **15**, 561–74.

Smith, A.G., Cabral, J.R.P., Carthew, P., Francis, J.E. and Manson, M.M. (1989) Carcinogenicity of iron in conjunction with a chlorinated environmental chemical, hexachlorobenzene, in C57BL/105ScSn mice. *Int. J. Cancer*, **43**, 492–6.

Smith, A.G., Francis, J.E., Green, J.A., Greig, J.B., Wolf, C.R. and Manson, M.M. (1990) Sex-linked hepatic uroporphyria and the induction of cytochromes P450IA in rats caused by hexachlorobenzene and polyhalogenated biphenyls. *Biochem. Pharmacol.*, **40**, 2059–68.

Snedecor, G.W. and Cochran, W.G. (1980) *Statistical Methods.* 7th Edition. Iowa State Univ. Press. 507 pp.

Snyder, C.A. (1989) The neuroendocrine system, an opportunity for immunotoxicologists. *Environ. Health Persp.*, **81**, 165–6.

Snyder, C.A., Goldstein, B.D., Sellakumar, A.R., Bromberg, I., Laskin, S. and Albert, R.E. (1980) The inhalation toxicology of benzene: Incidence of haematopoietic neoplasms and haematotoxicity in AKR/J and C57B1/6J mice. *Toxicol. Appl. Pharmacol.*, **54**, 323–31.

Sobokta, T.J., Brodie, R.E. and Cook, M.P. (1975) Psychophysiologic effects of early lead exposure. *Toxicology*, **5**, 175–91.

Sokal, R.R. and Rohlf, F.J. (1981) *Biometry. The Principles and Practice of Statistics in Biological Research.* Freeman and Co., New York. 839 pp.

Somlyay, I.M. and Várnagy, L.E. (1989) Changes in blood plasma biochemistry of chicken embryos exposed to various pesticide formulations. *Acta Vet. Hung.*, **37**, 179–84.

Somlyay, I.M., Várnagy, L.E. and Paviliscak, C. (1989) Effect of various pesticides on plasma biochemistry of chicken embryos. *Meded. Fac. Landbouwwet. Rijksuniv. Gent.*, **54**, 181–6.

Sonstegard, R.A. (1977) Environmental carcinogenesis studies in fishes of the Great Lakes of North America. *Ann. N.Y. Acad. Sci.*, **298**, 261–9.

Sonstegard, R.A. and Leatherland, J.F. (1979) Hypothyroidism in rats fed Great Lakes Coho Salmon. *Bull. Environ. Contam. Toxicol.*, **22**, 779–84.

Soper, K.A., Stolley, P.D., Galloway, S.M., Smith, J.G., Nichols, W.W. and Wolman, S.R. (1984) Sister-chromatid exchange (SCE) report on control subjects in a study of occupationally exposed workers. *Mutat. Res.*, **129**, 77–88.

Spear, P.A., Bourbonnais, D.H., Norstrom, R.J. and Moon, T.W. (1990) Yolk retinoids (vitamin A) in eggs of the Herring Gull and correlations with polychlorinated dibenzo-p-dioxins and dibenzofurans. *Toxicol. Environ. Chem.*, **9**, 1053–61.

Spear, P.A., Bourbonnais, D.H., Peakall, D.B. and Moon, T.W. (1989) Dove reproduction and retinoid (vitamin A) dynamics in adult females and their eggs following exposure to 3,3′,4,4′-tetrachlorobiphenyl. *Can. J. Zool.*, **67**, 908–13.

Spear, P.A., Garcin, H. and Narbonne, J.-F. (1988) Increased retinoic acid metabolism following 3,3′,4,4′,5,5′-hexabromobiphenyl injection. *Can. J. Physiol. Pharmacol.*, **66**, 1181–6.

Spear, P.A., Higueret, P. and Garcin, H. (1990). Increased thyroxine turnover after 3,3′,4,4′,5,5′-hexabromobiphenyl injection and lack of effect on peripheral triiodothyronine production. *Can. J. Physiol. Pharmacol.* **68**, 1079–84.

Spear, P.A. and Moon, T.W. (1985) Low dietary iodine and thyroid anomalies in ring doves, *Streptopelia risoria*, exposed to 3,4,3′,4′-tetrachlorobiphenyl. *Arch. Environ. Contam. Toxicol.*, **14**, 547–53.

Spear, P.A. and Moon, T.W. (1986) Thyroid-vitamin A interactions in chicks exposed to 3,4,3′,4′-tetrachlorobiphenyl: Influence of low dietary vitamin A and iodine. *Environ. Res.*, **40**, 188–98.

Spear, P.A., Moon, T.W. and Peakall, D.B. (1986) Liver retinoid concentrations in natural populations of herring gulls (*Larus argentatus*) contaminated by 2,3,7,8-tetrachlorodibenzo-p-dioxin and in ring doves (*Streptopelia risoria*) injected with a dioxin analogue. *Can. J. Zool*, **64**, 204–8.

Spiegelberg, T., Kordel, W. and Hochrainer, D. (1984) Effect of NiO inhalation on alveolar macrophages and the humoral immune systems in rats. *Ecotoxicol. Environ. Safety*, **8**, 516–25.

Spies, R.B., Felton, J.S. and Dillard, L. (1982) Hepatic mixed-function oxidases in California flatfishes are increased in contaminated environments and by oil and PCB ingestion. *Mar. Biol.*, **70**, 117–27.

Sprague, J.B. (1981) Ethologists are environmental drop-outs. In: *Ecology and Ethology of Fishes* (Eds D.L.G. Naokes and J.A. Ward), p. 138. Dr. W. Junk Publ., The Hague.

Springfield, A.C., Carlson, G.P. and DeFeo, J.J. (1973) Liver enlargement and modification of hepatic microsomal drug metabolism in rats by pyrethrum. *Toxicol. Appl. Pharmacol.*, **24**, 298–308.

Stanley, P.I. and Bunyan, P.J. (1979) Hazards to wintering geese and other wildlife from the use of dieldrin, chlorfenvinphos and carbophenothion as wheat seed treatments. *Proc. Roy. Soc. London, Ser. B*, **205**, 31–45.

Stanworth, D.R. (1985) Current concepts of hypersensitivity. In: *Immunotoxicology and Immunopharmacology* (Eds J.H. Dean, M.I. Luster, A.E. Munson, and H. Amos), pp. 91–8. Raven Press, New York.

Stanworth, D.R. (1987) Mechanisms of pseudo-allergy. In: *Immunotoxicology* (Eds A. Berlin, J. Dean, M.H. Draper, E.M.B. Smith and F. Spreafico), pp. 426–42. Martinus Nijhoff Publishers, Dordrecht.

Stark, D.M., Shopsis, C., Borenfreund, E. and Babich, H. (1986) Progress and problems in evaluating and validating alternative assays in toxicology. *Food Chem. Toxicol.*, **24**, 449–55.

Stegeman, J.J. (1989) Cytochrome P450 forms in fish: Catalytic, immunological and sequence similarities. *Xenobiotica*, **19**, 1093–1110.

Stegeman, J.J. and Chevion, M. (1980) Sex differences in cytochrome P-450 and mixed-function oxygenase activity in gonadally mature trout. *Biochem. Pharmacol.*, **29**, 553–8.

Stegeman, J.J., Kloepper-Sams, P.J. and Farrington, J.W. (1986) Monooxygenase induction and chlorobiphenyls in the deep-sea fish *Coryphanoides armatus*. *Science*, **231**, 1287–9.

Stegeman, J.J., Teng, F.Y. and Snowberger, E.A. (1987) Induced cytochrome P450 in winter flounder (*Pseudopleuronectes americanus*) from coastal Massachusetts evaluated by catalytic assay and monoclonal antibody probes. *Can. J. Fish Aquat. Sci.*, **44**, 1270–7.

Stein, J.E., Reichert, W.L., Nishimoto, M. and Varanasi, U. (1990) Overview of studies on liver carcinogenesis in English sole from Puget Sound; evidence for a xenobiotic chemical etiology II: Biochemical studies. *Sci. Total Environ.*, **94**, 51–69.

Stewart, R.E. and Aldrich, J.W. (1951) Removal and repopulation of breeding birds in a spruce-fir forest community. *Auk*, **68**, 471–82.

Stich, H.F., Leung, H.W. and Roberts, J.R. (Eds) (1982) *Workshop on the Combined Effects of Xenobiotics*. ACSCEQ Document NRCC 18978. National Research Council of Canada, Ottawa. pp. 254.

Stickel, L. and Stickel, W. (1969) Distribution of DDT residues in tissues of birds in relation to mortality, body condition and time. *Ind. Med. Surg.*, **38**, 44–53.

Stickel, W.H., Stickel, L.F., Dyrland, R.A. and Hughes, D.L. (1984) Aroclor 1254 residues in birds: Lethal residues and loss rates. *Arch. Environ. Contam. Toxicol.*, **13**, 7–13.

Stone, C.L. and Fox, M.R.S. (1984) Effects of low levels of dietary lead and iron on hepatic RNA, protein, and minerals in young Japanese quail. *Environ. Res.*, **33**, 322–32.

Story, D.L., Gee, S.J., Tyson, C.A. and Gould, D.H. (1983) Response of isolated hepatocytes to organic and inorganic cytotoxins. *J. Toxicol. Environ. Health*, **11**, 483–501.

Street J.C. and Chadwick, R.W. (1967) Stimulation of dieldrin metabolism by DDT. *Toxicol. Appl. Pharmacol.*, **11**, 68–71.

Stromborg, K.L., Grue, C.E., Nichols, J.D., Hepp, G.R., Hines, J.E. and Bourne,

H.C. (1988) Postfledging survival of European Starlings exposed as nestlings to an organophosphorus insecticide. *Ecology*, **69**, 590–601.

Subba, R. and Glick, B. (1977) Pesticide effects on the immune response and metabolic activity of chicken lymphocytes. *Proc. Soc. Exp. Biol. Med.*, **154**, 27–9.

Sutcliffe, W.H., Jr. (1965) Growth estimates from ribonucleic acid content in some small organisms. *Limnol. Oceanogr.*, **10**(Suppl.), r253–8.

Swartz, W.J. (1984) Effects of 1,1-bis(p-chlorophenyl)-2,2,2-trichloroethane (DDT) on gonadal development in the chick embryo: A histological and histochemical study. *Environ. Res.*, **35**, 333–45.

Szaro, R.C., Dieter, M.P., Heinz, G.H. and Ferrell, J.F. (1978) Effects of chronic ingestion of South Louisiana crude oil on Mallard ducklings. *Environ. Res.*, **17**, 426–36.

Talcott, P.A., Exon, J.H. and Koller, L.D. (1984) Alteration of natural cell-mediated cytotoxicity in rats treated with selenium, diethylnitrosamine and ethylnitrosourea. *Cancer Letters*, **23**, 313–22.

Temple, S.A. (1987) Do predators always capture substandard individuals disproportionately from prey populations? *Ecology*, **68**, 669–74.

Thaller, C. and Eichele, G. (1990) Isolation of 3, 4-didehydroretinoic acid, a novel morphogenetic signal in the chick wing bud. *Nature*, **345**, 815–19.

Thomas, P.T. and Faith, R.E. (1985) Adult and perinatal immunotoxicity induced by halogenated aromatic hydrocarbons. In: *Immunotoxicology and Immunopharmacology* (Eds J.H. Dean, M.I. Luster, A.E. Munson and H. Amos), pp. 305–13. Raven Press, New York.

Thomas, P.T. and Hinsdill, R.D. (1978) Effect of polychlorinated biphenyls on the immune responses of rhesus monkeys and mice. *Toxicol. Appl. Pharmacol.*, **44**, 41–51.

Thomas, P.T. and Hinsdill, R.D. (1979) The effect of perinatal exposure to tetrachlorodibenzo-p-dioxin on the immune response of young mice. *Drug Chem. Toxicol.*, **2** (1 & 2), 77–98.

Thomas, P.T., Ratajczak, H.V., Aranyi, C., Gibbons, R. and Fenters, J.D. (1985) Evaluation of host resistance and immune function in cadmium-exposed mice. *Toxicol. Appl. Pharmacol.*, **80**, 446–56.

Thompson, H.M., Walker, C.H. and Hardy, A.R. (1988) Avian esterases as indicators of exposure to insecticides – the factor of diurnal variation. *Bull. Environ. Contam. Toxicol.*, **41**, 4–11.

Thompson, J.N. (1976) Fat-soluble vitamins. *Comp. Anim. Nutr.*, **1**, 99–135.

Thompson, R.A., Schroder, G.D. and Connor, T.H. (1988) Chromosomal aberrations in the cotton rat, *Sigmodon hispidus*, exposed to hazardous waste. *Environ. Mol. Mutagen.*, **11**, 359–67.

Thunberg, T., Ahlborg, U.G., Håkansson, H., Krantz, C. and Monier, M. (1980) Effect of 2,3,7,8-tetrachlorodibenzo-p-dioxin on the hepatic storage of retinol in rats with different dietary supplies of vitamin A (retinol). *Arch. Toxicol.*, **45**, 273–85.

Thunberg, T., Ahlborg, U.G. and Wahlström, B. (1984) Comparison between the effects of 2,3,7,8-tetrachlorodibenzo-p-dioxin and six other compounds on the vitamin A storage, the UDP-glucuronosyltransferase and the aryl hydrocarbon hydroxylase activity in the rat liver. *Arch. Toxicol.*, **55**, 16–19.

Thurmond, L.M., Lauer, L.D., House, R.V., Cook, J.C. and Dean, J.H. (1987)

Immunosuppression following exposure to 7,12-dimethylbenz[a]anthracene (DMBA) in Ah-responsive and Ah-nonresponsive mice. *Toxicol. Appl. Pharmacol.*, **91**, 450–60.

Thurmond, L.M., House, R.V., Lauer, L.D. and Dean, J.H. (1988) Suppression of splenic lymphocyte function by 7,12-dimethylbenz[a]anthracene (DMBA) in vitro. *Toxicol. Appl. Pharmacol.*, **93**, 269–77.

Tillitt, D.E., Ankley, G.T. and Giesy, J.P. (1989) Planar chlorinated hydrocarbons (PCHs) in colonial fish-eating water bird eggs from the Great Lakes. *Mar. Environ. Res.*, **28**, 505–8.

Tillitt, D.E., Ankley, G.T., Giesy, J.P. and Kevern, N.R. (1988) H4IIE hepatoma cell extract bioassays-derived 2,3,7,8-tetrachlorodibenzo-p-dioxin-equivalents (TCDD-EQ) from Michigan water bird colony eggs, 1986 and 1987. *Report* 63 pp.

Tilson, H.A., Hudson, P.M. and Hong, J.S. (1986) 5,5-diphenylhydantoin antagonizes neurochemical and behavioral effects of p,p′-DDT but not of chlordecone. *J. Neurochem.*, **47**, 1870–8.

Tinsley, I.J. (1965) DDT ingestion and liver glucose 6-phosphate dehydrogenase activity – II. *Biochem. Pharmacol.*, **14**, 847–51.

Trizio, D., Basketter, D.A., Botham, P.A., Graepel, P.H., Lambre, C., Magda, S.J., Pal, T.M., Riley, A.J., Ronneberger, H., Van Sittert, N.J. and Bontinck, W.J. (1988) Identification of immunotoxic effects of chemicals and assessment of their relevance to man. *Fd. Chem. Toxic.*, **26**, 527–39.

Truscott, B., Walsh, J.M., Burton, M.P., Payne, J.F. and Idler, D.R. (1983) Effect of acute exposure to crude petroleum on some reproductive hormones in Salmon and Flounder. *Comp. Biochem. Physiol.*, **75C**, 121–30.

Trust, K.A., Miller, H.W., Ringelman, J.K. and Orme, I.M. (1990) Effects of ingested lead on antibody production in mallards (Anas platyrhynchos). *J. Wild. Diseases*, **26**, 316–22.

Tucker, R.K. and Haegele, M.A. (1971) Comparative acute oral toxicity of pesticides to six species of birds. *Toxicol. Appl. Pharmacol.*, **20**, 57–65.

Tucker, R.K. and Leitzke, J.S. (1979) Comparative toxicology of insecticides for vertebrate wildlife and fish. *Pharmacol. Ther.*, **6**, 167–220.

Tyson, C.A. and Green, C.E. (1987) Cytotoxicity measures: Choices and methods. In: *The Isolated Hepatocyte. Use in Toxicology and Xenobiotic Biotransformations* (Eds E.J. Raukman and G.M. Padilla), pp. 119–58. Academic Press, New York.

USDI (1987) Injury to fish and wildlife species. Type B Technical Information Document. *CERCLA 301 Project.* US Dept. Interior.

Umbriet, T.H. and Gallo, M.A. (1988) Physiological implications of estrogen receptor modulation by 2,3,7,8-tetrachlorodibenzo-p-dioxin. *Toxicol. Lett.*, **42**, 5–14.

Urnab, T. and Jarstrand, C. (1986) Selenium effects on human neutrophilic granulocyte function in vitro. *Immunopharmacol.*, **12**, 167–72.

Vanat, S.V. and Vanat, I.M. (1971) Contributions to the toxic-allergic reaction induced by DDT. *Klin. Med.*, **49**, 126–7.

Van der Gaag, M.A., Van de Kerkhoff, J.F.J., Van der Klift, H.W. and Poels, C.L.M. (1983) Toxicological assessment of river water quality in bioassays with fish. *Environ. Monit. Assess.*, **3**, 247–55.

Varanasi, U., Reichert, W.L., Le Eberhart, B.-T. and Stein, J.E. (1989a) Formation and persistence of benzo[a]pyrene-diolepoxide-DNA adducts in liver of English sole (*Paraphrys vetulus*). *Chem.-Biol. Interact.*, **69**, 203–16.

Varanasi, U., Reichert, W.L. and Stein, J.E. (1989b) ^{32}P-Postlabeling analysis of DNA adducts in liver of wild English sole (*Parophys vetulus*) and winter flounder (*Pseudopleuronectes americanus*). *Cancer Res.*, **49**, 1171–7.

Varney, J.R. and Ellis, D.H. (1974) Telemetering egg for use in incubation and nesting studies. *J. Wildl. Manage.*, **38**, 142–8.

Vecchi, A. (1987) Some aspects of immune alterations induced by chloro-dibenzo-p-dioxins and chloro-dibenzofurans. In: *Immunotoxicology* (Eds A. Berlin, J. Dean., M.H. Draper, E.M.B. Smith., F. Spreafico), pp. 308–17. Martinus Nijhoff Publishers, Dordrecht.

Vecchi, A., Sironi, M., Canegrati, M.A., Recchia, M. and Garattini, S. (1983) Immunosuppressive effects of 2,3,7,8-tetrachlorodibenzo-p-dioxin in strains of mice with different susceptibility to induction of aryl hydrocarbon hydroxylase. *Toxicol. Appl. Pharmacol.*, **68**, 434–41.

van Veld, P.A., Westbrook, D.J., Woodin, B.R., Hale, R.C. Smith, C.L., Huggett R.J. and Stegeman, J.J. (1990) Induced cytochrome P-450 in intestine and liver of spot (*Leiostromus xanthurus*) from a polycyclic aromatic hydrocarbon contaminated environment. *Aquat. Toxicol.*, **17**, 119–31.

Verma, S.R. and Chand, R. (1986) Toxicity effects of $HgCl_2$ on a few enzymes of carbohydrate metabolism of *Notopterus notopterus*. *Indian J. Environ. Health*, **28**, 1–7.

Verma, S.R., Rani, S. and Dalela, R.C. (1981) Isolated and combined effects of pesticides on serum transaminases in *Mystus vittatus* (African catfish). *Toxicol. Lett.*, **8**, 67–71.

Verschuuren, H.G., Ruitenberg, E.J., Peetoom, F., Helleman, P.W. and Van Esch, G.J. (1970) Influence of triphenyltin acetate on lymphatic tissue and immune responses in guinea pigs. *Toxicol. Appl. Pharmacol.*, **16**, 400–10.

Villeneuve, D.C., Grant, D.L., Phillips, W.E.J., Clark, M.L. and Clegg, D.J. (1971) Effect of PCB administration on microsomal enzyme activity in pregnant rabbits. *Bull. Environ. Contam. Toxicol.*, **6**, 120–8.

Villeneuve, D.C., Valli, V.E., Norstrom, R.J., Freeman, H., Sanglang, G.B., Ritter, L. and Becking, G.C. (1981) Toxicological response of rats fed Lake Ontario or Pacific Coho salmon for 28 days. *J. Environ. Sci. Health*, **B16**, 649–89.

Vos, J.G. and Koeman, J.H. (1970) Comparative toxicological study with polychlorinated biphenyls in chickens with special reference to porphyria, edema formation, liver necrosis, and tissue residues. *Toxicol. Appl. Pharmacol.*, **17**, 656–68.

Vos, J.G. (1977) Immune suppression as related to toxicology. *CRC Crit. Rev. Toxicol.*, **5**, 67–101.

Vos, J.G. and Moore, J.A. (1974) Suppression of cellular immunity in rats and mice by maternal treatment with 2,3,7,8-tetrachlorodibenzo-p-dioxin. *Int. Arch. Allergy*, **47**, 777–94.

Vos, J.G., Van Logten, M.J., Kreeftenberg, J.G., Steerenberg, P.A. and Kruizinga, W. (1979) Effect of hexachlorobenzene on the immune system of rats following combined pre- and post-natal exposure. *Drug Chem. Toxicol.*, **2**, 61–76.

Wachsmuth, E.D. (1983) Evaluating immunopathological effects of new drugs. In: *Immunotoxicology* (Eds G.G. Gibson, R. Hubbard and D.V. Parke), pp. 237–50. Academic Press, London.

Waggoner, J.P., III and Zeeman, M.G. (1975) DDT: Short-term effects on osmoregulation in black surfperch (*Embiotoca jacksoni*). *Bull. Environ. Contam. Toxicol.*, **13**, 297–300.

Wagner, M., Thaller, C., Jessell, T. and Eichele, G. (1990) Polarizing activity and retinoid synthesis in the floor plate of the neural tube. *Nature*, **345**, 819–22.

Wagner, S.R. and Greene, F.E. (1978) Dieldrin-induced alterations in biogenic amine content of rat brain. *Toxicol. Appl. Pharmacol.*, **43**, 45–55.

Wahlberg, J.E. and Boman, A. (1978) Sensitization and testing of guinea-pigs with cobalt chloride. *Contact Derm.*, **4**, 128–32.

Waksvik, H., Magnus, P. and Berg, K. (1981) Effects of age, sex and genes on sister chromatid exchanges. *Clin. Genet.*, **20**, 449–54.

Walker, C.H. (1978) Species differences in microsomal monooxygenase activity and their relationship to biological half-lives. *Drug Metabol. Rev.*, **7**, 295–323.

Walker, C.H. (1980) Species variations in some hepatic microsomal enzymes that metabolize xenobiotics. *Prog. Drug Metabol.*, **5**, 113–64.

Walker, C.H. (1989) The development of an improved system of nomenclature and classification of esterases. In: *Enzymes Hydrolysing Organophosphorous Compounds* (Ed. E. Reiner, W.H. Aldridge and F.C.G. Hoskin), pp. 236–45. Halsted Press, New York.

Walker, C.H. and Knight, G.C. (1981) The hepatic microsomal enzymes of sea birds and their interaction with liposoluble pollutants. *Aquat. Toxicol.*, **1**, 343–54.

Walker, C.H. and Mackness, M.I. (1983) Esterases: Problems of identification and classification. *Biochem. Pharmacol.*, **32**, 3265–9.

Wallach, D.F.H. (1987) *Fundamentals of Receptor Molecular Biology*. Chapter 11. Steroid hormones. Marcel Dekker, Inc., New York. pp. 398.

Walters, P., Khan, S., O'Brien, P.J., Payne, J.F. and Rahimtula, A.D. (1987) Effectiveness of a Prudhoe Bay crude oil and its aliphatic aromatic and heterocyclic fractions in inducing mortality and aryl hydrocarbon hydroxylase in chick embryo in ovo. *Arch. Toxicol.*, **60**, 454–9.

Wanek, N., Gardiner, D.M., Muneoka, K. and Bryant, S.V. (1991) Conversion of retinoic acid of anterior cells into ZPA cells in the chick wing bud. *Nature*, **350**, 81–3.

Wang, C. and Murphy, S.D. (1982) Kinetic analysis of species difference in acetylcholinesterase sensitivity to organophosphate insecticides. *Toxicol. Appl. Pharmacol.*, **66**, 409–19.

Ward, D.V., Howes, B.L. and Ludwig, D.F. (1976) Interactive effects of predation pressure and insecticide (Temefos) toxicity on populations of the marsh fiddler crab *Uca pugnax*. *Mar. Biol.*, **35**, 119–26.

Ward, E.C., Murray, M.J. and Dean, J.H. (1985) Immunotoxicity of nonhalogenated polycyclic aromatic hydrocarbons. In: *Immunotoxicology and Immunopharmacology* (Eds J.H. Dean, M.I. Luster, A.E. Munson, and H. Amos), pp. 291–303. Raven Press, New York.

Ward, E.C., Murray, M.J., Lauer, L.D., House, R.V., Irons, R. and Dean, J.H. (1984) Immunosuppression following 7,12-dimethylbenz[a]anthracene exposure in B6C3F1 mice. I. Effects on humoral immunity and host resistance. *Toxicol. Appl. Pharmacol.*, **75**, 299–308.

Warner, R.E., Peterson, K.K. and Borgman, L. (1966) Behavioural pathology in fish: A quantitative study of sublethal pesticide toxication. *J. Appl. Ecol.*, **3**(Suppl.), 223–47.

Weis, J.S. and Weis, P. (1987) Pollutants as development toxicants in aquatic organisms. *Environ. Health Perspect.*, **71**, 77–85.

Weisbart, M. and Feiner, D. (1974) Sublethal effect of DDT on osmotic and ionic regulation by the goldfish *Carassius auratus. Can. J. Zool.*, **52**, 739–44.

Welch, R.M., Rosenberg, P. and Coon, J.M. (1959) Inhibition of hexbarbital metabolism by chlorothion (p-nitro m-chloro phenyl dimethyl thionophosphate). *Pharmacologist*, **1**, 64.

Weseloh, D.V., Teeple, S.M. and Gilbertson, M. (1983) Double-crested cormorants of the Great Lakes: Egg-laying parameters, reproductive failure and contaminant residues in eggs, Lake Huron 1972–1973. *Can. J. Zool.*, **61**, 427–36.

Wesén, C., Carlberg, G.E. and Martinsen, K. (1990) On the identity of chlorinated organic substances in aquatic organisms and sediments. *Ambio*, **19**, 36–8.

Westlake, G.E., Bunyan, P.J., Martin, A.D., Stanley, P.I. and Steed, L.C. (1981a) Organophosphate poisoning. Effects of selected organophosphate pesticides on plasma enzymes and brain esterases of Japanese quail (*Coturnix coturnix japonica*). *J. Agric. Food Chem.*, **29**, 772–8.

Westlake, G.E., Bunyan, P.J., Martin, A.D., Stanley, P.I. and Steed, L.C. (1981b) Carbamate poisoning. Effects of selected carbamate pesticides in plasma enzymes and brain esterases of Japanese quail (*Coturnix coturnix japonica*). *J. Agric. Food Chem.*, **29**, 779–85.

Westlake, G.E., Bunyan, P.J. and Stanley, P.I. (1978) Variation in the response of plasma enzyme activities in avian species dosed with carbophenothion. *Ecotoxicol. Environ. Saf.*, **2**, 151–9.

Westlake, G.E., Martin, A.D., Stanley, P.I. and Walker, C.H. (1983) Control enzyme levels in the plasma, brain and liver from wild birds and mammals in Britain. *Comp. Biochem. Physiol.*, **76C**, 15–24.

Wetmore, A. (1919) Lead poisoning in waterfowl. *US Dept. Agric. Bull.*, **792**, 12 pp.

White, D.H., King, K.A., Mitchell, C.A., Hill, E.F. and Lamont, T.G. (1979) Parathion causes secondary poisoning in a laughing gull breeding colony. *Bull. Environ. Contam. Toxicol.*, **23**, 281–4.

White, K.L. and Holsapple, M.P. (1984) Direct suppression of in vitro antibody production by mouse spleen cells by the carcinogen benzo(e)pyrene but not by the noncarcinogenic congener benzo(e)pyrene. *Cancer Res.*, **44**, 3388–93.

White, D.H., Mitchell, C.A. and Hill, E.F. (1983) Parathion alters incubation behavior of laughing gulls. *Bull. Environ. Contam. Toxicol.*, **31**, 93–7.

Widdows, J., Moore, S.L., Clarke, K.R. and Donkin, P. (1983) Uptake, tissue distribution and elimination of [1-^{14}C] naphthalene in the mussel *Mytilus edulis. Mar. Biol.*, **76**, 109–14.

Wiemeyer, S.N. and Porter, R.D. (1970) DDE thins eggshells of captive American Kestrels. *Nature*, **227**, 737–8.

Wiemeyer, S.N., Spitzer, P.R., Krantz, W.C., Lamont, T.G. and Cromartie, E. (1975) Effects of environmental pollutants on Connecticut and Maryland ospreys. *J. Wildl. Manage.*, **39**, 124–39.

Wierda, D., Irons, R.D. and Greenlee, W.F. (1981) Immunotoxicity in C57B1/6 mice exposed to benzene and Aroclor 1254. *Toxicol. Appl. Pharmacol.*, **60**, 410–17.

Wieser, W. and Hinterleitner, S. (1980) Serum enzymes in rainbow trout as tools in the diagnosis of water quality. *Bull. Environ. Contam. Toxicol.*, **25**, 188–93.

Wilder, I.B. and Stanley, J.G. (1983) RNA–DNA ratio as an index to growth in salmonid fishes in the laboratory and in streams containing carbaryl. *J. Fish Biol.*, **22**, 165–72.

Wilkinson, C.F. (1968) Detoxification of pesticides and the mechanism of synergism. In: *The Enzymatic Oxidation of Toxicants* (Ed. E. Hodgson), pp. 113–49. North Carolina State University, Raleigh, NC.

Willhite, C. and Sharma, R.P. (1978) Acute dieldrin exposure in relation to brain monoamine oxidase activity and concentration of brain serotonin and 5-hydroxyindoleacetic acid. *Toxicol. Lett.*, **2**, 71–5.

Wistar, R. and Hildemann, W. (1960) The effect of stress on skin transplantation immunity in mice. *Science*, **131**, 159–60.

Wojdani, A. and Alfred, L.J. (1983) Alterations in cell-mediated immune functions induced in mouse splenic lymphocytes by polycyclic aromatic hydrocarbons. *Cancer Res.*, **44**, 942–5.

Woodwell, G.M., Wurster, C.F., Jr. and Isaacson, P.A. (1967) DDT residues in an east coast estuary: A case of biological concentration of a persistent insecticide. *Science*, **156**, 821–4.

Wurster, D.H., Wurster, C.F., Jr. and Strickland, W.N. (1965) Bird mortality following DDT spraying for Dutch elm disease. *Ecology*, **46**, 488–99.

Yamanaka, S., Hashimoto, M., Tobe, M., Kobayashi, K., Sekizawa, J. and Nishimura, M. (1990) A simple method for screening assessment of acute toxicity of chemicals. *Arch. Toxicol.*, **64**, 262–8.

Yang, K.H., Kim, B.S., Munson, A.E. and Holsapple, M.P. (1986) Immunosuppression induced by chemicals requiring metabolic activation in mixed cultures of rat hepatocytes and murine splenocytes. *Toxicol. Appl. Pharmacol.*, **83**, 420–9.

Young, D.R., Heesen, T.C. and McDermott, D.J. (1976) An offshore biomonitoring system for chlorinated hydrocarbons. *Mar. Pollut. Bull.*, **7**, 156–9.

Zbinden, G. (1984) Statistical considerations and protocols using small numbers of animals. Introductory remarks. In: *Acute Toxicity Testing: Alternative Approaches* (Ed. A.M. Goldberg), pp. 197–205. Mary Ann Liebert, New York.

Zinkl, J.G., Henny, C.J. and Shea, P.J. (1979) Brain cholinesterase activities of passerine birds in forests sprayed with cholinesterase inhibiting insecticides. In: *Animals as Monitors of Environmental Pollutants. Nat. Acad. Sci.*, 356–65.

Zitko, V., Choi, P.M.K. Wildlish, D.J., Monghan, C.F. and Lister N.A. (1974) Distribution of PCB and p,p'-DDE residues in Atlantic herring (*Clupea harengus harengus*) and yellow perch (*Perca flavescens*) in eastern Canada – 1972. *Pesticide Monitor J.*, **8**, 105–9.

Index

Note: as there are comprehensive appendices, many items are indexed under their acronyms. All references in italics are figures and those in bold type represent tables.

Parathion 29, 30, 32, 57, 64, 89, **159**, 176, 183
Partridge 23
  grey 64
  red-legged 18, 23
Passeriformes 22, 23, **145**, **152**
PBB 170, 171, 181–2, 187, 188
PBCO (Prudhoe Bay crude oil) 57, 95, 208, *210*
PCB 1, 2, 4, *5*, 6, 8, *9*, 10, 11, 25, 40, *41*, **52**, 53, 61–2, **63**, 64, 66, 71, 85, 95, 96, 97, 98–9, *100*, 101, 102, 104, 106, 109–10, 111, 112, 113, 119, 127–8, 128–9, 130, 131, 132, 134, **145**, *146*, *147*, 170, 171, 181–2, 187, 199, 219, 222
PCDD 1, 96, 97, 99, 102, 123, 125, 221
PCDF 1, 96, 97, 98, 99, 102, 123, 125, 221
p-cresol 71
Pelican, brown 60
Pentachlorophenol 163–4, 182
Perch *103*
  yellow 11
Permethrin 43
Pesticides 16, 18, 19, 26, 27, 28–9, 35–6, 88, 89, 156
  see also named pesticides
Petrel
  Leach's storm 68
  storm 11
Petroleum products 56–7, 77, 102
  see also oil
PHAH 2, 3, 4, 6, 16, 19, 38–41, 45, 50, 51, 52, 59–64, 65, 66, 68, 84–5, 95, 96, 98, 99, 102, 107, 108, 109, 110, 111, 112, 113, 114, 116, 120, 125–6, 127, 130, 131, 132, 134, 139, **140**, 143–4, **145**, 147, 148, 149, 157, 182, 205, *206*, **206**, 211, **213**, 218–9, 221–2
Phasianadae 22, 23
  see also chicken, quail, pheasant
Pheasant 23, 98, **192**
Phenobarbital 122
Phosphatases 150–1
Physiological injuries **215–7**

Phytoplankton 71, 148
Pig *93*
  miniature 22
Pigeon 92, 94, 96, 108–9, 113, 136, **140**
Pike 84, 104, 132
Plankton 4, *5*
Plants **12**
Ploceidae 23
Plover, piping 3–4
Poisoning 25–7
Pontoporeia 5
Porphyrin 17, 96, 125–7, *126*, **195**, 196, **216**
  altered by HCB 127–8
  altered by PCBs *100*, 128–9, 199
  relation to other biomarkers 123, 129–32
  use of biomarker 132–4, *133*, 205–6, **206**, **213**, 218
  relationship with other biomarkers 129–32
  use as biomarker 132–4, *133*
  see also HCP
Post-labelling techniques 72–3, 74, 75
Predation pressure 159–60
Pseudoallergy *171*, 172–3
Puffin 10, 53, 92
  Atlantic 149
Pyrethroids 64, 89

Quail 22, 23, 24, 25, 32, 34, 50–1, 62, 72, 93, 96, 99, *100*, 113, 119, 120, 128, 129, 130, **140**, **141**, **142**, 143, 144, **145**, *146*, 192, 211
  bobwhite 25, 29–30, 32, 57, 64, 151, **152**, **159**
  Californian **192**
Quelea, red-billed 23
Quokka *93*

Rabbit 22, 66, *93*, 119, 139, **175**, 181, **192**
Rat 12–14, *13*, 16, 20, 24, 25, 30, 32, 37, 39, 42–3, 44, 51, **52**, 61, 64, 66, 77, 82, 88–9, 90, 91, 92, *93*, 95, 96, 98, 99, 101, 105, 111–12, 113, 116, 119–20, 121, 122, 127–8, 129, 132, 134, 135, 136, **140**, **145**, 147, 148, 149, 159,